ALLOGRAFT PROCUREMENT, PROCESSING AND TRANSPLANTATION

A Comprehensive Guide for Tissue Banks

ALLOGRAFT PROCUREMENT, PROCESSING AND TRANSPLANTATION

A Comprehensive Guide for Tissue Banks

Aziz Nather

National University of Singapore, Singapore

Norimah Yusof

Malaysian Nuclear Agency, Malaysia

Nazly Hilmy

BATAN Research Tissue Bank, Indonesia

World Scientific

NEW JERSEY · LONDON · SINGAPORE · BEIJING · SHANGHAI · HONG KONG · TAIPEI · CHENNAI

Published by

World Scientific Publishing Co. Pte. Ltd.

5 Toh Tuck Link, Singapore 596224

USA office: 27 Warren Street, Suite 401-402, Hackensack, NJ 07601

UK office: 57 Shelton Street, Covent Garden, London WC2H 9HE

British Library Cataloguing-in-Publication Data
A catalogue record for this book is available from the British Library.

ISBN-13 978-981-4291-18-7
ISBN-10 981-4291-18-8

Typeset by Stallion Press
Email: enquiries@stallionpress.com

Printed in Singapore by Mainland Press Pte Ltd.

This book is dedicated to Professor Glyn Owen Phillips, Technical Advisor to DDG, IAEA (Mr Qian) for his vision and mission to development of tissue banking (RAS 7/008) in the Asia-Pacific region (1985 to 2008) and to the development and success of IAEA/NUS Diploma Course and IAEA/NUS Training Centre in Singapore for training tissue bank operators (since November 1997).

Professor Phillips (left), Mrs. Rhiain Phillips (center) and Mr. Ramen Mukherjee (Technical Officer for RAS 7/008: right)

Professor Phillips and Mr. Mukheerjee received Honorary Membership of APASTB during the 8th Scientific Conference of Asia Pacific Association of Surgical Tissue Banking (APASTB) in Bali, 2000.

Contents

Preface

Singapore has been actively involved with the International Atomic
Energy Agency (IAEA) in training tissue bank operators from the Asia-
Pacific region, Latin America, Eastern Europe and Africa, since NUH
Tissue Bank was appointed to be the IAEA/NUS Regional Training
Centre for RCA member states (Asia Pacific) in 1997. In November 1997,
NUS held the first IAEA/NUS Diploma Course in Tissue Banking, the
first of its kind in the world. Since then, ten batches have successfully
participated — the 11th batch, at the time of writing, has started as an
online diploma course in August 2009. The course has been conducted as
an online course since 2004 (starting from the 6th batch). To date a total
of 141 tissue bank operators (from 27 countries in the Asia-Pacific region,
Latin America, Eastern Europe, Africa and Australia) have convocated
with an NUS Diploma in Tissue Banking.

Singapore has been instrumental in development of the IAEA/NUS
Multi-Media Curriculum on Tissue Banking, which was produced in April
1998. This curriculum served as the textbook for all diploma courses run
by IAEA jointly with NUS. This curriculum was translated into Spanish
and used as textbook for all IAEA courses held in Latin America, with
Argentina as the Regional Training Centre since 2000. It has also been
translated into Korean language in 2003 by IAEA for use in its Korean
National Training Courses run annually since 2004.

This book is the latest update on all topics in the previous Multi-
Media Curriculum ranging from donor selection criteria, procurement
techniques, laboratory testing, processing techniques for deep-frozen and

freeze-dried allografts, quality control issues, radiation sterilization techniques, clinical transplantation of bones and soft tissues and principles of sterile techniques in the operating room. Incorporated in the book are the basic science topics such as anatomy, biomechanics, microbiology, and immunology.

This book will be useful for all tissue bank operators, doctors, nurses, technicians and allied healthcare professionals dealing with the practice of tissue procurement, processing and clinical transplantation.

Aziz Nather

Editor

Director, National University Hospital Tissue Bank

Director, IAEA/NUS International Training Centre for Tissue Bank Operators (Singapore)

Director, IAEA/NUS Diploma Course in Tissue Banking (Singapore)

Senior Consultant

Department of Orthopaedic Surgery

Yong Loo Lin School of Medicine

National University of Singapore

About the Editors

Associate Professor Abdul Aziz Nather
Director, National University Hospital Tissue Bank
Director, IAEA/NUS Training Centre for Tissue Bank Operators
Director, IAEA/NUS Diploma Course in Tissue Banking
Senior Consultant, Department of Orthopaedic Surgery
Yong Loo Lin School of Medicine
National University of Singapore
Singapore

Dr Norimah Yusof
Head, Malaysian Nuclear Agency Research Tissue Bank
Radiation Biologist
Director, Division of Agriculture and Biosciences
Head, Division of Biological Applications
Malaysian Nuclear Agency
Bangi, Selangor, Malaysia

Dr Nazly Hilmy
Senior Consultant
BATAN Research Tissue Bank (BRTB)
Radiation Biologist
Center for Research and Development of Isotopes
    and Radiation Technology
BATAN, Jakarta, Indonesia

List of Contributors

Basril Abbas
BATAN Research Tissue Bank (BRTB)
Center for the Application of Isotopes and Radiation Technology
BATAN, Jakarta 12070
Indonesia

Noriah Mod Ali
Malaysian Nuclear Agency (NM)
Bangi 43000 Kajang, Selangor
Malaysia

Zameer Aziz
NUH Tissue Bank
Department of Orthopaedic Surgery
Yong Loo Lin School of Medicine
National University of Singapore
Singapore

Chan Poh Lin
NUH Tissue Bank
Department of Orthopaedic Surgery
Yong Loo Lin School of Medicine
National University of Singapore
Singapore

Chan Xin Bei
NUH Tissue Bank
Department of Orthopaedic Surgery
Yong Loo Lin School of Medicine
National University of Singapore
Singapore

Chong Cui Lian
NUH Tissue Bank
Department of Orthopaedic Surgery
Yong Loo Lin School of Medicine
National University of Singapore
Singapore

Han Yan Yi
NUH Tissue Bank
Department of Orthopaedic Surgery
Yong Loo Lin School of Medicine
National University of Singapore
Singapore

Asnah Hassan
Malaysian Nuclear Agency (NM)
Bangi 43000 Kajang, Selangor
Malaysia

Nazly Hilmy
BATAN Research Tissue Bank (BRTB)
Center for the Application of Isotopes and Radiation Technology
BATAN, Jakarta 12070
Indonesia

Awang Hazmi Awang Junaidi
Faculty of Veterinary Medicine
Universiti Putra Malaysia
43400 UPM Serdang, Selangor
Malaysia

Koh Si Qi
NUH Tissue Bank
Department of Orthopaedic Surgery
Yong Loo Lin School of Medicine
National University of Singapore
Singapore

Lee Choon Wei
NUH Tissue Bank
Department of Orthopaedic Surgery
Yong Loo Lin School of Medicine
National University of Singapore
Singapore

Low Jia Ming
NUH Tissue Bank
Department of Orthopaedic Surgery
Yong Loo Lin School of Medicine
National University of Singapore
Singapore

Menkher Manjas
Dr. M. Djamil Hospital Tissue Bank
Department of Surgery
Faculty of Medicine
Andalas University
Padang, Indonesia

Hasim Mohamad
School of Medical Science,
University of Science, Malaysia

Department of Surgery,
Hospital Raja Perempuan Zainab II
Kota Bharu, Malaysia

Siti Zubaidah Mordiffi
Evidence Based Nursing Unit
Nursing Department
National University Hospital
Singapore

Aziz Nather
NUH Tissue Bank
IAEA/NUS Training Centre for Tissue Bank Operators (Singapore)
Department of Orthopaedic Surgery
Yong Loo Lin School of Medicine
National University of Singapore
Singapore

Neo Shu Hui
NUH Tissue Bank
Department of Orthopaedic Surgery
Yong Loo Lin School of Medicine
National University of Singapore
Singapore

Paramita Pandansari
BATAN Research Tissue Bank (BRTB)
Center for the Application of Isotopes and Radiation Technology
BATAN, Jakarta 12070
Indonesia

Sim Yi Lin
NUH Tissue Bank
Department of Orthopaedic Surgery
Yong Loo Lin School of Medicine
National University of Singapore
Singapore

Mary Tan
Nursing Education Unit
Nursing Department
National University Hospital
Singapore

Retno Dwijartini Tantin
BATAN Dental Clinic
Center for Technology of Radiation Safety and Metrology
BATAN, Jakarta 12440
Indonesia

Tay Li Min
NUH Tissue Bank
Department of Orthopaedic Surgery
Yong Loo Lin School of Medicine
National University of Singapore
Singapore

Petrus Tarusaraya
Sitinala Leprosy Hospital
Tangerang, Indonesia

Teng Wen Hui
NUH Tissue Bank
Department of Orthopaedic Surgery
Yong Loo Lin School of Medicine
National University of Singapore
Singapore

Norimah Yusof
Malaysian Nuclear Agency (NM)
Bangi 43000 Kajang, Selangor
Malaysia

Zheng Shushan
NUH Tissue Bank
Department of Orthopaedic Surgery
Yong Loo Lin School of Medicine
National University of Singapore
Singapore

Ahmad Hafiz Zulkify
Department of Orthopaedics, Traumatology and Rehabilitation
Kulliyah of Medicine
International Islamic University Malaysia
Malaysia

Part I

History

Chapter 1

Evolution of Allograft Transplantation

Aziz Nather and Shushan Zheng

Introduction

To everyone who reads this book and endeavours to master the operation of a tissue bank, it is important to first understand the history behind bone and skin allograft transplantation.

The history of bone and skin allograft is replete with controversies due to the radical nature of the procedure in earlier times. Understandably, the act of transferring bone or skin from one person to another invited much public contention on social, ethical and religious grounds. The many works to advance the field of allograft transplantation were likewise, often frowned upon by the public and the medical authorities (Bradley and Hamilton, 2001). Furthermore, considering the dearth of knowledge on the subject in the past, failure of a study was as much an outcome as success. It is therefore not beyond our imagination to understand how these factors would have deterred the progress of allograft transplantation in earlier times.

At other times, however, events served as impetus to the progress of allograft transplantation. World War II, for example, resulted in large numbers of casualties. Losses of bone, fractures or burn wounds in victims of the War compelled surgeons of their time to come up with methods to repair these defects (bone-grafting and bone-transplantation; Lancet, 1918). Indeed, necessity is the mother of invention.

NUH Tissue Bank, Department of Orthopaedic Surgery, Yong Loo Lin School of Medicine, National University of Singapore, Singapore.

Regardless of circumstances and public opinion, the pioneers have undertaken research to shed light on allograft techniques and their significance in the realm of medicine. Their motivation for doing so would perhaps be aptly expressed by Gaspare Tagliacozzi of Bologna (1546–1599), pioneer of the autologous procedure known as the "Forearm Flap". He wrote, "We reconstruct and complete parts which nature had given but which were destroyed by fate, and we do so, not so much for the enjoyment of an eye, as for psychic comfort to the afflicted" (Ben-hur and Converse, 1980; Ang, 2005).

The Early Miracles of Transplantation

The early history of transplantation was marked by accounts of transplantation that carried with them an element of wonder. One such famous account was that of the legend of Saints Cosmas and Damian. Born in the third century as sons of a physician, these two Christian Arab twins became physicians and practiced the art of healing across Turkey, Rome and Greece, all the while providing their services for free. They were subsequently martyred in 287 AD during the persecution of Christians by the Emperor Diocletian and were buried in their hometown of Egea. Two centuries later, in ancient Rome, the Deacon Justinian was so exhausted by the pain from his ulcerated leg that he fell asleep during his prayers. In his dream, the twin physicians came to him and replaced his diseased leg with that from a recently deceased Moor (Julien *et al.*, 1987; Rinaldi, 1987; Mankin, 2002; Mankin *et al.*, 2005). The successful transplantation, later called the "black leg miracle", has since been depicted in many paintings (Fig. 1) and decorations in religious manuscripts.

History of Skin Allograft Transplantation

The practice of skin grafting originated in India — as described in the *Sushruta Samhita* (ca. 600 BCE) — for the purpose of nasal reconstruction. The mutilation of the nose was rampant as a form of punishment in the past and this necessitated skin grafting for the repair of the mutilated nose (Sushruta, 1907; Nichter *et al.*, 1983; Ang, 2005). With time, the applications of skin grafting evolved to include reconstruction after

Fig. 1. Saints Cosmas and Damian performing the "miracle of the black leg". This painting is attributed to Master of Los Balbases, Burgos, Spain, c. 1495. Reproduced from the Wellcome Institute Library, London (reference L0014276).

surgery, treating patients with burn wounds and patients with epidermolysis bullosa, treatment of chronic ulcers, and hair restoration to areas of hair loss (Herman, 2002).

The early recorded attempts at skin transplantation were made by Branca de Branca, an Italian, who in 1442 AD, employed the binding of the patient's arm to the site of skin graft to transplant a slave's nose to his master's. While Branca pioneered the use of surgical flap for nasal reconstruction, credit went to his compatriot, Gaspare Tagliacozzi, instead. (Davis, 1941; Herman, 2002).

Tagliacozzi's work came over a hundred years later, and was described in his publication *De Curtorum Chirurgia per Institionem* (*Surgery of the Mutilated by Grafting*) in 1597. The procedure, called the "Forearm Flap", was likewise used for nasal reconstruction using a graft harvested from the inner arm. The difference between the work of Tagliacozzi and Branca, however, was that the former's technique was autologous, whereas the latter's was allogenic. Unfortunately, this difference was not apparent to the many intellectuals of the 17th and 18th century who despised Tagliacozzi's work. Many satirical stories about the use of noses from slaves were propagated. The Church was also against such work and hence exhumed Tagliacozzi's body from its burial site lest it desecrate the holy ground (Koch, 1941; Gina, 2005). It is therefore ironic when we learn that Tagliacozzi was himself against the idea of using allografts because of considerations about the "force and power" of the individual. It would be another 400 years before this "force and power" was to be recognised as a major biological phenomenon (Phillips, 1998).

The First Skin Allograft

In 1869, Jacques-Louis Reverdin (1842–1929) of Switzerland performed the first skin allograft transplantation with the use of epidermal grafts, otherwise known as split thickness skin grafts (Reverdin, 1869). He described this technique during a meeting of the Société Impériale de Chirurgie during the same year (Freshwater and Krizek, 1978; Ang, 2005) and was hailed as the father of skin transplantation due to his work in this area (Hauben, 1985; Ang, 2005).

In 1871, the Englishman George Pollock (1897–1917) introduced the idea of treating burns patients with epidermal grafts. He purportedly donated small pieces of his own skin and used it together with skin from the patient in the treatment process (Pollock, 1871; Freshwater and Krizek, 1978; Herman, 2002). His first patient was an 8-year-old called Anne T, who had severe burns on her thighs from having caught fire on her dress (Freshwater and Krizek, 1971; Ang, 2005).

A year after, in France, Louis Xavier Édouard Léopold Ollier (1830–1900) reported the successful transplant of skin using the entire epidermis and a portion of the dermis (Ollier, 1872). In the United States, John Girdner (1881) described the first allograft skin transplantation using skin from a human cadaver (Herman, 2002). Girdner procured skin from the inner thigh of a young German boy within 6 hours of his death and transplanted the skin onto the shoulder blade of a 10-year-old boy who had been struck by lightning (Obeng *et al.*, 2001). And in 1886, following Thiersch's method of removing split-thickness grafts, the use of new and improved skin grafting methods spread (Phillips, 1998).

The Discovery of Immunogenicity

The progress of skin allograft techniques and applications made a significant impact on the treatment of skin and burns patients. However, it was observed that all allografts, despite starting out well initially, would ultimately fail to integrate into the recipient's body. The concepts that explained the loss of grafted tissue were still unknown at that time. It was not until 1920 that the experiments of Emile Holman in Baltimore hinted at the first signs of what would become the concept of rejection. Holman, a surgeon at John Hopkins, transplanted the skin of a mother onto a badly burnt child. The subsequent grafting of more skin onto the child a few days later resulted in the inflammation of both the mother and the child's own skin. The implications of immunogenicity from this experiment, though noted by Holman, were not fully explored at that point of time (Holman, 1924; Hakim, 1997).

In 1927, the German Karl Bauer performed a successful skin allograft between identical twins and the skin on the twins stayed on indefinitely. With our current knowledge of the mediation of the body's immune system, such a result is hardly surprising (Bauer, 1927; Lytton, 2005).

Yet it was only in 1943 that the physiology of graft survival was unravelled. World War II had resulted in many victims of the war, amongst them the sufferers of massive burns. After seeing a seriously burnt patient in an Oxford hospital, a young zoologist, Peter Medawar, became interested in the problems of skin grafting. Following the establishment of the Burns Unit at Glasgow, he was asked by the British Medical Research Council to collaborate with Glasgow surgeon Thomas Gibson to perfect skin grafting, and in doing so, discovered the "Second Set Response". When transplanting skin allograft onto a burnt patient, they observed the hastening of the rejection of a second set of allograft. Medawar returned to Oxford and experimented on rabbits (Tilney, 2000). He observed that when skin was transplanted from animal A onto animal B, the graft survived about 7 days. However, when the same transplantation was carried out a second time, the graft was rejected in half that period. From this discovery, Medawar drew the link between the body's immune system and the rejection of grafted tissue (Gibson and Medawar, 1943; Hakim, 1997) and termed the rejection mechanism as "actively acquired immunity".

Further Developments

In 1944, Jerome Pierce Webster (1888–1974), a Professor of Surgery at Columbia University Hospital, described the use of refrigerated skin as a temporary dressing to treat burns patients (Webster, 1944; Herman, 2002). A few years later, in 1949, the first proper skin bank, the US Navy Skin Bank, was established (Trier and Sell, 1968; Herman, 2002). In that same year in England, a doctoral student Christopher Polge and his mentors Parkes and Smith discovered the way to preserve tissues for transplantation with cryopreservatives (Polge *et al.*, 1949; Herman, 2002).

The second half of the twentieth century marked even more developments in the application and understanding of skin allografts. Skin from human cadavers was used as biological dressings in burn patients. In 1958, Eade discovered the use of skin grafts in controlling surface wound infection (Eade, 1958; Herman, 2002). By 1968, cryopreserved skin was successfully used as skin allografts. (Cochrane, 1968; Herman, 2002) In 1971, O'Donoghue and Zarem described the stimulation of neovascularisation by skin grafts (O'Donoghue and Zarem, 1971; Herman, 2002).

In 1975, Rheinwald and Green developed a method of producing cultured epithelial sheets from human keratinocytes (Rheinwald and Green, 1975; Herman, 2002). Following this, in 1981, O'Connor reported the successful use of cultured epithelial autografts (CEA) (O'Connor *et al.*, 1981; Brychta *et al.*, 1994). However, as cultured epithelial autograft took a long time for culture, and cultured epithelial allograft was, in contrast, a readily available source for recipients, the latter became widely used as well (Matouskova *et al.*, 2002).

More recently, in 1993, Kirsner described the ability of skin allografts to release growth factors as well as to act as pharmacologic agents (Kirsner *et al.*, 1993; Herman, 2002). Fortunato Benaim and Alberto Bolgiani, both leaders in the realm of burn care in Latin America, also reported in 1993 that skin allografts provided by tissue banks were routinely used in burns centre as a temporary cover for serious burn wounds (Bolgiani and Benaim, 1993; Bourroul, 2002).

From repairing mutilated noses to becoming an important means of wound care in burn patients, the field of skin allograft has witnessed a giant leap since Branca's time, and it will no doubt continue to be invigorated by the continuing research and development in this field.

History of Bone Allograft

The earliest work that demonstrated the viability of bone allograft was that by the Scottish surgeon and anatomist, John Hunter. In 1770, Hunter transplanted the spur of a cock into its comb. The spur, being an outgrowth of the tarsometatarsus, contained a solid mass of bone within it (Lancet, 1918). In 1771, Hunter described in his book, the *Treatise on the Natural History of the Human Teeth,* the successful reimplantation of a premolar that was lost through trauma. Thereafter, he conducted his famous experiments on the transplantation of human tooth into the comb of a cock. Hunter's work had intrigued others into conducting experiments on tooth transplantation. Unfortunately, those experiments often exploited the poor as donors and placed their lives at the mercy of infections, and hence abated in the latter half of the eighteeth century (Tilney, 2000).

The Problem of Osteogenesis

In the eighteenth century, the fate of allografts and their role in bone formation became of interest to many orthopaedic surgeons. A controversy over the science of osteogenesis — the formation of bone — had emerged following the opposing views of Duhamel and von Haller.

In 1739, Duhamel performed an experiment in which he implanted silver wires subperiosteally. Weeks later, he found that the wires were buried in bone and concluded that the periosteum had led to new bone formation (Duhamel, 1739). Duhamel went on in 1742 (Duhamel, 1742) and 1743 (Duhamel, 1743) to repeat and extend the madder feeding experiments of Belchier (Belchier, 1736a; Belchier, 1736b). He noted that madder (the root of rubbia tinctoria giving a red dye) stained only growing bone, and distinguished between two layers of the periosteum — a superficial supporting layer and a deep osteogenic layer, which he termed the "cambium layer" (cambium meaning a layer of cells between bark and wood) (Chase and Herndon, 1955; Bassett, 1962).

However, von Haller (1763), the professor of John Hunter, claimed the opposite: the periosteum was not osteogenic. According to von Haller, the periosteum merely acted as the support for blood vessels, and it was the exudation from arteries that caused osteogenesis (Chase and Herndon, 1955; Bassett, 1962).

The views of these two men soon led to two opposing schools of thought and formed the basis of the aptly named "Duhamel-Haller Controversy". Hunter joined in the conflict and performed a number of experiments to substantiate his professor's claim (Bassett, 1962). On the other hand, Heine (1836) described findings that lent support to Duhamel's claim — he removed ribs subperiosteally and found that they grew back. However, it was Flourens (1842) who went a long way in settling the controversy when he conclusively showed that periosteum was osteogenic and was the chief agent in the healing of bone defects (Chase and Herndon, 1955).

In 1858, Ollier took what was arguably the first really scientific approach to tackle the problem of osteogenesis (Ollier, 1858; Chase and Herndon, 1955). Despite the lack of modern histological techniques or aseptic surgery, he performed comprehensive studies on the periosteum. Ollier's experiments were published in two volumes entitled *"Traite Experimental et clinique de la regeneration des os"* in 1867. His conclusion was that transplanted periosteum and bone survived and could become osteogenic under proper circumstances. Additionally, Ollier believed that periosteum-covered transplants were the best bone grafts for use, and that the contents of the Haversian canals and the endosteum were also osteogenic (Chase and Herndon, 1955).

Owing to the thoroughness of Ollier's work, his views remained almost indisputable for the following decades. Then in 1893, Barth, a student of Marchand, began to challenge those views with his numerous papers. By basing his work almost solely on replanted discs removed by trephine from the skull, Barth claimed that all transplanted bone, marrow and periosteum would eventually die and be replaced by surrounding tissue, and that bone grafts were but a form of passive scaffolding. In response to this challenge, scores of papers were written by others to defend each side of the argument (Chase and Herndon, 1955).

Axhausen entered the debate with his thorough and scientifically rooted studies on osteogenesis and bone transplantation (Axhausen, 1907; Axhausen, 1909a; Axhausen, 1909b; Axhausen, 1909c). He showed that the survival and osteogenic property of the periosteum varied between different types of graft: they were highest in autografts; significantly less so in allografts and null in xenografts. He also believed that most of the periosteum would survive and lead to osteogenesis while the transplanted bone would die. Hence, like Ollier, Axhausen preferred bone grafts with attached periosteum. His evidence was so convincing that in 1908, Barth, influenced by Axhausen's findings and similar findings by others, rescinded his views and accepted Axhausen's principles (Chase and Herndon, 1955).

Then in 1912, Macewen in his work "The Growth of Bone" contradicted Axhausen's views. He denied that periosteum was osteogenic and regarded it as only a limiting membrane. Macewen ascribed all the phenomenon of bone graft repair to osteoblasts. The cause of this contradiction was due to Macewen's failure to recognise the cambium layer as part of the periosteum (Chase and Herndon, 1955). Ollier had shown that periosteum without the deep layer did not produce bone. Axhausen had also showed that osteogenesis in the graft occurred beneath the periosteum. The confusion arose due to the anatomical description of the periosteum. Hey Groves (1917) suggested using the term "epiosteal" to denote structures on the surface of the bone but beneath the periosteum. Thus, Axhausen showed it was the epiosteal layer which was osteogenic. In fact, Macewen's work showing that periosteum was useless, unless the underlying surface of the cortex was attached, formed the basis for the present day concept of the osteoperiosteal graft.

In 1914, Phemister performed a series of experiments in dogs to further investigate osteogenesis. His findings showed that other than the periosteum, the endosteum, and the contents of the Haversian canals also had the capacity for osteogenesis. He explained that the surface location of the periosteum and endosteum allowed it to receive sufficient nutrition for survival and proliferation. However, the great mass of bone cells that were separated from the surface by an impermeable calcified matrix would eventually be necrotised and absorbed. A few bone cells lying about the periphery and lining the larger vascular spaces as well as the fibrous elements of the latter might survive and proliferate. Blood-forming cells of the marrow, despite their favourable nutrition, would necrotise because of their higher degree of specialisation (Phemister, 1914). Hence, Phemister proved that Axhausen's claim that osteogenesis did not occur from transplanted bone devoid of periosteum and endoteum was incorrect. The old view of Barth, later advocated by Murphy (1913), that there is no osteogenesis from a transplant, and that substitution occurs entirely from the ingrowth of new bone from the host fragment ends, was also proven wrong.

Still, the basic principles advocated by Axhausen have largely been corroborated by later investigators. The wide acceptance of Axhausen's views has thus brought an end to the debate over the role of the periosteum in osteogenesis.

The First Bone Allograft

Other than his work on osteogenesis, the Scottish surgeon William Macewen was also the first to perform a bone allograft. In 1879, using the tibia of a child with rickets, he transplanted the allograft onto the humeral shaft of a young boy whose humerus was lost through osteomyelitis. This work was later described in 1881 in a paper called "Observations concerning transplantation of bone. Illustrated by a case of inter-human osseous transplantation, whereby over two-thirds of the shaft of a humerus was restored" (Jones, 1952; Macewen, 1881). The allograft was a success because Macewen operated under antiseptic conditions as introduced by Lister. Furthermore, the transplanted tibia was, by chance, subjected to muscular stress and this encouraged the osteoblasts present to form bones (Burwell, 1994).

Developments in the Use of Bone Allograft

In 1908, Erich Lexer described the first massive allograft, in which tissue harvested from amputees were used to restore motion to joints following osteomyelitis and to reconstruct defects following resection of bone tumours (Lexer, 1908a; Lexer, 1908b; Burwell, 1994; Enneking, 2005). A year later, in 1909, Judet reported a whole-joint transplantation — femur, tibia and patella — in the knee joint of man. The procedure, as he described, could be carried out to treat the trauma of infective arthritis or tuberculosis (Judet, 1909; Burwell, 1994).

Use of Fresh Bone Allograft

Before advancements in refrigeration techniques, fresh bones were used for most allografts. In 1915, Trout reported the successful bone transplantation between father and child to treat spina bifida, with the use of fresh bone allograft (Trout, 1915; Burwell, 1994). The trend of using fresh bone allografts continued through the 1930s and 1940s, mainly in cases of bone transplants from parent to child to treat pseudarthrosis, cysts and tumours. In 1948, Alldredge described the treatment of nonunion in adults using fresh bone allograft (Alldredge, 1948). However, the development in the use of fresh bone allografts took a turn in the 1950s. It was reported that fresh allografts produced a strong immune response within the body (Bonfiglio *et al.*, 1955; De Boer, 1988). The use of fresh bone allografts has since dwindled to a stop with the increased understanding of immunogenicity (Burwell, 1994).

Preservation of Bone Allografts

In 1910, Bauer demonstrated the successful transplantation of bones stored by refrigeration for as long as three weeks (Bauer, 1910; Burwell, 1994). A year later, Tuffier described the use of thin bone slices refrigerated for as long as five days in the treatment of patients (Tuffier, 1911; Burwell, 1994). Then in 1912, Albee, an American orthopaedic surgeon, recommended that all tissues — bones included — be stored at 4–5°C (Albee, 1912; Tomford, 1994). In 1942, Inclan reported the successful storing of autologous bones by refrigeration. Following the report, the idea of bone banking received much mentioned in various publications (Inclan, 1942; De Boer, 1988; Donati, 2007).

Influenced by the success of Inclan, Bush and Wilson independently described the preservation of bone grafts at −20°C and the building of a bone bank for small fragments (Bush, 1947; Wilson, 1947; Donati, 2007). After Wilson's report, many orthopaedic surgeons across the United States followed suit. They stored femoral heads collected from their patients in freezers in their own hospitals, and used these small bones on an individual basis (Tomford, 1994), much like the practice of a cottage industry.

Establishment of Tissue Banks

Spurred by the needs of the wounded following World War II and encouraged by the climate of interest surrounding bone banking, George Hyatt, an orthopaedic surgeon, founded the US Navy Tissue Bank in Bethesda, Maryland in 1949. Hyatt developed a system for procuring tissue, with a focus on bones, from cadavers in operating theatres, and employed freeze-drying to store bones. These helped to increase the availability of allogenic bones. (Hyatt, 1950; Hyatt *et al.*, 1952; Tomford, 2000).

Many other banks were also founded in Europe during this period. In 1952, Rudolph Klen set up the Hradec Kralove Tissue Bank in the old Czechoslovakia. In the UK, the Leeds Tissue Bank was founded in the city of Leeds, Yorkshire, in 1955 with Frank Dexter managing the day to day running of the bank (Kearney, 2006). In 1956, the Central German Tissue Bank was set up in Charité University Hospital in Berlin, the old East Germany. The bank has since been re-established as the German Institute for Cell and Tissue Replacement in 1994 by Rudiger von Versen. In Athens, Greece, the Demokritos Human Tissue bank was set up by Nicholas Triantafyllou (Phillips, 1998).

In Poland, the Central Tissue Bank was set up at the Medical University of Warsaw in 1963. Under the leadership of Janus Komender and Kazimierz Ostrowski, followed by Anna Dziedzic-Goclawska, the tissue bank went on to become the oldest tissue bank in the world to employ radiation technology in the sterilisation of tissue grafts.

In 1965, the activities in Leeds Tissue Bank were shifted following a decision to centralise all tissue banking activity within the Yorkshire region. The Yorkshire Regional Tissue Bank was thus established at the Pinderfields Hospital in Wakefield, with Frank Dexter appointed as the head of the bank (Kearney, 2006).

Allograft Immunogenicity

Since the 1950s, there has been much research into the antigenicity of the various types of allograft bones. Herndon and Chase noted that freezing allogenic bones could reduce immune response to it (Chase and Herndon, 1955; Hubble, 2001) Curtiss made similar observations in his animal studies and agreed that freezing cadaveric bones would reduce the immune response and hence also reduce the rejection rate (Curtiss *et al.*, 1959; Mankin *et al.*, 2005).

In the 1960s, Geoffrey Burwell, an orthopaedic surgeon in Leeds, embarked on a series of experiments on bone transplantation which led to his discoveries in allograft immunogenicity. He showed that the bone marrow was responsible for the immune response to fresh allogenic bone and that frozen bones performed better compared to fresh allogenic bones (Phillips, 2008). More significantly, his experiments shed light on the science of bone preservation and led to the development of the protocol for bone preservation we use today (Burwell, 1962; Burwell, 1963; Burwell, 1964a; Burwell, 1964b; Burwell, 1965; Burwell, 1966; Burwell and Gowland, 1961; Burwell and Gowland, 1962; Burwell *et al.*, 1963; Kearney, 2006).

Many other investigators sought to understand the immunology behind allografts and most reached the same conclusion, as summarised here by Friedlaender: fresh bone allografts were the most immunogenic; freeze-drying dramatically reduced the immune response and frozen bones had an immunogenicity that was between that of the former two (Friedlaender, 1976; Friedlaender *et al.*, 1983; Burwell, 1994).

The growth in knowledge of allograft immunogenicity proved to be instrumental in influencing the decline in the use of fresh bone allograft and the development of freezing and freeze-drying techniques for use in bone banks.

The Discovery of the BMP

The journey that led to the discovery of bone morphogenetic protein (BMP) started in 1965, when Marshall Urist made the discovery that demineralised bone matrices stimulated ectopic bone induction in his experiment using rats (Urist, 1965; Glowacki, 1992). The discovery prompted Urist to study the mechanism behind the stimuli, and finally led to his identification of BMP as the protein responsible for bone induction activity in 1971 (Urist and Strates, 1971; Meikle, 2007).

The significance of this discovery lies in the field of possibilities BMP opens up for the study of bone development through all stages, from beginning to the end. BMP can be manipulated to initiate bone development (Urist, 1994). However, the amino acid sequence of BMP needed identification before BMP can be easily manipulated. It was in the late 1980s that the purification and sequencing of bovine BMP and human BMP was finally achieved (Wozney *et al.*, 1988; Luyten *et al.*, 1989; Wozney, 1992; Chen *et al.*, 2004). Since then, 20 BMPs have been identified. The clinical applications of BMP have also been widely studied.

Further Developments in the Use of Bone Allograft

Following his classic work on massive bone allograft transplantation in 1908, Lexer described in 1925 the results of using fresh cadaveric tissue on 11 half joints and 23 whole joints and reported a reasonable success rate (Lexer, 1925; Mankin *et al.*, 2005).

Operations using massive bone allograft were far and few between until the 1970s, when Volkov in Russia expanded on Lexer's work with a large series on transplanting whole joints and articular surfaces (Volkov, 1970). At around the same time, Carlos Ottolenghi in Buenos Aires, Argentina, reported his use of deep frozen allografts for mostly bone tumours (Ottolenghi, 1966). Together with his successor, Luis Muscolo, he later presented the results of a series of long-term follow-ups on massive bone allografts transplanted (Ottolenghi *et al.*, 1982). In Houston, Texas, Frank Parrish, upon learning of Volkov's reported success, performed a series of experiments on the use of frozen massive osteoarticular allografts in reconstructing defects following the removal of tumour (Parrish, 1973; Enneking, 2005). The efforts of these surgeons across the region inspired other groups to look towards bone allograft as an alternative to metallic implants. One of these group, led by Henry Mankin at the Massachusetts General Hospital, reported the largest clinical series on the use of bone allografts in bone tumour surgery and showed success in three quarters of the patients (Mankin *et al.*, 1976; Tomford, 2000).

With the development of chemotherapy, imaging and surgical reconstruction techniques, the use of allograft in limb-sparing procedures became adopted in the early 1980s (Enneking, 2005). Since then, the procedure has offered patients with bone tumours a much-needed alternative to amputation.

Development of the Living Bone

While the use of massive bone allografts gained impetus in the 1980s, it was during this same period that the problems with allografts were reported. Enneking, with the help of Burchardt, performed a series of experiments to investigate the fate of large bone allografts and found that the allografts were often poorly incorporated into the host bone and were vulnerable to infection and fatigue fractures (Enneking *et al.*, 1975; Burchardt *et al.*, 1983; Tomford, 2000). Solutions were needed to overcome the histoincompatibility of bone allografts and to improve allograft incorporation. One such solution came in the form of a living bone.

The possibility of a living bone was first raised by Curtis in 1893, when he described the ideal "living bone which will exactly fill the gap and will continue to live without absorption" (Curtis, 1893; De Boer, 1988). This ideal living bone inched towards reality with the birth of vascular surgery through the work of Alexis Carrel (Carrel, 1908), then with further developments in microsurgery in the 1960s. In 1975, the first free vascularised graft was performed by Ostrup (Ostrup and Frederickson, 1974; De Boer, 1988). Following that, the technique of revascularising bone with anastamosis became adopted in transplants for treating bone tumours in the 1980s (Weiland *et al.*, 1983; Wood and Cooney, 1984; Wood *et al.*, 1984; Wood *et al.*, 1985; De Boer, 1988). During the same period, the technique of transplanting both allograft and vascularised graft was also introduced to improve the success of transplantation (Gross *et al.*, 1984; Bieber and Wood, 1986; De Boer, 1988). The living bone proved successful in reducing graft fractures, nonunion between the graft and host bone, resorption of bone and infection (De Boer, 1988).

The Issue of Disease Transmission

The first case of AIDS surfaced in 1981 (Centers for Disease Control, 1981; Tomford, 2000) and the first case of HIV-1 transmission in bones was reported in 1984, followed by a second case in 1985 (Centers for Disease Control, 1988; Tomford, 2000). The transmission of disease came at a time when proper screening methods were yet to be developed. Two cases of hepatitis C transmission were also reported in the 1990s, the second case occurring despite the existence of a first-generation screening test (Eggen and Nordbo, 1992; Conrad *et al.*, 1995; Tomford, 2000).

The occurrence of disease transmission prompted the development of proper donor-screening and bone preparation and processing methods. When Kenneth Sell became the head of the US Navy Tissue Bank, he started a programme in which a group of fellows researched on graft technology. Many prominent investigators, including Andrew Bassett, Gary Friedlaender, Theodore Malinin, William Tomford, and Michael Strong, were recruited. The findings of the programme helped immensely in advancing the knowledge in ensuring the safety of allografts and the prevention of disease transmission in the field of tissue banking (Lord *et al.*, 1988; Buck *et al.*, 1989; Tomford *et al.*, 1990; Strong *et al.*, 1991; Lietman *et al.*, 2000; Quinn *et al.*, 2001; Mankin *et al.*, 2005).

Developments in Tissue Banking

With the development in the use of bone and other allografts, it became imperative to move away from the "cottage industry" model that had continued well into the latter half of the twentieth century. Not only was the practice unreliable for preservation of bone, it was unsafe for surgeons to simply retrieve the bones they stored in their private freezers for transplantation purposes. However, there were exceptions to the model, of which includes the US Navy Tissue Bank (1949). From those first yawnings of activity, tissue banks began its slow sprouting across the world.

In the United States, the Boston group, led by Henry Mankin, researched extensively on tissue banking. It seems as though there was a gestation period, which in due course exploded. In 1976, in a meeting of a group of 26 individuals with backgrounds ranging from the sciences to the clinical to tissue banking, the American Association of Tissue Banks (AATB) was established (Phillips, 1998). The AATB was to "facilitate the provision of transplantable tissues of uniform high quality in quantities sufficient to meet national needs". Kenneth Sell was elected as the first president of AATB in 1977 (Joyce, 2000). In 1990, there were 30 tissue banks in the United States (Phillips, 2008).

The situation was similar in Europe. There had been a period during which tissue banks had very little contact with one another. Europe was divided into East and West. The East, with its more liberal tissue donation laws, had generally progressed faster than the West. In the UK, for example, while Geoffrey Burwell collaborated with Frank Dexter and supplied bones to orthopaedic surgeons in the North of England, there was no movement to extend the progress to the remainder of the country. The principal reason was that orthopaedic surgeons had not been convinced of the value of allografts for their procedures. Then in 1991, an International Conference in Tissue Banking was held in Berlin, and only then did the various centres realise the full extent of participation. The following year, in June 1992, the European Association of Tissue Banks (EATB) was set up in Marseilles, France, with Rudiger von Versen elected as the first president (Phillips, 1998).

In the Asia-Pacific region, tissue banking began in a few countries as early as the 1980s. In 1981, the Burma Tissue Bank was set up by Dr U Pe Khin at the Rangoon Orthopaedic Hospital in Rangoon, Burma. In 1984, Dr Yongyudh Vajaradul set up the Bangkok Biomaterial Centre at Siriraj Hospital in Bangkok, Thailand. Tissue banks sprouted across the Asia-Pacific region over the next few years. In October 1988, the Asia Pacific Association of Surgical Tissue Banks (APASTB) was established in Bangkok, Thailand, with Vajaradul as its first president and Aziz Nather as its first vice president (Nather *et al.*, 2005).

A New Beginning

Five associations have now come together once every three years to run a World Congress. They include the AATB, EATB, APASTB, ATBF (Australasian Tissue Banking Forum) and ALABAT (Latin America

Association of Tissue Banks). At the Fifth World Congress, held in Kuala Lumpur, Malaysia from 2–6 June 2008, a World Union of Tissue Banks was set up on 4 June 2008 with representatives from all five associations. The Sixth World Congress will be held in 2011 in Portugal by the EATB.

References

Albee FH (1912). Discussion of preservation of tissues and application in surgery by Alexis Carrel. *J Am Med Assoc* **59**: 527–536.

Alldredge RH (1948). In discussion after the paper of Henry (1948).

Ang GC (2005). History of skin transplantation. *Clin Dermatol* **25**: 320–324.

Axhausen G (1907). Histologische untersuchungen uber knochentransplantation am menschen. *Duetsche Zeitschr F Chir* **91**: 388–428.

Axhausen G (1909a). Die histologischen und klinischen gesetze der freien osteoplastik auf grund von thierversuchen. *Arch F Klin Chir* **88**: 23–145.

Axhausen G (1909b). Zur frage der freien osteoplastik. *Zentralbl Chir* **36** (Beilage: 133–134).

Axhausen G (1909c). Ueber den vorgang partieller sequestrirung transplantierten knochengewebes nebst neuen histologischen untersuchungen uber knochentransplantation am menschen. *Arch F Klin Chir* **89**: 281–302.

Barth A (1893). Ueber histologische befunde nach knochenimplantationen. *Arch F Klin Chir* **46**: 409–417.

Bassett CAL (1962). Current concepts of bone formation. *J Bone Joint Surg* **44A**: 1217–1244.

Bauer H (1910). Ueber knochentransplantation. *Zentralbl Chir* **37**: 20–21.

Bauer KH (1927). Homoiotransplantation von epidermis bei eineiigen zwillingen. *Beitr Klin Chir* **141**: 442–446.

Belchier J (1736a). An account of the bones of animals being changed to a red colour by aliment only. *Philos Trans R Soc* **39**: 287–288.

Belchier J (1736b). A further account of the bones of animals being made red by aliment only. *Philos Trans R Soc* **39**: 299–300.

Ben-hur N and Converse JM (1980). The impact of plastic surgery on transplantation from skin graft to microsurgery. *Transplant Proc* **12**: 616–620.

Bieber EJ and Wood MB (1986). Bone reconstruction. *Clin Plast Surg* **13**: 645–655.

Bolgiani A and Benaim F (1993). Primer banco de piel de argentina, organización y funcionamiento. *Actualización en el tratamiento de las quemaduras. VI Congresso de Quemaduras*, Argentina, pp. 90–99.

Bone-grafting and bone-transplantation (1918, February 16). *Lancet*: 264–265.

Bonfiglio M, Jeter WS, and Smith CL (1955). The immune concept: its relation to bone transplantation. *Ann NY Acad Sci* **59**: 417–433.

Bourroul SC, Pino ES, Herson MR, and Mathor MB (2002). Study of irradiated skin allografts and its use on burnt patients. In *American Nuclear Energy Symposium* 2002, Miami, FL.

Bradley JA and Hamilton DNH (2001). Organ transplantation: an historical perspective. In Hakim NS and Danovitch GM (eds.), *Transplantation Surgery,* Springer, London, pp. 1–22.

Brychta R, Adler J, Rihovd R, Suchdnek V, and Komdrkovd J (1994). Cultured skin cells for treatment of burns. *Ann Medit Burns Club* **71**: 206–208.

Buck RE, Malinin TI, and Brown MD (1989). Bone transplantation and human immunodeficiency virus. An estimate of risk of acquired immunodeficiency syndrome (AIDS). *Clin Orthop* **240**: 129–136.

Burchardt H, Glowczewski FP, and Enneking WF (1983). The effect of adriamycin and methotrexate on the repair of segmental cortical autografts in dogs. *J Bone Joint Surg* **65A**: 103–108.

Burwell RG (1962). Studies in the transplantation of bone. IV. The immune response to lymph nodes draining second-set homografts of fresh cancellous bone. *J Bone Joint Surg* **44B**: 688–710.

Burwell RG (1963). Studies in the transplantation of bone. V. The capacity of fresh and treated homografts of bone to evoke transplantation immunity. *J Bone Joint Surg Br* **45B**: 386–401.

Burwell RG (1964a). Studies in the transplantation of bone. VII. The fresh composite homograft-autografts of cancellous bone. An analysis of factors leading to osteogenesis in marrow transplants and in marrow containing bone grafts. *J Bone Joint Surg* **46B**: 110–140.

Burwell RG (1964b). Biological mechanisms in foreign bone transplantation. In Clark JMP (ed.), *Modern Trends in Orthopaedics 4, Science of Fractures*, Butterworths, London, pp. 138–190.

Burwell RG (1965). Osteogenesis in cancellous bone grafts considered in terms of cellular changes, basic mechanisms and the perspective of growth-control and its possible aberrations. *Clin Orthop* **40**: 35–47.

Burwell RG (1966). Studies in the transplantation of bone. VIII. Treated composite homograft-autografts of cancellous bone: an analysis of inductive mechanisms in bone transplantation. *J Bone Joint Surg* **48B**: 532–566.

Burwell RG (1994). History of bone grafting and bone substitutes with special reference to osteogenic induction. In Urist MR, O'Connor BT and Burwell RG (eds.), *Bone Grafts, Derivatives and Substitute,* Butterworth-Heinemann, London, pp. 3–102.

Burwell RG and Gowland G (1961). Studies in the transplantation of bone. II. The changes occurring the lymphoid tissue after homografts and autografts of fresh cancellous bone. *J Bone Joint Surg* **43B**: 820–843.

Burwell RG and Gowland G (1962). Studies in the transplantation of bone. III. The immune responses of lymph nodes draining components of fresh homologous cancellous bone and homologous bone treated by different methods. *J Bone Joint Surg* **45B**: 131–148.

Burwell RG, Gowland G, and Dexter F (1963). Studies in the transplantation of bone. VI. Further observations concerning the antigenicity of homologous cortical and cancellous bone. *J Bone Joint Surg* **45B**: 597–608.

Bush LF (1947). The use of homogenous bone grafts. *J Bone Joint Surg* **29**: 620–628.

Carrel A (1908). Results of the transplantation of blood vessels, organ and limbs. *J Am Med Assoc* **51**: 1662–1667.

Centers for Disease Control (1981). Pneumocystis pneumonia — Los Angeles. *Morb Mortal Wkly Rep* **30**: 250–252.

Centers for Disease Control (1988). Transmission of HIV through bone transplantation: case report and public health recommendations. *Morb Mortal Wkly Rep* **37**: 597–599.

Chase SW and Herndon CH (1955). The fate of autogenous and homogenous bone grafts. A historical review. *J Bone Joint Surg Am* 37A: 809–841.

Chen D, Zhao M, and Mundy GR (2004). Bone morphogenetic proteins. *Growth Factors* **22**(4): 233–241.

Cochrane T (1968). The low temperature storage of skin: a preliminary report. *Br J Plast Surg* **21**: 118–125.

Conrad EU, Gretch DR, and Obermeyei KR (1995). Transmission of the hepatitis C virus by tissue transplantation. *J Bone Joint Surg* **77A**: 214–224.

Curtis BF (1893). Cases of bone implantation and transplantation for cyst of tibia, osteomyelitis cavities and ununited fractures. *Am J Med Sci* **106**: 30–43.

Curtiss PH, Powell AE, and Herndon CH (1959). Immunological factors in homogeneous bone transplantation. III. The inability of homogeneous rabbit bone to induce circulating antibodies in rabbits. *J Bone Joint Surg* **41A**: 1482–1488.

Davis JS (1941). The story of plastic surgery. *Ann Surg* **113**: 641–656.

De Boer HH (1988). The history of bone grafts. *Clin Orthop* **226**: 293–298.

Donati D, Zolezzi C, Tomba P, and Vigano A (2007). Bone grafting: historical and conceptual review, starting with an old manuscript by Vittorio Putti. *Acta Orthop* **78**(1): 19–25.

Duhamel HL (1739). Sur une racine qui a la faculte de triendre en rouge les os des animaux vivants. *Mem Acad Roy des Sciences* Paris **52**: 1–13.

Duhamel HL (1742). Sur le developpment et la crue des os des animaux. *Mem Acad Roy des Sciences* Paris **55**: 354–370.

Duhamel HL (1743). Quatrieme memoire sur les os dans lequel on se propose de rapporter de nouvelles preuves qui etablissent que les os croissent en grosseur par l'addition de couches osseuses qui tirent leur origine du perioste. *Communication a L'Acad Roy des Sciences* Paris **56**: 87–111.

Eade GG (1958). Relationship between granulation tissue, bacterial and skin grafts in burned patients. *Plast Reconstr Surg* **22**: 42–55.

Eggen BM and Nordbo SA (1992). Transmission of HCV by organ transplantation (letter). *N Engl J Med* **326**: 411.

Enneking WF (2005). Transplanting allografts. *J Am Coll Surg* **201**(1): 5–6.

Enneking WF, Burchardt H, Puhl JL, and Piotrowski G (1975). Physical and biological aspects of repair in dog cortical-bone transplants. *J Bone Joint Surg* **57A**: 237–246.

Flourens P (1842). Recherches sur le developpement des os et des dents. *Annales du Museum d'Histoire Naturelle*, Paris.

Freshwater MF and Krizek TJ (1971). Skin grafting of burns: a centennial. A tribute to George David Pollock. *J Trauma* **11**: 862–865.

Freshwater MF and Krizek TJ (1978). George David Pollock and the development of skin grafting. *Ann Plast Surg* **1**: 96–102.

Friedlaender GE (1976). The antigenicity of preserved allografts. *Transplant Proc* **8** (suppl 1): 195–200.

Friedlaender GE, Mankin HJ, and Langer F (1983). Immunology of osteochondral allografts: background and considerations. In Friedlaender GE, Mankin HJ, and KW Sell (eds.), *Osteochondral Allografts. Biology, Banking, and Clinical Applications*, Little, Brown & Co., Boston, pp. 133–140.

Gibson T and Medawar PB (1943). The fate of skin homografts in man. *J Anat* **11**: 299–309.

Girdner JH (1881). Skin-grafting with grafts taken from the dead subject. *Med Record NY* **20**: 119–120.

Glowacki J (1992). Tissue-response to bone-derived implants. In Habal MB and Reddi AH (eds.), *Bone Grafts & Bone Substitutes,* Saunders, Philadelphia, pp. 84–92.

Gross AE, McKee N, Farine I, Czitrom A, and Langer F (1984). Reconstruction of skeletal defects following en bloc excision of bone tumours. In Uhthoff HK and Stahl E (eds.), *Current Concepts of Diagnosis and Treatment of Bone and Soft Tissue Tumours,* Springer-Verlag, Berlin, pp. 163–174.

Groves EWH (1917). Methods and results of transplantation of bone in the repair of defects caused by injury or disease. *Br J Surg* **5**: 185–242.

Hakim NS (1997). History of transplantation. In Hakim NS (ed.), *Introduction to Organ Transplantation,* Imperial College Press, London, pp. 1–14.

Hauben DJ (1985). The history of free skin transplant operations. *Acta Chir Plast* **27**: 66–70.

Heine B (1836). Uber die wiedererzeugung neuer knochenmassen und bildung neuer knochen. *J Chir Augenheilk* **24**: 513–527.

Herman AR (2002). The history of skin grafts. *J Drugs Dermatol* **3**: 298–301.

Holman E (1924). Protein sensitization in isoskingrafting. Is the latter of practical value? *Surg Gynecol Obstet* **38**: 100–106.

Hubble MJW (2001). Bone transplantation. *Cur Orthop* **15**: 199–205.

Hyatt GW (1950). Fundamentals in the use and preservation of homogenous bone. *US Armed Forces Med J* **1**: 841–852.

Hyatt GW, Turner TC, Bassett CAL, Pate JW, and Sawyer PN (1952). New methods for preserving bone, skin and blood vessels. *Postgrad Med* **12**: 239–254.

Inclan A (1942). The use of preserved bone graft in orthopaedic surgery. *J Bone Joint Surg* **24**: 81–96.

Jones AR (1952). Sir William Macewen. *J Bone Joint Surg Br* **34B**(1): 123–128.

Joyce MJ (2000). American Association of Tissue Banks: a historical reflection upon entering the 21st century. *Cell Tissue Bank* **1**: 5–8.

Judet H (1909). La greffe des articulations. *Rev de chirurg*, Paris **40**: 1.

Julien P, Ledermann F, and Touwaide A (1993). *Cosma e damiano.* Antea Edizioni, Milano.

Kearney JN (2006). Yorkshire regional tissue bank — circa 50 years of tissue banking. *Cell Tissue Bank* **7**(4): 259–264.

Kirsner RS, Falanga V, and Eaglstein WH (1993). Biology of skin grafts: grafts as pharmacologic agents. *Arch Dermatol* **39**(6): 1007–1010.

Koch SL (1941). The transplantation of skin and subcutaneous tissue to the hand. *Surg Gynecol Obstet* **72**: 1–13.

Lexer E (1908a). Ueber glenktransplantation. *Med Klin* **4**: 817.

Lexer E (1908b). Die verwendung der freien knochenplastik nebst versuchen uber gelenkversteifung und gelenktransplantation. *Arch Klin Chir* **86**: 939–954.

Lexer (1925). Joint transplantations and arthroplasty. *Surg Gynecol Obstet* **40**: 782–809.

Lietman SA, Tomford WW, Gebhardt MC, Springfield DS, and Mankin HJ (2000). Complications of irradiated allografts in orthopaedic tumour surgery. *Clin Orthop* **375**: 214–217.

Lord CF, Gebhardt MC, Tomford WW, and Mankin HJ (1988). The incidence, nature and treatment of allograft infections. *J Bone Joint Surg* **70A**: 369–376.

Luyten FP, Cunningham NS, Ma S, Muthukumaran N, Hammonds RG, Nevins WB, Woods WI, and Reddi AH (1989). Purification and partial amino acid sequence of osteogenin, a protein initiating bone differentiation. *J Biol Chem* **264**: 13377–13380.

Lytton B (2005). The early history of kidney transplantation at Yale (1967–1985): a personal memoir. *Yale J Biol Med* **78**: 171–182.

Macewen W (1881). Observations concerning transplantation on bone. *Proc R Soc Lond* **32**: 232.

Macewen W (1912). *The Growth of Bone. Observations on Osteogenesis. The Experimental Inquiry into the Development and Reproduction of Diaphyseal Bone.* James Maclehose and Sons, Glasgow.

Mankin HJ, Vogelson FS, and Thrasher AZ (1976). Massive resection and allograft replacement in the treatment of malignant bone tumors. *N Engl J Med* **294**: 1247–1255.

Mankin HJ (2002). History of treatment of musculoskeletal tumors. In Klenerman L (ed.), *The Evolution of Orthopaedic Surgery*, Royal Society of Medicine Press, London, pp. 191–210.

Mankin HJ, Hornicel FJ, Gebhardt MC, and Tomford WW (2005). Bone allograft transplantation: theory and practice. In Lieberman R and Friedlander GE (eds.), *Bone Regeneration and Repair: Biology and Clinical Applications*, Humana Press, Totowa NJ, pp. 241–261.

Matouskova E, Broz L, Pokorna E, and Konigova R (2002). Prevention of burn wound conversion by allogeneic keratinocytes cultured on acellular xenodermis. *Cell Tissue Bank* **3**: 29–35.

Meikle MC (2007). On the transplantation, regeneration and induction of bone: the path to bone morphogenetic proteins and other skeletal growth factors. *Surgeon* **5**(4): 232–244.

Murphy JB (1913). Osteoplasty. *Surg Gynecol Obstet* **16**: 493–536.

Nather A, Ong HJC, Feng MCB, and Aziz Z (2005). Asia-Pacific Association of Surgical Tissue Banking — past, present and future. *ASEAN Orthop Assoc* **17**(1): 17–19.

Nichter LS, Morgan RF, and Nichter MA (1983). The impact of Indian methods for total nasal reconstruction. *Clin Plast Surg* **10**: 635–647.

Obeng MK, McCauley RL, Barnett JR, Heggers JP, Sheridan K, and Schutzler SS (2001). Cadaveric allograft discards as a result of positive skin cultures. *Burns* **27**: 267–271.

O'Connor NE, Mulliken JB, Banks-Schegel S, Kehinde O, and Green H (1981). Grafting of burns with cultured epithelium prepared from autologous epidermal cells. *Lancet* **1**: 75–78.

O'Donoghue MN and Zarem HA (1971). Stimulation of neovascularization — comparative efficiency of fresh and preserved skin grafts. *Plast Reconstr Surg* **48**: 474–477.

Ollier L (1858). De la production artificielle des os au moyen de la transplantation de perioste et des greffes osseux. *Comp Rend Soc de Biol* **5**: 145.

Ollier L (1867). Traite experimental et clinique de la regeneration des os et de la production artificielles du tissue osseux. *Victor Mason et Fils*, Paris.

Ollier L (1872). Sue les greffes cutanees ou autoplastiques. *Bull Acad Med Paris* **2**: 243.

Ostrup LT and Frederickson JM (1974). Distant transfer of a free living bone graft by microvascular anastomoses: an experimental study. *Plast Reconstr Surg* **54**: 274–285.

Ottolenghi CE (1966). Massive osteoarticular bone grafts. *J Bone Joint Surg* **48B**: 646–659.

Ottolenghi CE, Muscolo DL, and Maenza R (1982). Bone defect reconstruction by massive allograft: technique and results of 51 cases followed for 5 to 32 years.

In Straub LR and Wilson PD (eds.), *Clinical Trends in Orthopaedics*, Thieme-Stratton, New York, pp. 171–182.

Parrish FF (1973). Allograft replacement of all or part of the end of a long bone following excision of a tumor. *J Bone Joint Surg* **55A**: 1–22.

Phemister DB (1914). The fate of transplanted bone and regenerative power of its various constituents. *Surg Gynecol Obstet* **19**: 303–333.

Phillips GO (Coord. Ed.) (1998). *Multi-media Distance Learning Package on Tissue Banking. Module 0: Historical Background.* National University of Singapore, IAEA/NUS Regional Training Centre (RCA), IAEA/NUS Interregional Training Centre, Singapore.

Phillips GO (2008, June). The Hudson Silva lecture: global and future of tissue banking. In *5th World Congress on Tissue Banking*, Kuala Lumpur, Malaysia.

Polge C, Smith AU, and Parkes AS (1949). Revival of spermatozoa after vitrification and dehydration at low temperature. *Nature* **164**: 666.

Pollock GD (1871). Cases of skin grafting and skin transplantation. *Trans Clin Soc Lond* **4**: 37.

Quinn RH, Manking HJ, Springfield DS, and Gebhardt MC (2001). Management of infected bulk allografts with antibiotic-impregnated polymethylmethacrylate spacers. *Orthopedics* **24**: 971–975.

Rinaldi E (1987). The first homoplastic limb transplant according to the legend of saint Cosmas and saint Damian. *Ital J Orthop Traumatol* **13**(3): 393–406.

Reverdin JL (1869). Greffe epidermique, experience faite dans le service de M le docteur Guyon, a l'hopital necker. *Bull Imp Soc Chir Paris* **10**: 511–515.

Rheinwald J and Green H (1975). Serial cultivation of strains of human epidermal keratinocytes: formation of keratinizing colonies from single cells. *Cell* **6**: 331–344.

Strong DM, Sayers MH, and Conrad EU (1991). Screening tissue donors for infectious markers. In Friedlaender GE and Goldberg VM (eds.), *Bone and Cartilage Allografts*, American Academy of Orthopaedic Surgeons, Park Ridge IL, pp. 193–209.

Sushruta (1907). *Sushruta samhita* (K Bhishagratna, Trans.). Bose, Calcutta. (Original work published 600 BC.)

Tilney NL (2000). Transplantation and its biology: from fantasy to routine. *J Appl Physiol* **89**: 1681–1689.

Tomford WW (1994). A perspective in bone banking in the United States. In Urist MR, O'Connor BT, and Burwell RG (eds.), *Bone Grafts, Derivatives and Substitute,* Butterworth-Heinemann, London, pp. 193–195.

Tomford WW (2000). Bone allografts: past, present and future. *Cell Tissue Bank* **1**: 105–109.

Tomford WW, Thongphasuk J, Mankin HJ, and Ferraro MJ (1990). Musculoskeletal allografts. A study of the clinical incidence and causes of infection associated with their use. *J Bone Joint Surg* **72A**: 1137–1143.

Trier WC and Sell KW (1968). United states navy skin bank. *Plast Reconstr Surg* **48**: 543–548.

Trout HH (1915). Spina bifida, tibial transplant, father to child. *Surg Gynecol Obstet* **22**: 523.

Tuffier T (1911). Des greffes de cartilage et d'os humain dans les resections articulaires. *Bull et Mem Soc de Chir de Paris* **37**: 278.

Urist MR (1965). Bone: formation by autoinduction. *Science* **150**: 893–899.

Urist MR and Strates BS (1971). Bone morphogenetic protein. *J Dent Res* **50**: 1392–1406.

Urist MR (1994). The search for and the discovery of bone morphogenetic protein (BMP). In Urist MR, O'Connor BT and Burwell RG (eds.), *Bone Grafts, Derivatives and Substitute,* Butterworth-Heinemann, London, pp. 315–362.

Volkov M (1970). Allotransplantation of joints. *J Bone Joint Surg* **52B**: 49–53.

Von Haller A (1763). Experimenta de ossium formatione. *Opera minora*, Lausanne.

Webster JP (1944). Refrigerated skin grafts. *Ann Surg* **120**: 431–448.

Weiland AJ, Moore JR, and Daniel RK (1983). Vascularized bone autografts: experience with 41 cases. *Clin Orthop* **174**: 87–95.

Wilson PD (1947). Experiences with a bone bank. *Ann Surg* **126**: 932–946.

Wood MB and Cooney WP (1984). Vascularized bone segment transfers for management of chronic osteomyelitis. *Orthop Clin North Am* **15**: 461–472.

Wood MB, Cooney WP, and Irons GB (1984). Posttraumatic lower extremity reconstruction by vascularized bone graft transfer. *Orthopedics* **7**: 255–262.

Wood MB, Cooney WP, and Irons GB (1985). Skeletal reconstruction by vascularized bone transfer: indications and results. *Mayo Clin Proc* **60**: 729–734.

Wozney JM, Rosen V, Celeste AJ, Mitsock LM, Whitters MJ, Kriz RW, Hewick RM, and Wang EA (1988). Novel regulators of bone formation: molecular clones and activities. *Science* **242**: 1528–1534.

Wozney JM (1992). The bone morphogenetic protein family and osteogenesis. *Mol Reprod Dev* **32**: 160–167.

Chapter 2

Tissue Banking in the Asia-Pacific Region: Current Trends and Future Directions

Aziz Nather, Yi Lin Sim and Shushan Zheng

Introduction

While tissue banking is well established in USA, Europe and Canada, tissue transplantation gained impetus in the Asia-Pacific region only in the last decade (Nather, 1999a; Nather, 1999b). New tissue banks have been set up in countries such as Korea, Indonesia, Malaysia and India. This development runs in parallel with the increasing demand for tissue allografts for transplantation in the region in the last few years.

Two major driving forces have been responsible for the development of tissue banking in the Asia-Pacific region:

- Regional Cooperative Agreement (RCA) Project RAS 7/008: Radiation Sterilisation of Tissue Grafts (1985 to 2004) under the IAEA Programme on Tissue Banking, and
- the Asia Pacific Association of Surgical Tissue Banks (APASTB), which was established in October 1988 (Nather, 1999b; Nather *et al.*, 2005a)

RCA Project RAS 7/008: Radiation Sterilisation of Tissue Grafts

The Regional Cooperative Agreement (RCA) was an agreement among 16 member states in the Asia-Pacific region, namely Australia, Bangladesh,

NUH Tissue Bank, Department of Orthopaedic Surgery, Yong Loo Lin School of Medicine, National University of Singapore, Singapore.

China, India, Indonesia, Japan, Korea, Malaysia, Mongolia, Myanmar, Pakistan, the Philippines, Singapore, Sri Lanka, Thailand, and Vietnam.

The RCA Project RAS 7/008 (1985–2004) was responsible for setting up and developing 15 tissue banks in 13 member states, i.e. Bangladesh, China, India, Indonesia, Korea, Malaysia, Myanmar, Pakistan, the Philippines, Sri Lanka, Thailand, Vietnam, and Singapore. Financial support from IAEA was given to 12 of these member states (excluding Singapore) to provide the necessary equipment for establishing a National Tissue Bank in each state. Furthermore, fellowships and scientific visits were sponsored, and experts were sent to the National Tissue Bank in these countries.

IAEA Expert Missions

Numerous missions had been conducted by IAEA experts in the Asia-Pacific region, Latin America and Africa. The experts include: Dr Glyn Phillips (UK), Dr Rudiger von Versen (Germany), Dr Heinz Winkler (Austria), Dr Michael Strong (USA), Dr Aziz Nather (Singapore), Dr Norimah Yusof (Malaysia), Dr Nazly Hilmy (Indonesia), Dr Samuel Doppelt (USA), Dr Yongyudh Vajaradul (Thailand), Dr Martha Anderson (USA) and Dr Jan Koller (Slovakia). These missions were instrumental in the setting up of tissue banks and the development of tissue banking activities in several countries.

Dr Nather was contracted as an UN/IAEA expert to help set up tissue banks in 10 missions involving Malaysia (Kota Bahru and Kuala Lumpur), Vietnam (Ho Chi Minh City and Hanoi), Zambia, Myanmar, Argentina, Brazil, Cuba, Sri Lanka, and Korea. He was also invited to set up two tissue banks in Hong Kong by the Hong Kong Orthopaedic Association, namely at the Queen Mary Hospital and Prince of Wales Hospital in 1995; and to set up a bone bank in Kobe, Japan, by Professor Maruo Souji in 1996. By conducting such missions, the author has gained good insight into the local situation of these countries.

Myanmar

Tissue banking began in the Asia-Pacific region in Burma in 1981, when Dr U. Pe Khin, a Consultant Orthopaedic Surgeon, founded the Burma

Tissue Bank in Rangoon, Burma. Dr U. Pe Khin had been the Medical Superintendent at the Rangoon Orthopaedic Hospital when he became the Chairman of the Burma Tissue Bank. After his sudden death in 1987, Dr Myo Mint succeeded him in 1992 and reactivated the tissue bank project. However, his efforts were short-lived as he was transferred to Mandalay in 1995. Dr Myo Mint was succeeded by Dr Khin Maung Han.

There is no law for tissue donation in Myanmar, apart from the Eye Donation Law. Approximately 85% of the population are Buddhists, who favour tissue donation. Myanmar has a gamma irradiation facility, a gamma chamber belonging to the Department of Agriculture, that is available for use.

Thailand

Following the major setback suffered by Burma with the demise of its pioneer Dr U. Pe Khin, Thailand became the forerunner of the region. In December 1984, Dr Yongyudh Vajaradul, a Consultant Orthopaedic Surgeon, set up the Bangkok Biomaterial Centre at the Siriraj Hospital, Mahidol University, Thailand.

Like Myanmar, there is no tissue donation law in Thailand. Thailand is a predominantly Buddhist country favouring tissue donation. The radiation sterilisation of tissue grafts started in Thailand in 1986, when the bank acquired its own gamma chamber.

China

The China Institute for Radiation Protection (CIRP) Tissue Bank was set up in 1988 in Taiyuan, Shanxi Province, by Dr Sun Shiquan and Dr Li Youchen. This was quickly expanded to become the first provincial tissue bank in China, the Shanxi Provincial Tissue Bank, in July 1993. In 1994, it secured approval from the government to distribute its tissue grafts to Beijing and other cities in China. Although China does not have a human transplantation act, tissue banks in China follow the principle of obtaining written consent from donors before the procurement of tissues.

All tissue grafts are gamma-irradiated at the CIRP. The CIRP has been conducting national training courses for tissue bank operators in China for several years. In 2000, the Shanxi Provincial Tissue Bank was privatised by OsteoRad Biomaterial Co. Ltd. and fitted with modern up-to-date facilities. Dr Li Bao Xing became its Executive Director.

The number of tissue banks in China remains unknown. Indeed, China is a subcontinent in itself. With the modernisation of China, more tissue banks in China are being recognised. In time, China may exert a major influence on tissue banking in the Asia-Pacific region.

India

The late Dr N. M. Kavarana founded the Tata Memorial Hospital Tissue Bank in 1988 at the Tata Memorial Hospital, Mumbai, in collaboration with the IAEA. Tissue banking follows the Bombay Anatomy Act 1949, which covers the use of unclaimed or donated bodies for therapeutic purposes, medical education, or research; and the more recent Transplantation of Human Organs Act 1994, in which consent is required from the donor or next of kin. The irradiation of tissue grafts is performed in a Gamma Chamber 900 donated by the Atomic Energy Commission, Government of India. The current director of this tissue bank is Dr Astrid Lobo Gajiwala.

In 2006, there was much renewed interest to set up tissue banks in many parts of India, such as Chennai (Dr Mayil Natarajan), Coimbatore (Dr S Rajasekaran), New Delhi (Dr Surya Bahn), Kerala, Mumbai, Kolkata, etc. Dr Nather has also been invited to help set up tissue banks in these countries. Like China, India is a subcontinent that has recently sprung into activity. The development of tissue banking in India is also expected to have a considerable impact on the rest of the region.

Singapore

The National University of Singapore (NUS) Bone Bank was set up in October 1988 as a research tissue bank for the Department of Orthopaedic Surgery by Dr Nather using an NUS Research Grant, RP 880334 "Bridging of Large Bone Defects by Allografts" (Nather and Wang, 2002). The bank was pushed into clinical activity fairly quickly with the rapid rise in clinical demand for tissue allografts.

In 1994, the bank was awarded a S$239 965 grant by the Ministry of Education (Totalisator Board) to upgrade its clinical facilities and functions, and to start the production of lyophilised gamma-irradiated morsellised bone allografts. It acquired two sets of lyophiliser units (each lyophiliser unit included a band saw, a shaker bath, a lyophiliser, and a laminar flow cabinet) and two new freezers (in addition to its original two). That same year, the NUS Bone Bank became the national bone bank supplying bone and soft tissue allografts to all nine hospitals in the country.

Tissue donation follows the Medical (Therapy, Education and Research) Act of 1972, whereby any person of sound mind and 18 years of age or above may give all or any part of his or her body for education or transplantation. It is an "opting-in" law, requiring written consent from the donor or next of kin.

In September 1992, Singapore began irradiating deep-frozen long bones by sending the bones packed with dry ice in Polylite containers by air to the Malaysian Nuclear Agency (MINT) in Bangi, Selangor. They are irradiated in a cobalt-60 plant (Sinagama) by Dr Norimah Yusof. Since 1994, with the production of lyophilised, gamma-irradiated morsellised bones, these small grafts have been irradiated in a gamma chamber at the Department of Nuclear Medicine, Singapore General Hospital, by Dr Betty Xun Fei.

The National University Hospital (NUH) Tissue Bank was officially inaugurated in September 1995, and became a hospital cost facility in 1998. The bank has a wet processing laboratory, a dry processing laboratory, a documentation room, and a reception area with a total floor space of about 2000 square feet (Nather, 2004).

The bank also became the Regional Training Centre for the Asia-Pacific region in 1997 and the International Training Centre for tissue bank operators since 2002.

Indonesia

Indonesia's first surgical tissue bank, the BATAN Research Tissue Bank, at the Centre for Application of Isotopes and Radiation (CAIR), National Atomic Energy Agency (BATAN), was established by Dr Nazly Hilmy in Jakarta in 1990. Initially, due to the shortage of deceased donors, the bank processed both human and bovine bone grafts for clinical application in addition to processing human amnion. The Indonesia 1992 Health Regulation is an "opting-in" law that allows retrieval of tissues from living donors only.

In 1986, Dr Nazly Hilmy managed to enact a "Fatwa for Bone, Skin, and Amnion", thus permitting procurement from deceased donors as well. However, this religious breakthrough did not cause any significant change in the attitude of donors due to cultural factors. Hence, the shortage of deceased donors in Indonesia persists.

A second bone bank, the Dr M. Djamil Hospital Tissue Bank, has been set up in Padang, Sumatra, by Dr Menkher Manjas, a Consultant Orthopaedic Surgeon. In addition, the late Dr Abdurrahman, a Consultant Orthopaedic Surgeon and Musculoskeletal Oncologist, inaugurated the Dr Soetomo Hospital Tissue Bank in the Department of Orthopaedic Surgery, Air Langga University, Surabaya, during the 8th Scientific Meeting of the APASTB in Bali in 2000.

Dr Abdurrahman also set up the Indonesian Association of Tissue Banks and became its first president in 2001. Unfortunately, his demise due to an illness caused a major setback in the development of tissue banking in Surabaya. His leadership position at the Dr Soetomo Hospital Tissue Bank has since been succeeded by Dr Ferdiansyah, another Consultant Orthopaedic Surgeon and Musculoskeletal Oncologist.

Hong Kong

Two tissue banks have been set up at the two medical universities in Hong Kong. In 1990, Dr David Fang formalised a regional musculoskeletal tissue bank at the Queen Mary Hospital, University of Hong Kong. Dr T. L. Poon took over as Director in 1994. In 1992, Dr Shekhar Kumta formalised another regional musculoskeletal tissue bank at the Sir Y. K. Pao Centre for Cancer, Prince of Wales Hospital, Chinese University of Hong Kong.

Both banks are supported by the Hong Kong Government. Tissue banking in Hong Kong follows the Human Organ Transplantation Ordinance 1997, an "opting-in" law requiring consent from the donor or next of kin.

Korea

The Korea Biomaterial Research Institute was established in 1990 by Dr Chang Joon Yim, a Dental Surgeon, at the College of Dentistry, Dankook University, in Cheonan, Korea. The tissue grafts are irradiated by the Korea Atomic Energy Research Institute (KAERI) in Taejon, Korea.

Brain death was legally recognised in Korea in February 2000. Since then, there has been tremendous activity in setting up new tissue banks in Korea, notably at the St.Vincent's Hospital, Catholic Medical University, in Seoul, Korea, by Professor Yong Koo Kang. More than 20 new tissue banks have already been established.

Plans are underway to set up one or two regional tissue banks in Korea. Two private tissue banks, Bioland and Hans Biomed have also been set up with government approval.

Japan

The Kitasato University Hospital Bone Bank was set up in April 1991 by Dr Moritoshi Itoman. Tissue procurement is illegal in Japan, except for cornea as stated in the Law for Transplantation of Kidneys and Corneas 1979. Radiation sterilisation was introduced by Dr Itoman in 1994, with the cooperation of the Japan Atomic Energy Research Institute (JAERI).

Few organs have been procured despite the 1997 legislation recognising the concept of brain death. In a landmark event on March 1, 1999, a liver, heart, and both kidneys were procured from a brain-dead donor in Kochi and airflown for transplantation to four recipients, each in a different city (Reuters). While it was hoped that the event would spur tissue banking activity in Japan, this did not happen.

Urabe *et al.* (in press) noted that between the year 2000 and 2004, a total of 134 782 bone grafts have been performed in 2239 institution. Of this, 4% of the bone grafts utilised were banked bone grafts. Only 9% (≈200) of the 2239 institutions have their own bone banks. More importantly, only three regional bone banks could procure, process and preserve banked allografts from cadaveric donors and supply them to other institutions.

Progress has been made with the legislation of Advanced Medical Treatment in December 2004, which allows approved institutions to charge the expenses of authorised treatment directly to the patient. Prior to this legislation, bone banks had to bear all expenses, with one implantation costing 282 588 yen (1760 euro) (Urabe *et al.*, in press). In March 2007, cryopreserved allogenic tissue from cadaveric donors was approved as advanced medical treatment.

However, the number of institutional and regional bone banks in Japan remains low. Urabe *et al.* (in press) attributes this to the high expenditure borne by bone banks, and to the low number of regional banks, which is insufficient in supporting over 200 institutional banks.

The Philippines

In Manila, the University of Philippines General Hospital (UPGH) Tissue Bank was set up by the late Dr Norberto Agcaoili in May 1990 at the Department of Orthopaedic Surgery, University of Philippines College of Medicine. Tissue donation follows the Republic Act 7170, 1991, an "opting-in" law whereby the legacy or donation of all or part of a human body after death for specified purposes must be authorised by the donor or next of kin. While Filipinos are predominantly Catholic, their cultural attitudes do not favour donation. All tissues are gamma-irradiated at the Philippines Nuclear Research Institute (PNRI) in Manila.

Malaysia

In Malaysia, two tissue banks were set up in 1991:

1. The Malaysian National Tissue Bank by Dr Hasim Mohamad, a General Surgeon, at the University of Science Malaysia, Kota Bahru, Kelantan, and
2. The Malaysian Institute for Nuclear Technology Research (MINT) Tissue Bank by Dr Norimah Yusof at the Malaysian Institute for Nuclear Technology, Bangi, Selangor (later renamed as the Malaysian Nuclear Agency or MNA in 2006).

As with other countries, tissue banking in Malaysia follows an "opting-in" law, the Laws of Malaysia, Act 130, 1974.

In September 1995, the Malaysia Islamic Centre passed a "Fatwa on Bone, Skin, and Amnion" — a religious ruling allowing Muslims to donate. Despite this, there has been little change in the attitude of Muslims towards tissue donation. Hence, the shortage of donors in Malaysia persists.

Dr Hasim Mohamad helped inaugurate the Malaysian National Tissue Bank in Kota Bahru, Kelantan on November 5, 1994. A third tissue bank, the General Hospital Kuala Lumpur Bone Bank, was inaugurated by Dr Ruzlan in 1998 during the 7th Scientific Meeting of the APASTB held in Kuala Lumpur. Two more new tissue banks are being set up: one at the University of Malaya Medical Centre, Kuala Lumpur; and one at the International Islamic University of Malaysia, Kuantan. All tissue allografts are sent to Dr Norimah Yusof at the MNA for gamma irradiation.

Under the leadership of Dr Hasim Mohamad and Dr Norimah Yusof, the development of tissue banking in Malaysia has expanded tremendously. The Malaysian Association for Cell and Tissue Banking was established in 2005.

Vietnam

Dr Tran Bac Hai was a key person in driving the development of the Biomaterial Research Laboratory University Training Centre for Health Care Professionals in Ho Chi Minh City in January 1993. The bank procures amnion and lyophilised morsellised chip grafts. All the grafts are sent to Dalat for irradiation. During the same period, the Hanoi Tissue Bank — a skin bank — was set up at the Laboratory of Biomaterial Preparation (Vinatom) by Dr Pham Quang Ngoc. The bank processes skin and amnion grafts. All tissues are sent to the Hanoi Irradiation Centre for irradiation.

Tissue banking in Vietnam follows two laws: Article 32 of the Civil Code, Chapter 2, on tissue donation; and The People's Health Protection Code, Chapter 4, on tissue transplantation.

Sri Lanka

The Sri Lanka Model Human Tissue Bank was set up by the late Dr Hudson Silva and his wife. Dr Silva is renowned for having procured

43 000 corneas to be distributed to several countries in the world on humanitarian grounds. In 1993, an agreement was signed between the IAEA and the Ministry of Health in Sri Lanka, upon which the Government allotted a large piece of land for the development of the tissue bank. The bank was inaugurated on May 8, 1996, by the Prime Minister of Sri Lanka.

This project was conceptualised as a model project because there were many donors in this predominantly Buddhist country, and it was thought that they could supply the much-needed grafts to other countries in the region which have a great shortage of donors. Unfortunately, this plan did not succeed due to laws in the neighbouring countries that prevented tissues from crossing borders easily.

Sri Lanka follows an "opting-in" law, the Human Tissue Transplantation Act No. 48 of 1987, requiring consent from the donor or next of kin. The bank has its own gamma irradiation facility — a Gamma Cell 200 — in its own premises.

Currently, more professional education is needed in Sri Lanka to promote the utilisation of tissue grafts in the country.

Australia

Four tissue banks in Australia incorporate radiation processing for the terminal treatment of grafts by the Australian Nuclear Science and Technology Organisation (ANSTO). These include the following:

1. Queensland Bone Bank (led by Dr David Morgan),
2. Donor Tissue Bank of Victoria (led by Dr Lyn Ireland, succeeded by Dr Marissa Herson),
3. Perth Bone and Tissue Bank (led by Dr John Pearman), and
4. South Australia Tissue Bank (led by Dr Steven Nailer).

Tissue banks in Australia have to comply with the Australian Code of Good Manufacturing Practice for Therapeutic Goods — Human Tissues, September 1995. This Code adopts and applies basic quality system principles from the ISO 9000 series of standards to tissue banking.

The strength of tissue banking in Australia lies in the high standards these tissue banks are required to comply with, according to the Therapeutic Goods Administration (TGA) Act, to obtain yearly licensing. Annual audits of the banks are performed. In contrast, in the USA, while the American Association of Tissue Banks (AATB) has set very high standards, accreditation is not compulsory. In Europe, the European Council will issue a set of European Council Standards that all tissue banks in Europe must comply with. However, it will take some years before the implementation of common standards for the whole of Europe. Singapore follows the Australian system of compulsory auditing and licensing, with Singaporean auditors being trained by TGA authorities in Australia.

The New South Wales Bone Bank, a private tissue bank with two clean rooms has recently been developed in New South Wales at the cost of about A$1 million. The Queensland Bone Bank and Queensland Skin Bank costing A$12 million has also been set up in Brisbane by the Government of Queensland, with six clean rooms and state-of-the-art facilities. It is helmed by Dr David Morgan as Chairman and Ron Simard as Manager. Both of these banks will play a role in influencing tissue banking not only in Australia, but perhaps also in the Asia-Pacific region.

IAEA ERA: The "Golden Age" of Tissue Banking in the Asia-Pacific Region

The Tissue Banking Programme run by the IAEA ushered in the "golden age" of tissue banking in the Asia-Pacific region. Under the programme, the RCA Project RAS 7/008 was initiated in 1985 and ran for nearly 20 years until its completion in 2004. The programme has provided its 12 member states with experts and training for tissue bank operators, and successfully established a National Tissue Bank in each of the member states. Furthermore, IAEA has supported member states by sponsoring their attendance in the scientific meetings of the Asia Pacific Association of Surgical Tissue Banks (APASTB) from 1992 to 2002.

Post-IAEA ERA

The year 2004 was a challenging one for tissue banks in the Asia-Pacific region. The completion of the RCA Project RAS 7/008 meant the withdrawal of IAEA support to tissue banking activities in the region. As a result, some countries experienced difficulties with continuing their tissue banking efforts. However, a core group of countries, comprising Malaysia, Korea, Indonesia, Hong Kong, India, and Singapore, continued to be active in promoting tissue banking in the region and participating in the activities of the APASTB.

Future of Tissue Banking in the Asia-Pacific Region

Living in the Tissue Engineering Era

The emergence of tissue engineering in 2000 posed a strong challenge to tissue banking. For a while, hospital authorities and tissue engineers considered allograft transplantation as outmoded and expected the decline of tissue banks. Funding for tissue banks became scarce because tissue engineering was put on priority.

However, so far tissue banks have flourished despite the advances of tissue engineering. The failure to produce scaffolds capable of fulfilling both the biological and biomechanical functions of the tissues they are replacing is the major limiting factor to the progress of tissue engineering. Existing artificial scaffolds for bone — tricalcium phosphate, hydroxyapatite, and polycaprolactone — are only of cancellous strength (about 30 megapascals), but scaffolds of cortical strength (about 200 megapascals) are needed. In the absence of the latter, bone allografts (natural scaffolds) are currently the best scaffolds for tissue engineering. Until further improvement in technology occurs, natural scaffolds will remain the best scaffolds. The situation with scaffolds for ligaments is even worse, as poly(L-lactide) and polyglycolic acid are only of suture strength. Again, ligament allografts (natural scaffolds) are the best scaffolds at present.

The way forward may be to combine the use of bone and ligament allografts with adult mesenchymal stem cells (MSCs) and growth factors. Recent literature on the subject points in the same direction (Lucarelli *et al.*, 2005; Korda *et al.*, 2008 and Quarto *et al.*, 2001).

Nather observes that so far there has been no case report on reconstruction of large defects using allograft with MSCs. However, clinical studies are probably underway and clinical case studies of allograft with MSCs are expected to be reported soon.

In the near future, it is anticipated that the combination of MSCs using Good Manufacturing Practice (GMP) facilities with bone and ligament allograft as scaffolds would produce the best engineered tissues for massive tissue reconstruction — allograft engineering.

References

Korda M, Blunn G, Goodship A, and Hua J (2008). Use of mesenchymal stem cells to enhance bone formation around revision hip replacements. *J Orthop Res* **26**(6): 880–885.

Lucarelli E, Fini M, Beccheroni A, Giavaresi G *et al.* (2005). Stromal stem cells and platelet-rich plasma improve bone allograft integration. *Clin Orthop* **435**: 62–68.

Nather A (1999a). Tissue banking in the Asia Pacific region: current status and future developments. *J Orthop Surg* **7**(2): 89–93.

Nather A (1999b). Tissue banking in the Asia Pacific region — the Asia Pacific Association of Surgical Tissue Banking. In Phillips GO, Strong DM, von Versen R, and Nather A (eds.) (1999c), *Advances in Tissue Banking*, Vol. 3, World Scientific, Singapore, pp. 419–425.

Nather A (2000a). Tissue banking in Asia Pacific region — ethical, legal, religious, cultural and other regulatory aspects. *ASEAN Orthop Assoc* **13**(1): 60–63.

Nather A (2000b). Diploma training for technologists in tissue banking. *Cell Tiss Bank* **1**(1): 41–44.

Nather A (ed.) (2001). *The Scientific Basis of Tissue Transplantation*. World Scientific, Singapore.

Nather A (2004). Musculoskeletal tissue banking in Singapore — Fifteen years of experience (1988–2003). *J Orthop Surg* **12**: 184–190.

Nather A and Wang LH (2002). Bone banking in Singapore — Fourteen years of experience. *ASEAN Orthop Assoc* **15**(1): 20–29.

Nather A, Ong HJC, Feng MCB, and Aziz Z (2005a). Asia-Pacific Association of Surgical Tissue Banking — past, present and future. *ASEAN Orthop Assoc* **17**(1):17–19.

Nather A, Teo WY, and Wang LH (2005b). Diploma course training of tissue bank operators: seven years of experience. In Nather A (ed.), *Bone Grafts and Bone Substitutions: Basic Science and Clinical Applications*, World Scientific, Singapore, pp. 213–226.

Phillips GO (ed.) (2000). *Radiation and Tissue Banking*. World Scientific, Singapore.

Quarto R, Mastrogiacomo M, Cancedda R, Kutepov SM, Mukhachev V, Lavroukov A, Kon E, and Marcacci M (2001). Repair of large bone defects with the use of autologous bone marrow stromal cells. *N Engl J Med* **344**(5): 385–386.

Reuters AFP (1999). Media frenzy shocks heart-donors family. Breaking of taboo. The Straits Times, Tuesday, March.

Chapter 3

Asia Pacific Association of Surgical Tissue Banks — Past, Present and Future

Aziz Nather*, Zameer Aziz* and Shushan Zheng*

Introduction

The development of Regional Cooperative Agreement (RCA) Project RAS 7/008: Radiation Sterilisation of Tissue Grafts by the International Atomic Energy Agency (IAEA) since 1985 (Nather *et al.*, 2001; Phillips and Pedraza, 2003) played an important role in the development of the Asia Pacific Association of Surgical Tissue Banks (APASTB).

Historical Background

Pan-Asiatic Tissue Banking Association

The Pan-Asiatic Tissue Banking Association was set up in 1985 in Bangkok, Thailand, by Dr Yongyudh Vajaradul (Thailand), Dr Alain Patel (France), and the late Dr Norberto Agcaoili (Philipines) with its Secretariat set up at the Bangkok Biomaterial Centre in Siriraj Hospital.

Asia Pacific Association of Surgical Tissue Banks

The Asia Pacific Association of Surgical Tissue Banks (APASTB) was founded in October 1988 at the 3rd International Symposium on Locomotor

*NUH Tissue Bank, Department of Orthopaedic Surgery, Yong Loo Lin School of Medicine, National University of Singapore, Singapore.

Tissue Banking in Bangkok, Thailand. Dr Yongyudh Vajaradul was appointed as the first president of the association and Dr Aziz Nather as the first vice president. Its secretariat was set up in the Bangkok Biomaterial Centre in Thailand. The association was an extension of the earlier Pan-Asiatic Tissue Banking Association first proposed in 1985 in Bangkok, Thailand by Dr Yongyudh Vajaradul. All members of the latter association became members of APASTB.

The membership of APASTB has increased steadily over the years to include 200 members from 16 countries in the Asia-Pacific region — Thailand, Singapore, Japan, Philippines, China, Australia, Malaysia, Indonesia, Vietnam, Korea, Hong Kong, India, Pakistan, Sri Lanka, Bangladesh, and Myanmar.

APASTB Scientific Meetings

	Location	Year	Organised by
Inaugural	Thailand	1989	Dr. Yongyudh Vajaradul
Second	Singapore	1990	Dr. Aziz Nather
Third	Japan	1991	Dr. Moritoshi Itoman
Fourth	The Philippines	1992	Dr. Norberto Agcaoili
Fifth	Suzhou, China	1994	Dr. Tang Zhong Yi
Sixth	Gold Coast, Australia	1996	Dr. David Morgan
Seventh	Malaysia	1998	Dr. Hasim Mohamad
Eighth	Bali, Indonesia	2000	Dr. Abdurrahman
Ninth	Seoul, Korea	2002	Dr. Chang Joon Yim
Tenth	Hong Kong	2004	Dr. Shekhar Kumta
Eleventh	Bombay, India	2006	Dr. Astrid Lobo Gajiwala
Twelfth	Malaysia	2008	Dr. Norimah Yusof
Thirteenth	Padang, Indonesia	2010	Dr. Menkher Manjas

Initially, the meetings were held annually. In 1992 during the 4th meeting in Manila, the Board unanimously decided to hold subsequent meetings every two years. IAEA helped to sponsor council members from RCA member states to attend the scientific meetings. In this way, council members from all RCA member states were able to attend APASTB meetings from the fourth one organised in Manila to the 9th APASTB meeting held in Seoul, Korea, in 2002. This synergy between IAEA and APASTB helped considerably towards the development of APASTB. Unfortunately, IAEA stopped sponsoring members from RCA member states to the 10th meeting held in Hong Kong in 2004 as the RCA Programme for Tissue Banking (RAS 7/008) came to an end in 2004.

APASTB Presidents

Dr. Yongyudh Vadarajul	Thailand	1988–1990
Dr. Aziz Nather	Singapore	1990–1992
Dr. Moritoshi Itoman	Japan	1992–1994
Dr. Norberto Agcaoili	Philippines	1994–1996
Dr. Sun Shi Quan	China	1996–1998
Dr. David Morgan	Australia	1998–2000
Dr. Hasim Mohamad	Malaysia	2000–2002
Dr. Abdurrahman	Indonesia	2002–2004
Dr. Chang Joon Yim	Korea	2004–2006
Dr. Shekhar Kumta	Hong Kong	2006–2007
Dr. Astrid Lobo Gajiwala	India	2007–2010

Current Office Bearers of APASTB, 2008–2010 (Fig. 1)

Immediate Past President	Shekhar Kumta	Hong Kong
President	Astrid Lobo Gajiwala	India
First Vice President	Yong Koo Kang	Korea
Second Vice President	Menkher Manjas	Indonesia
Third Vice President	Sharon Bryce	Australia
Secretary-General	Ken Urabe	Japan
Assistant Secretary-General	Viwat Chuntrasatic	Thailand
Treasurer	Suzina Sheikh Ab. Hamid	Malaysia
Auditors	Moritoshi Itoman	Japan
	Hasim Mohamad	Malaysia
Editors	Aziz Nather	Singapore
	Norimah Yusof	Malaysia

Fig. 1. Current office bearers of APASTB (from left) Suzina, Kang, Menkher, Astrid, Urabe, Nather and Yusof.

Current Status

Networking of Member States

Members of APASTB are able to meet every two years to participate actively in the Scientific Meetings. In addition, APASTB produces a yearly newsletter. This serves to highlight important events on tissue banking held by member states, forecast important future events on tissue banking and also acts as a forum for members to discuss issues on tissue banking.

General Standards

The APASTB Standards Subcommittee started work on the APASTB's own standards during the 9th APASTB Scientific Meeting in Korea, 2002. The Standards were drafted according to the template of the "IAEA International Standards in Tissue Banking", developed under the IAEA INT/6/052 Interregional Programme, and follows the standards adopted by American Association of Tissue Banks (AATB) and European Association of Tissue Banks (EATB).

After two years of deliberations, the first draft was finalised during the 10th APASTB Scientific Meeting in Hong Kong, 2004. It was during this meeting that APASTB made the recommendation for all member states to conform to these standards. The first edition of the APASTB General Standards was printed in January 2007 in the textbook "Radiation in Tissue Banking", edited by A. Nather, N. Yusof, and N. Hilmy.

International Journal on Tissue Banking

APASTB has adopted an international journal on tissue banking entitled *Cell and Tissue Banking* as its official journal, with Dr Rudiger von Versen as the Chief Editor. The journal is published four times a year and provides an avenue where new knowledge regarding tissue banking and transplantation can be circulated. The first issue of the journal was published in 2000. This journal has proven to be a very useful channel for tissue bankers in APASTB to publish their work in both basic and clinical research.

World Congresses on Tissue Banking

			Hosted by	Chairperson
First	1996	Gold Coast, Australia	APASTB	Dr. David Morgan
Second	1999	Warsaw, Poland	EATB	Dr. Janus Komender
Third	2002	Boston, USA	AATB	Dr. Samuel Doppelt
Fourth	2005	Brazil, Latin America	ALABAT	Dr. Marissa Herson
Fifth	2008	Kuala Lumpur, Malaysia	APASTB	Dr. Hasim Mohamad

The 1st World Congress of Tissue Banks involving APASTB, EATB and AATB was hosted by APASTB during its 6th Scientific Meeting held in Gold Coast, Australia, in October 1996. APASTB is proud to say that the 5th World Congress on Tissue Banking, in conjuction with held its 12th APASTB meeting in Kuala Lumpur in 2008, with Dr. Hasim Mohamad as Chairperson and Dr Norimah Yusof as Secretary, was an outstanding success with more than 250 delegates.

Future Role

APASTB will continue to play an important role promoting tissue banking activities in the Asia-Pacific region. It will continue to maintain and upgrade its General Standards periodically. It will also play role in the World Union on Cell and Tissue Banking.

References

Nather A (1999). Tissue banking in the Asia Pacific region — the Asia Pacific Association of Surgical Tissue Banking. In Phillips GO, Strong DM, von Versen R, and Nather A (eds.), *Advances in Tissue Banking*, Vol. 3, World Scientific, Singapore, pp. 419–425.

Nather A, Phillips GO, Chang HF, and Lim MYL (2001). Development of IAEA/NUS internet diploma course in tissue banking. *ASEAN Orthop Assoc* **14**: 5–7.

Nather A, Phillips GO, and Morales J (2003). IAEA/NUS distance learning diploma training course for tissue bank operators — past, present and future. *Cell Tissue Bank* **4**: 77–84.

Nather A, Yusof N, and Hilmy N (2007). *Radiation in Tissue Banking. Basic Sciences and Clinical Applications of Irradiated Tissue Allografts*, World Scientific, Singapore.

Phillips GO and Pedraza JM (2003). The International Atomic Energy Agency (IAEA) Programme in radiation and tissue banking: past, present and future. *Cell Tissue Bank* **4**: 69–76.

Chapter 4

Distance Learning Diploma Course for Training Tissue Bank Operators — 12 Years of Experience by IAEA/NUS Training Centre

Aziz Nather, Poh Lin Chan and Shushan Zheng

Introduction

In many regions of the world — Asia Pacific, Latin America, Africa, and Eastern Europe — the directors of tissue banks are mostly part-time volunteers (Nather, 2000). Only the technologists are employed as full-time staff. This is in contrast to large banks in the USA and Europe, many of which are run by large corporations as business ventures.

Furthermore, the responsibility of performing the day-to-day activities of a tissue bank lies with the technologists. Therefore, it is of utmost importance that they are well trained to perform all the duties required of them. These include the following:

- Screening of potential donors (both living and deceased)
- Serological investigations
- Procurement of tissues
- Processing of tissues
- Documentation

NUH Tissue Bank, Department of Orthopaedic Surgery, Yong Loo Lin School of Medicine, National University of Singapore, Singapore.

- Distribution of tissues
- Promotion of public awareness of tissue banking and transplantation
- Promotion of professional awareness of tissue banking and transplantation.

In the past, the available training programmes were limited to a short two-week course conducted by the American Association of Tissue Banks (AATB). There was therefore demand for a structured year-long training programme with a comprehensive curriculum leading to diploma certification by an internationally recognised university.

RAS 7/008: IAEA/RCA Programme on "Radiation Sterilisation of Tissue Grafts"

Since 1985, the International Atomic Energy Agency (IAEA) — under the Regional Cooperative Agreement (RCA) for member states in the Asia-Pacific region — began running a programme on the "Radiation Sterilisation of Tissue Grafts" (RAS 7/008) involving tissue banks in 13 countries, namely Bangladesh, China, India, Indonesia, Korea, Malaysia, Myanmar, Pakistan, the Philippines, Singapore, Sri Lanka, Thailand, and Vietnam (Nather, 1999; Nather, 2000; Nather *et al.*, 2003).

RAS 7/008 was refined to a thematic model project under the leadership of Professor Glyn Phillips, who is technical advisor to the Deputy Director-General, Department of Technical Cooperation, IAEA (Mr Qian Jihui). The objective was to raise the quality standard of tissue banking to an international level. Efforts were directed to harmonise the quality standards of tissue banks in the region so as to facilitate the exchange of grafts from one country to another in the long run.

National coordinators from each member state spent several years developing and writing an IAEA/RCA draft curriculum on tissue banking, with Professor Phillips as the coordinating editor. The first draft curriculum was successfully assembled during the RCA Workshop in Suzhou, China, in 1994, and was the first of its kind in the world.

The curriculum was piloted in Singapore during the IAEA/RCA Regional Workshop on the "Dissemination of Information on Procedures

Fig. 1. The IAEA/RCA Regional Workshop on "Dissemination of Information on Procedures for Production and Radiation Sterilisation of Tissue Allografts" in 1995.

for Production and Radiation Sterilisation of Tissue Allografts" in September 1995 (Fig. 1). Twenty-one trainers used the curriculum to teach 35 trainees. This was the largest workshop ever held for curriculum training and it was deemed to be successful, as the curriculum was found to be effective and very suitable for training tissue bank operators. The NUH Tissue Bank was inaugurated as a hospital tissue bank during the opening ceremony of this workshop (Nather, 2000).

Development of Singapore as the Regional Training Centre (RTC) for the Asia-Pacific Region in Singapore

In September 1996, the NUH Tissue Bank was appointed by the IAEA to become the IAEA/NUS Regional Training Centre (RTC) for training tissue bank operators in the Asia-Pacific region (Nather, 1999; Nather, 2000). The government of Singapore (represented by the Ministry of Environment), with the National Science and Technology Board (NSTB) as the funding agency, awarded a S$225 500 grant to build a new purpose-built tissue bank cum Regional Training Centre. The 2000 square feet centre is located on the second level of the National University Hospital and has separate wet and dry processing laboratories, a documentation/distribution room, and a reception area.

The IAEA/NUS Regional Training Centre was launched during the IAEA/RCA Regional Training Course for the "Delivery of Curriculum to Tissue Bank Operators" on November 3, 1997. It was inaugurated by the Deputy Vice Chancellor of NUS, Professor Chong Chi Tat. The IAEA was represented by Mr Thomas Tisue, special advisor to the Deputy Director-General (Mr Qian Jihui). At the same time, the first ever IAEA/NUS Diploma Course on Tissue Banking was launched — another first in the world.

NUS Diploma Course in Tissue Banking

The NUS Tissue Banking Course is a one-year distance learning diploma course. The minimum criteria for admission are at least five passes in the GCE O-Level Examination (or its equivalent), experience in working in a tissue bank or association with a tissue bank for at least a year, and proficiency in English. The course fee is only US$100.

The curriculum for the NUS Diploma course includes the following:

- The conversion of the IAEA draft curriculum on tissue banking into a multi-media curriculum, which consists of eight modules, accompanying sets of slides, seven video demonstrations, and an audio cassette (Fig. 2). The components of each module are contained in specially designed box containers (Nather, 2000; Nather *et al.*, 2003). The production costs of this curriculum (about S$100 000) are borne by the NSTB. The eight modules comprise the following:

  Module: Guide to Curriculum
  Module 0: Historical Background
  Module 1: Rules and Regulations
  Module 2: Organisation
  Module 3: Quality Assurance
  Module 4: Procurement
  Module 5: Processing
  Module 6: Distribution and Utilisation
  Module 7: Future Developments in Tissue Banking

- Lectures on basic sciences. The basic science subjects include basic anatomy, basic microbiology, introduction to transmissible diseases,

Fig. 2. The IAEA/NUS Multi-Media Curriculum produced by Singapore.

basic immunology, principles of sterile technique, basic radiation science, biology of healing of tissue transplantation, and biomechanics of tissue transplantation.

* Recommended textbook, *The Scientific Basis of Tissue Transplantation, Advances in Tissue Banking*, Vol. 5, (A. Nather, ed., World Scientific, Singapore, 2002).

The course structure consists of three components:

* Two-week foundation course at the RTC, Singapore, with lectures and practical demonstrations, ending with a theory and practical (OSPE) examination (Phase I)
* Three assignments given at quarterly intervals, the last assignment being a practical assignment
* A terminal examination conducted by NUS over one week at the RTC (Phase II)

The marking allocation scheme for the diploma course is as follows:

* Foundation course exam (theory, practical)   20%
* Assignments   40%

- Terminal NUS Exam (theory, practical, viva)   40%

  The NUS Diploma in Tissue Banking is awarded in three categories:

- Distinction >80 marks
- Credit   70–79 marks
- Pass   50–69 marks

Unsuccessful candidates are allowed to resit for the examination up to a maximum of three attempts for the main or supplementary examinations.

Curriculum Update

The curriculum has been updated in four phases (Nather *et al.*, 2001):

- Phase I — Seven video tape demonstrations on the procurement, processing, and transplantation of tissues were converted into two compact discs in March 2000 (Fig. 3).
- Phase II — Text booklets for Modules 0 to 7 were updated with a companion book, *Radiation and Tissue Banking* (G.O. Phillips, ed., World Scientific, Singapore, 2000), in July 2000 (Fig. 3).
- Phase III — A text on basic sciences was produced for the first time as a textbook, *The Scientific Basis of Tissue Transplantation*, in January 2002 (Fig. 3).

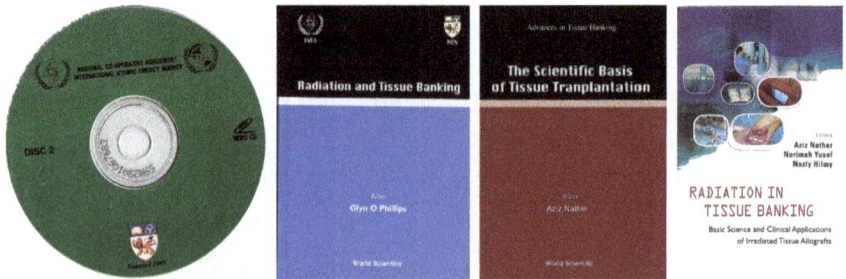

Fig. 3. The curriculum updates. (From left) compact discs, *Radiation and Tissue Banking, The Scientific Basis of Tissue Transplantation and Radiation in Tissue Banking.*

- Phase IV — A textbook on *Radiation in Tissue Banking* was produced in April 2007 (Fig. 3).

IAEA/NUS Diploma Courses Conducted (1997–2003)

The first diploma course was launched on November 3, 1997, with 17 candidates; and the first NUS diploma examination was held in October 1998. Overall, 12 candidates graduated; of these, 4 passed with distinction, 5 with credit, and 3 with pass only (Nather, 2000, Nather *et al.*, 2003). Diploma courses have since been held every consecutive year.

Development of Singapore as the International Training Centre (ITC)

A memorandum of understanding (MOU) was signed between the NUS (represented by the Dean of Faculty of Medicine) and the IAEA (represented by the Deputy Director-General) on July 4, 2002 (Nather *et al.*, 2003). With this memorandum, Singapore was appointed as the IAEA/NUS Interregional Training Centre (ITC) for four regions: the Asia-Pacific, Latin America, Africa, and Europe (Fig. 4). All of the local costs for the development of the ITC were borne by a local grant obtained from the Lee Foundation (S$85 000) in Singapore.

Transformation of Diploma Course into an Internet Course

The demand for training has increased tremendously over the years not only for technologists in the Asia-Pacific region, but also for tissue bank operators in other regions such as Africa and parts of Eastern Europe (Nather *et al.*, 2001; Nather *et al.*, 2003). The financial costs borne by the IAEA were very large. For each foundation course (Phase I), the cost incurred for sponsoring 6 overseas lecturers and 20 students from 13 member states (Asia-Pacific region) was about US$100 000. In addition, the cost incurred for holding the Phase II one-week terminal examination at the end of the year (including sponsoring the same students plus three overseas examiners) was about US$40 000. From 1997 to 2003, the total

Fig. 4. The network of the IAEA Training Programme, with Singapore as the Interregional Training Centre for the Asia-Pacific region, Latin America, and Korea.

cost incurred for five batches was approximately US$700 000, a staggering amount indeed.

For the IAEA to continue sponsoring similar courses in the future, these costs had to be substantially reduced. Also, it was not possible for the IAEA to continue sponsoring such courses indefinitely. Thus, plans were made by the RTC in Singapore to continue running such courses on its own, without the financial support of the IAEA. The RTC aimed to become self-sufficient. Likewise, member states were to start paying for the training costs of their own tissue bank operators.

The conversion of the diploma course into an internet course proved to be an effective solution. In October 2000, Singapore was approached by the IAEA to consider such a conversion (Nather *et al.*, 2001; Nather *et al.*, 2003). The funding for the cost of this development, which began in 2001, came from the IAEA. The introduction of this internet course eliminated the need for a foundation course. However, for the NUS to confer a diploma, the terminal examination still had to be held in the RTC, Singapore; instead of one week and three examiners, it could be held for just three days and involve only one external examiner. The estimated cost for the exam would then be only about US$15 000. Had the training course been designed as a distance learning internet course from the start, the cost incurred by the IAEA for the first five batches (1997–2003) would have been only US$75 000 plus the cost of registration fees (US$500 × 101 = US$50 500) — i.e. a total of US$125 000 instead of US$700 000, a five-fold decrease in the expenditure that has been spent.

Instruction Materials for Internet Delivery

The instruction materials that have been converted for internet delivery include the IAEA/NUS Multi-Media Curriculum with eight modules (text booklets), and two compact discs. In addition, textbooks were recommended to supplement the online curriculum. These include *Scientific Basis of Tissue Transplantation* and *Radiation in Tissue Banking*.

Internet Diploma Course in Tissue Banking (2004 onwards)

The internet course was piloted with the fourth batch of diploma students in April 2001, and with the fifth batch in 2002. The first internet course was launched with the sixth batch on February 9, 2004, with 16 students sponsored by the IAEA. The IAEA funded the costs of the new registration fees (US$500 × 16 = US$8000). In addition, there were three other self-sponsored students. The terminal examination for this first internet course was held in February 2005. Two other batches have since graduated.

Results of the IAEA/NUS Diploma Training Course

Between 1997 and 2009, ten courses were conducted by the RTC with a total of 225 tissue bank operators. Of these, 193 were from the Asia-Pacific region (13 countries), 9 from Latin America (Brazil, Chile, Cuba, Peru, Uruguay), 11 from Europe (Greece, Slovakia, Poland, Ukraine), 11 from Africa (Zambia, Libya, Egypt, Algeria) and 3 from Australia.

Even though sponsorship from IAEA stopped since 2004, the training centre in Singapore has continued to run the internet diploma course. Students since the seventh batch have been paying for their own enrollment. The results so far have been encouraging.

The last (eleventh) batch involved 18 students who registered in August 2009 and are due to sit for the terminal examination in August

2010. Currently, ten batches have completed the diploma training. A total of 141 tissue bank operators have convocated with an NUS Diploma in Tissue Banking; of these, 29 have completed the course with distinction, 68 with credit, and 44 with pass only (Table 1). Forty-three students did not complete the diploma course. Increased participation from regions outside the Asia-Pacific was seen from the fourth batch onwards (Table 2).

Transfer of Technology to Latin America

In October 1998, an IAEA Interregional Trainers Workshop on the "Distant Learning Use of the Curriculum Package on Tissue Banking" was conducted by Dr Phillips and Dr Nather in Singapore with participant trainers from Argentina, Brazil, Chile, Cuba, Mexico, and Peru (Nather, 2000). One copy of the Multi-Media Curriculum (English version) was presented to each trainer from Latin America with compliments of the Singapore government. The curriculum was subsequently translated into

Table 1. Results of the IAEA/NUS Diploma Courses conducted (1997–2009).

Training courses	No. of students registered	No. of students convocated	Results			
			Distinction	Credit	Pass	Fail
First course (Nov 97–Oct 98)	18	12	4	5	3	6
Second course (Apr 99–Mar 2000)	17	15	2	5	8	2
Third course (Apr 2000–Mar 2001)	21	17	1	5	11	4
Fourth course (Apr 2001–Mar 2002)	24	19	1	11	7	5
Fifth course (Apr 2002–Aug 2003)	21	14	3	8	3	7
Sixth course (Feb 2004–Feb 2005)	19	12	4	5	3	7
Seventh course (Mar 2005–Mar 2006)	18	13	5	6	2	5
Eighth course (Apr 2006–Apr 2007)	22	18	3	10	5	4
Ninth course (Apr 2007–Aug 2008)	20	6	3	3	0	0
Tenth course (Aug 2008–Aug 2009)	27	15	3	10	2	3
Eleventh course (Aug 2009–Aug 2010)	18					
Total	225	141	29	68	44	43

Table 2. Regional distribution of tissue bank operators registered (1997–2009).

Batch no.	Asia Pacific	Latin America	Africa	Europe	Australia
		No. of students registered			
First batch (Nov 97–Oct 98)	18	0	0	0	0
Second batch (Apr 99–Mar 2000)	16	0	0	1 (Slovakia)	0
Third batch (Apr 2000–Mar 2001)	19	2 (1 Brazil, 1 Chile)	0	0	0
Fourth batch (Apr 2001–Mar 2002)	19	0	3 (2 Zambia, 1 Algeria)	2 (1 Greece, 1 Slovakia)	0
Fifth batch (Apr 2002–Aug 2003)	14	3 (1 Cuba, 1 Peru, 1 Uruguay)	3 (I Egypt, 1 Libya, 1 Zambia)	1 (Poland)	0
Sixth batch (Feb 2004–Feb 2005)	12	0	2 (l Libya, 1 Algeria)	5 (3 Slovakia, 1 Ukraine, 1 Poland)	0
Seventh batch (Mar 2005–Mar 2006)	17	1 (USA)	0	0	0
Eighth batch (Apr 2006–Apr 2007)	18 2 (Iran)	0	2 (South Africa)	0	2
Ninth batch (Apr 2007–Aug 2008)	20	0	0	0	0
Tenth course (Aug 2008–Aug 2009)	24	1	1	0	1
Eleventh course (Aug 2009–Aug 2010)	16	2	0	0	0
Total	193	9	11	9	3

Spanish for use by Latin American countries, with Argentina established as the RTC for Latin America.

Transfer of Technology to Africa

From 3–7 June 1999, Dr Phillips and Dr Nather conducted the "IAEA Regional Training Course on Tissue Banking" in Algiers, Algeria.

The participants included representatives from Algeria, Egypt, Ghana, Libya, Zambia, and Nigeria. However, no further development occurred following the workshop.

National Training Programmes

The Korean National Training Centre has been set up in St. Vincent's Hospital, Catholic Medical University, Seoul, Korea, with Professor Yong Koo Kang as the Director. This centre is jointly run by the Korean Association of Tissue Banks (KATB) and the Korean Musculoskeletal Transplantation Society (KMTS). The centre uses the IAEA Multi-Media Curriculum translated into the Korean language with funds from the IAEA, largely due to the efforts of Dr Chang Joon Kim, Dr Glyn Phillips, and Mr Jorge Morales. The first Korean National Training Course (KNTC) was launched in 2003 with support from the IAEA. Eleven students participated using the Multi-Media Curriculum printed in Korean.

The structure of the course (also a one-year distance learning programme) is similar to the IAEA/NUS Diploma Course run in Singapore, and consists of the following:

- A one-week foundation course (optional)
- Three assignments
- Three weekend courses (with lectures, assignments, and practical demonstrations)
- A terminal face-to-face examination.

The KNTC is run in collaboration with the IAEA/NUS Training Centre in Singapore, which serves all Korean students with the internet curriculum in English.

The first examination was held at St. Vincent's Hospital in November 2004, with Dr Nather as the IAEA consultant and external examiner. All 11 students passed. The second KNTC commenced in November 2004 with 12 participants. The examination was held in November 2005, (Dr Nather as the external examiner with IAEA support) 10 students passed. At the same time, 22 students enrolled in the third KNTC without

IAEA involvement. The examination was conducted in December 2006 (Dr Nather as the external examiner).

Conclusion

Singapore has played a vital role in the global training of tissue bank operators over the last 12 years, providing training not only to the Asia-Pacific region, but also to Latin America, Africa, and Europe. Its Regional Training Centre has grown and, as of February 2004, now functions as an Interregional Training Centre.

Despite the termination of the IAEA programme in 2004, the centre continued the training courses on its own. In order to succeed, it has forged partnerships with key countries in the Asia-Pacific region, namely Korea, Malaysia, and Indonesia. These countries have indicated their interest to run National Training Programmes in collaboration with Singapore. So far, only Korea has started its own National Training Course in the Korean language, supported by resources from the ITC in Singapore. Malaysia and Indonesia are planning to run similar National Training Courses, with Singapore providing the internet curriculum.

References

Nather A (1999). Tissue banking in the Asia Pacific region — the Asia Pacific Association of Surgical Tissue Banking. In Phillips GO, Strong DM, von Versen R, and Nather A (eds.), *Advances in Tissue Banking*, Vol. 3, World Scientific, Singapore, pp. 419–425.

Nather A (2000). Diploma training for the technologists in tissue banking. *Cell Tissue Bank* **1**: 41–44.

A Nather (ed.) (2002). *Advances in Tissue Banking*, Vol. 5, World Scientific, Singapore.

Nather A, Phillip GO, Cheong HF, and Ling MY (2001). Development of IAEA/NUS internet diploma course in tissue banking. *ASEAN Orthop Assoc* **14**: 5–7.

Nather A, Phillips GO, and Morales J (2003). IAEA/NUS distance learning diploma training course for tissue bank operators — past, present and future. *Cell Tissue Bank* **4**: 77–84.

Part II

Organisation Systems

Chapter 5

Manpower Organisation

Poh Lin Chan and Aziz Nather

Introduction

A proper management structure is important for a tissue bank to be successful and attain good productivity (i.e. high number of good quality tissues being procured, processed and utilised) (Nather and Lee, 2007).

It is pertinent for a tissue bank to adopt a management structure which will complement its local condition and ensure proper running of the bank in its own country. All successful tissue banks have good management structures which are fairly similar to each other.

Basic Management Structure

Countries under the Regional Cooperative Agreement (RCA) have small manpower within their tissue banks (i.e. there may only be two to three technical staff working in the bank). Under such circumstances, an elaborate organisation structure which consists of quality control manager, processing manager and safety officer, etc., is not always feasible. Whilst the functions of these personnel are not omitted and must be fulfilled by the technical staff within the tissue bank, a simpler organisation would be a more practical choice in allowing the bank to be more effective.

A basic management structure that may be adopted to run a tissue bank is shown in Fig. 1.

NUH Tissue Bank, Department of Orthopaedic Surgery, Yong Loo Lin School of Medicine, National University of Singapore, Singapore.

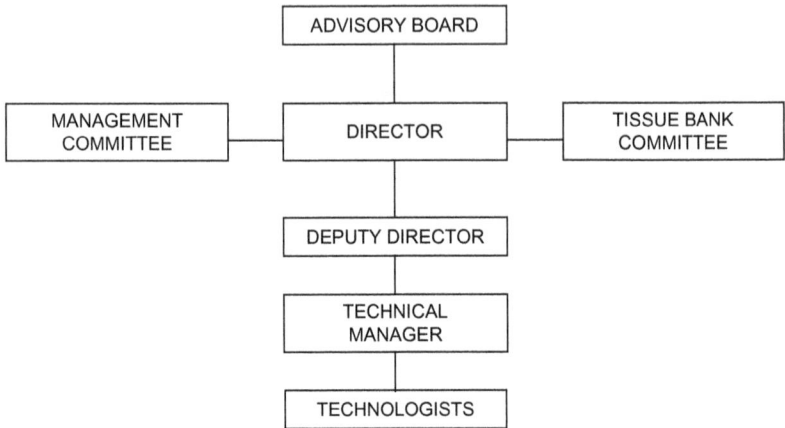

Fig. 1.　Basic management structure that can be adopted by tissue banks.

The key person or central figure in this basic management structure is the Director, who is committed to the bank and responsible for its overall functions. The Director reports to an Advisory Board and chairs two other committees: Tissue Bank Committee and Management Committee. It is important to obtain approval for involvement in the bank from all key personnel during the initial stages of planning, in order for them to function effectively in the Advisory Board, Tissue Bank Committee and Management Committee.

Roles of Key Personnel

The key personnel of a tissue bank include:

- Director
- Deputy Director
- Technical Manager
- Technologists

We shall now examine the various roles of each abovementioned individual in detail.

Director

The Director is a medical/dental officer or a radiation biologist/pathologist/microbiologist. In the Asia-Pacific region, Directors are mostly part-time surgeons or radiation scientists (Nather, 2000). The Director is responsible for:

- Determining overall policies, purpose, and direction of the bank
- Implementing appropriate policies and procedures for tissue procurement, such as donor selection criteria
- Promoting the use of tissue grafts processed by the bank
- Distributing and monitoring the Procedure and Quality Manuals of the bank
- Seeking approval from relevant ethical bodies for the procurement of donor tissues for the bank
- Administration of operations of the bank
- Authorising tissue grafts, which satisfy the quality standards adopted by the tissue bank, for utilisation and distribution
- Training tissue bank staff
- Taking care of the welfare of all personnel in the bank.

Deputy Director

In the absence of the Director, the Deputy Director will assume the duties of the Director.

Technical Manager

The Technical Manager is a technologist appointed by the Director to perform and oversee the following duties:

- Managing the day-to-day operations of the bank, which includes tissue procurement, processing, storage and distribution
- Ensuring all tissue bank staff follow strictly every guideline with regards to donor screening criteria, procurement techniques, and processing methods as laid down in the bank's procedure manual

- Ensuring all tissue bank staff follow rigorously all policies laid down in the tissue bank's Quality Manual, in order to achieve high standards of quality assurance of tissue graft products
- Ensuring all tissue bank staff maintain the bank equipment in good, functioning condition
- Taking up the role of Quality Control Manager in the tissue bank, in charge of Quality System Coordination and ensuring all technologists adhere strictly to all guidelines in the Quality Manual
- Assuring all tissue graft requests are met
- Assisting the Director in selecting the best matched tissue graft to meet the needs of a specific recipient

Technologist

Technologists work in collaboration with the Technical Manager, performing the day-to-day activities of the bank which includes:

- Procurement
- Processing
- Storage
- Distribution

The duties of a technologist include:

Living Donor Procurement

- Providing double jars to the operating theatre
- Ensuring all scrub nurses follow all guidelines stated in the Procedure Manual for collection of specimens
- Filling up the Living Donor Form
- Obtaining consent from donor using Consent Form
- Sending tissue specimen for aerobic and anaerobic culture and sensitivity tests
- Obtaining blood samples from donor to perform the following tests:

    (a) AIDS (anti-HIV1, anti-HIV2)
    (b) Hepatitis B (HbsAg)

(c) Hepatitis C (anti-HCV)

(d) Syphilis (RPR, and when positive, TPHA)

- Appropriate labelling and storing of tissue grafts in the quarantine freezer at $-80°C$
- Proper filing of the following:

  (a) Living Donor Form
  (b) Test reports for culture, anti-HIV_1, anti-HIV_2, HBsAg, anti-HCV, RPR, TPHA
  (c) Consent Form

Deceased Donor Procurement

- Sterilising the major cleansing set and surgical instrument set
- Autoclaving all linen wraps and glass bottles
- Preparing all necessary supplies required for deceased donor procurement
- Acting as a circulating nurse by providing consumables in a sterile manner to surgeons performing the procurement of tissues
- Scrubbing up as an assistant to procurement surgeons during tissue procurement
- Labelling all specimens with donor particulars and recording the types of tissue grafts procured
- Appropriate storage of tissue grafts in the quarantine freezer
- Completing the Deceased Donor Form
- Proper filing of the following:

  (a) Deceased Donor Form
  (b) Test reports for culture, anti-HIV_1, anti-HIV_2, HBsAg, anti-HCV, RPR, TPHA
  (c) Consent Form

Despatch

- Recording details of a request for a tissue graft, such as patient's particulars, diagnosis, type of operation, and type of tissue graft required

- Notifying the Director of requests for tissue grafts
- Assisting the Director in performing the matching of grafts to provide the most appropriate tissue graft available to meet the needs of a specific recipient
- Delivering tissue graft to a hospital operating theatre for internal despatch
- Scheduling tissue collection for long-distance and overseas despatch
- Procuring dry ice for the packaging of tissue grafts for long-distance and overseas despatch
- Completing the following forms

  (a) Recipient Form
  (b) Consent Form for recipient

Record Keeping

- Keeping clear and precise records of the following:

  (a) Freezer log book
  (b) Lyophiliser log book
  (c) Living donor book
  (d) Deceased donor book
  (e) Recipient book
  (f) Lyophilisation log book
  (g) Record book of bones lyophilised
  (h) Graft despatch book

Stock Checking

- Ensuring a sufficient supply of consumables within the tissue bank; items used up must be replaced to avoid shortfalls

Processing

- Processing bone grafts
- Recording all processing in the Wet Processing Form and Dry Processing Form

- Issuing surgeons with package inserts, along with all lyophilised tissue grafts, which provide necessary instructions on proper graft usage
- Recording the date and time of operation of lyophiliser in the lyophiliser log book
- Ensuring that Radiation Sterilisation Forms are completed and certified by radiation officers

Environmental Cleanliness

- Ensuring cleanliness of laboratories at all times for safe and clean processing of tissue grafts

Role of Boards/Committees for a Tissue Bank

Having understood the roles of the Director, Technical Manager and Technologist, it is pertinent to understand the role of the following important boards/committees:

- Advisory Board
- Tissue Bank Committee
- Management Committee

Advisory Board

Hospital Tissue Bank

In a hospital tissue bank, the board members include:

- Chairman of the hospital's medical board
- Chief Executive Officer
- Head of Department (in which the Director of the bank belongs to), etc.

The following chart illustrates the organisation structure of the NUH Tissue Bank in Singapore (Fig. 2).

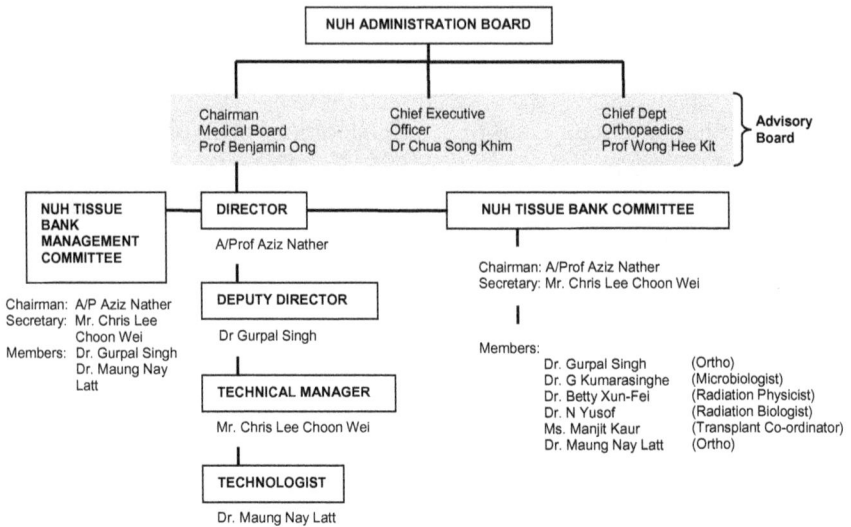

Fig. 2. Organisation structure for the NUH Tissue Bank in Singapore.

University Tissue Bank

In a university tissue bank, the board members include:

- Dean
- Vice-Dean
- Head of Department, etc.

The Malaysian National Tissue Bank in the Department of Orthopaedic Surgery, University Science Malaysia located in Kota Bahru, Kelantan, Malaysia, is an example of a university tissue bank with the abovementioned organisation chart (Fig. 3).

Radiation Institution Tissue Bank

In a radiation institution tissue bank, the Advisory Board members include:

- Director-General of Institute
- Deputy Director-General of Institute
- Head of Department, etc.

ORGANISATION CHART
NATIONAL TISSUE BANK UNIVERSITY SCIENCE OF MALAYSIA

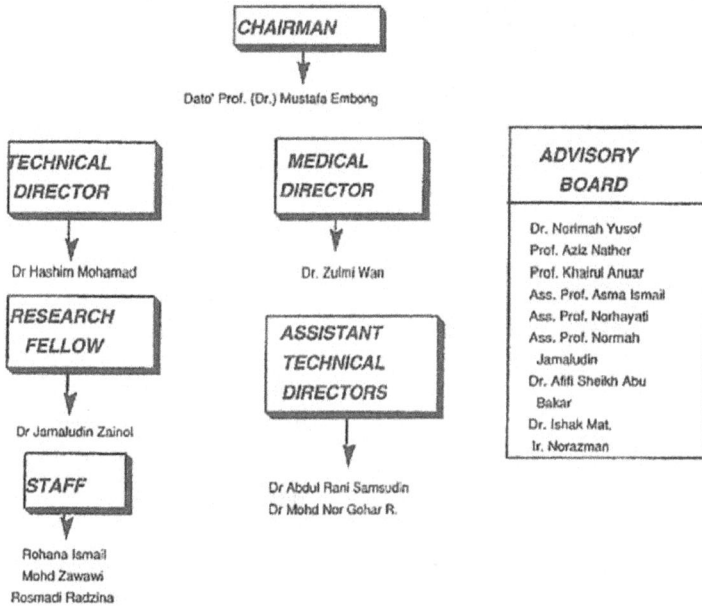

Fig. 3. Organisational chart for the Malaysian National Tissue Bank in Kelantan, Malaysia.

Voluntary Welfare Organisation Tissue Bank

In the case of a tissue bank run by a voluntary welfare organisation, with the support of the Ministry of Health, the Advisory Board will consist of:

- Appointees from the Ministry of Health
- Trustees of the voluntary organisation

The Advisory Board for the Sri Lanka Model Human Tissue Bank, as illustrated in the organisational chart below (Fig. 4) is one such example.

Role of the Advisory Board (Nather and Wang, 2005)

- Supervise the progress and development of the tissue bank
- Evaluate annual tissue bank reports

```
        ┌─────────────────────┐
        │   ADVISORY BOARD    │
        └─────────────────────┘
        (appointed by Ministry of Health)
                 │
        ┌─────────────────────┐
        │  EXECUTIVE DIRECTOR │
        └─────────────────────┘
           Dr. Hudson Silva
                 │                          ┌─────────────────────────┐
                 │                          │  TISSUE BANK COMMITTEE   │
        ┌─────────────────────┐            └─────────────────────────┘
        │  MEDICAL DIRECTOR   │
        └─────────────────────┘
          Dr. L.M. Amarasinghe
```

Chairman   : Dr. L.M. Amarasinghe (M. Director)
Vice Chairmen : Dr. Vasantha Perera (Dy. M. Director)
                         Dr. S.D. Karunaratne (Dy. M. Director)
Secretary   : Mr. Hemaka de Mel
Members   : 1. Professor C. Niriella (Prof. of Forensic Medical,
                             Korapitaya Hospital)
                         2. Dr. S.D. Atukorala (Head, Dept of Bacteriology,
                             Colombo National Hospital)
                         3. Dr. Upali Mendis (Director, Colombo Eye
                             Hospital)
                         4. Dr. M.H.S. Casim (Dy. Director, Colombo Eye
                             Hospital)

```
  ┌──────────────────────────────┐
  │  DEPUTY MEDICAL DIRECTOR     │
  └──────────────────────────────┘
       Dr. Vasantha Perera
```
                         5. Dr. D.K. Dias (Maxillo-Facial Surgeon, General
       Dr. D.D. Karunaratne   Hospital Matara)
                         6. Dr. M. Hennayaka (Paediatric Surgeon,
                             Children's Hospital, Colombo)

```
  ┌──────────────────────────────┐
  │     LABORATORY MANAGER       │
  └──────────────────────────────┘
       Mr. Hemaka de Mel
```
                         7. Dr. Banagala (Head, Dept of Orthopaedics,
                             Colombo National Hospital)
                         8. Dr. A. Aluwihare (Professor of Surgery, Kandy
                             General Hospital)
                         9. Dr. Ellawala (Orthopaedic Surgeon Teaching
                             Hospital, Karapitiya)

Quality   Procurement   Processing   Radiation

Control   Mr. Prabhanaverdra   Mr. Surandra   Mr. Hemaka de Mel

Mr. Jayathilaka   (Mr. Aruna J)   (Mr. Aruna J)

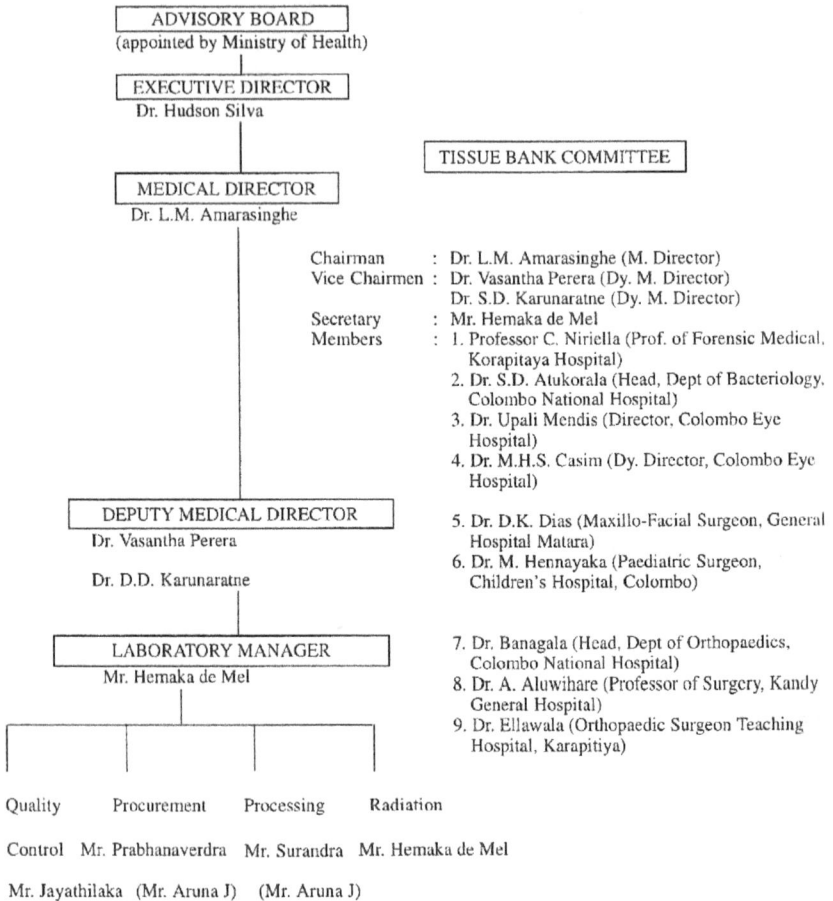

Fig. 4.   Organisational chart for the Model Human Tissue Bank in Sri Lanka.

- Review recommendations on all aspects of tissue banking put forth by the Tissue Bank Committee
- Endorse all guidelines written in the Procedure and Quality Manuals produced by the Tissue Bank Committee
- Approve new amendments made by the Tissue Bank Committee on these guidelines

Tissue Bank Committee

The Tissue Bank Committee is chaired by the Tissue Bank Director and is responsible for promoting the activities of tissue banking (Nather and Wang, 2005).

The following personnel may be included in this committee:

- Orthopaedic surgeon(s): promote bone procurement and utilisation
- Maxillofacial surgeon(s): promote bone utilisation
- Obstetrician(s): promote amnion procurement
- Plastic surgeon(s)/burns surgeon(s): promote amnion utilisation
- Eye surgeon(s): promote cornea procurement and utilisation and amnion utilisation
- ENT surgeon(s): promote soft tissue utilisation
- Pathologist(s)/Microbiologist(s): facilitate serological testing of donors
- Radiation biologist(s)/Radiation licensing authority: facilitate gamma sterilisation of tissues
- Transplant coordinator(s): increase donor pool
- Prominent community worker(s): raise public awareness

Role of Tissue Bank Committee

- Formulate guidelines, as spelt out in the Procedure Manual, regarding procurement, processing, packaging and distribution which will be adopted by the bank
- Create quality assurance policies, as spelt out in the Quality Manual, which will be adopted by the bank
- Monitor the bank's progress
- Raise public awareness on tissue banking
- Raise professional awareness on tissue procurement and transplantation
- Report all tissue bank activities to the Administration Board

Management Committee

The Management Committee is chaired by the Director of Tissue Bank to oversee the day-to-day running of the tissue bank (Nather and Lee, 2007).

The committee includes:

- Director
- Deputy Director
- Technical Manager
- Technologists

The committee has a meeting at least once a month to discuss the following issues:

- Tissues procured in the month
- Tissues utilised in the month
- Administrative matters
- Personnel matters
- Other matters

It is important that the committee functions regularly and efficiently, in order for the tissue bank to run smoothly and effectively to become successful.

References

Nather A (2000). Diploma training for technologists in tissue banking. *Cell Tissue Banking* **1**: 41–44.

Nather A and Lee CCW (2007). Setting up a tissue bank. In Nather A, Yusof N, and Hilmy N (eds.), *Radiation in Tissue Banking*, World Scientific, Singapore, p. 67.

Nather A and Wang LH (2005). Setting up a tissue bank. In Nather A (ed.), *Bone Grafts and Bone Substitutes*, World Scientific, Singapore, p. 155.

Chapter 6

Design of Tissue Banks

Aziz Nather* and Poh Lin Chan*

Introduction

To ensure that a human tissue bank functions smoothly and efficiently, it is important for its philosophy of design to be understood clearly and its physical layout well-planned.

This chapter will discuss the following factors that should be considered when designing a tissue bank:

- Location
- Layout and composition
- Flow of procedures

Location

The proper functioning of any tissue bank is highly dependent on both donor tissue procurement (input) and the distribution of processed graft (output). It is therefore ideal for a tissue bank to be located in vicinity where both procedures can be approached easily, such as near a main district hospital in which procurement and distribution of tissues for application can be done efficiently.

---

* NUH Tissue Bank, Department of Orthopaedic Surgery, Yong Loo Lin School of Medicine, National University of Singapore, Singapore.

Layout and Composition

The design of a tissue bank should achieve the objective of housing and facilitating all tissue banking activities. It would be ideal to contain the following rooms:

- Wet and dry processing laboratories
- Storage room
- Microbiology laboratory
- Office space
- Library
- Meeting room
- Lead-lined room for gamma irradiation (if available, should be housed in a corner section of the bank with limited access)
- Wash room

Unfortunately, almost all countries within the Asia-Pacific region face space and budgetary constraints. As such, good planning is important to ensure that high quality tissue processing can still be achieved despite working in less than optimal physical conditions (Nather and Lee, 2007). Tissue banks facing these limiting factors need not house all of the above-mentioned rooms, but should be able to accommodate the following minimum requirements:

- Two separate rooms for tissue processing by authorised personnel:

  1. Wet processing room
     A wet processing room or an isolation room is for the reception of donor tissues. It allows for wet processing of lyophilised tissue, which includes:

     o Dissection
     o Cutting
     o Washing
     o Pasteurisation

These activities can result in contamination of the environment with water or bone particles and thus should be best restricted in this wet processing area.

2. Dry processing room

    A dry processing room or a clean room should ideally be laminar flow rooms. In the Asia-Pacific region, laminar flow cabinets are used in place of laminar flow rooms. This is due to budgetary concerns with the cost of constructing a room with laminar flow between S$500 000 to S$1 000 000. Besides, with the recommended practice of end-sterilisation of all tissue graft products with 25 kGy of gamma radiation, there is no need for such laminar flow room (Nather and Wang, 2005). This room allows for dry processing of lyophilised tissue which includes:

    o   Freeze-drying
    o   Packaging
    o   Labelling
    o   Sealing

• Documentation room

  A documentation room allows proper filing of all documentation records while maintaining strict confidentiality.

As cleanliness is a crucial aspect of good tissue manufacturing practice, the layout and composition of the bank must allow for easy maintenance and cleaning. For example, surfaces of flooring and fixations should be non-porous so they do not absorb any contamination and require a lesser amount of bactericidal agents. In addition, coverings should be rolled up onto the walls to hide cracks which may harbour contamination.

Flow of Procedures

The layout of the tissue bank should be designed to enable the process of tissue procurement and processing to be conducted smoothly and efficiently.

There are two separate flowcharts for tissue graft production (the two differing mainly in terms of space requirements) that can be adopted. The flowchart selected will play a role in determining the design of the bank.

If sufficient space is available for the construction of the tissue bank, a design which allows the following ideal flow chart for tissue processing (Fig. 1) should be adopted, with separate distribution area and reception/ documentation area.

Tissues received by the bank must first be properly documented in the Documentation Area, prior to storage at −80°C in an electrical freezer in the Wet Processing Laboratory. The tissues may then undergo wet processing as the first part of production of lyophilised tissues. The tissues are then transported to a separate room known as the Dry Processing Laboratory for final processes in the production of freeze-dried tissues. Activities performed here have a very low bioburden and it is therefore vital for them to be conducted separately from those in the Wet Processing Laboratory, where bioburden is high. Only authorised personnel should be allowed into both the Wet and Dry Processing Laboratories. To enforce this, a logbook should be maintained which will allow source tracing should contamination occur. The tissue graft products, packed and sealed, are then transported to

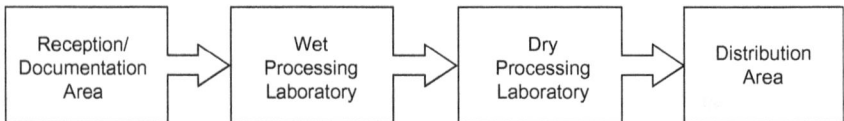

Fig. 1. Ideal flow chart for tissue processing if there is sufficient space.

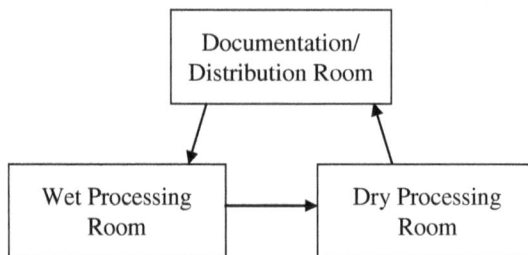

Fig. 2. Modified flow chart for tissue processing where there is limited space.

a gamma irradiation facility for gamma irradiation at a dose of 25 kGy to sterilise the grafts. The sterilised end-products are now ready for use and can be returned to the Distribution Area to be despatched for utilisation.

In cases where space is limited, reception, documentation and distribution work can be performed in the same area. As a result, the following modified flow chart for tissue processing (Fig. 2) should be adopted.

Examples of Building Plans of Tissue Banks in the Asia-Pacific Region

The Model Human Tissue Bank, Sri Lanka (Fig. 3), Malaysian National Tissue Bank, University Science Malaysia, Kota Bahru, Kelantan,

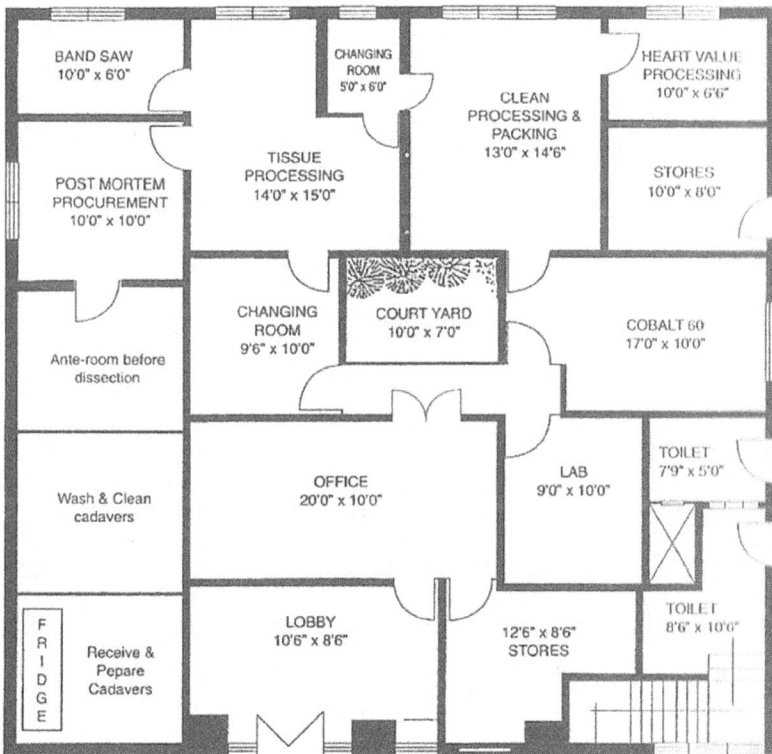

Fig. 3. Building plan of the Model Human Tissue Bank in Sri Lanka.

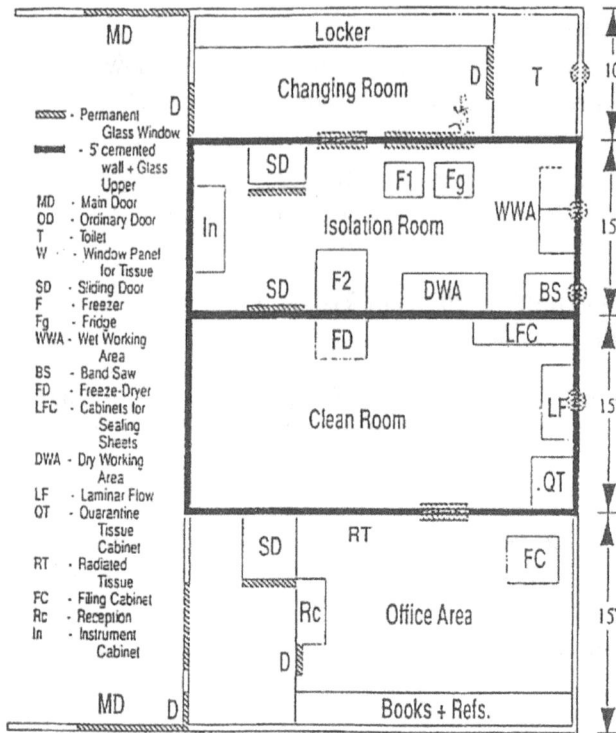

The two rooms; Isolation Room and Clean Room
are minimum requirements and could represent the initial provision

Fig. 4. Building plan of the Malaysian National Tissue Bank in Kelantan, Malaysia.

Malaysia (Fig. 4) and Tata Memorial Hospital Tissue Bank, Mumbai, India (Fig. 5) are examples of tissue banks that have met the minimum requirements for building design of a tissue bank.

The Model Human Tissue Bank is a purpose-built facility which demonstrates all the ideal principles and practices. However, there are also many other tissue banks that function effectively, despite fewer resources and physical space. Nevertheless, the minimum requirements mentioned earlier should be maintained.

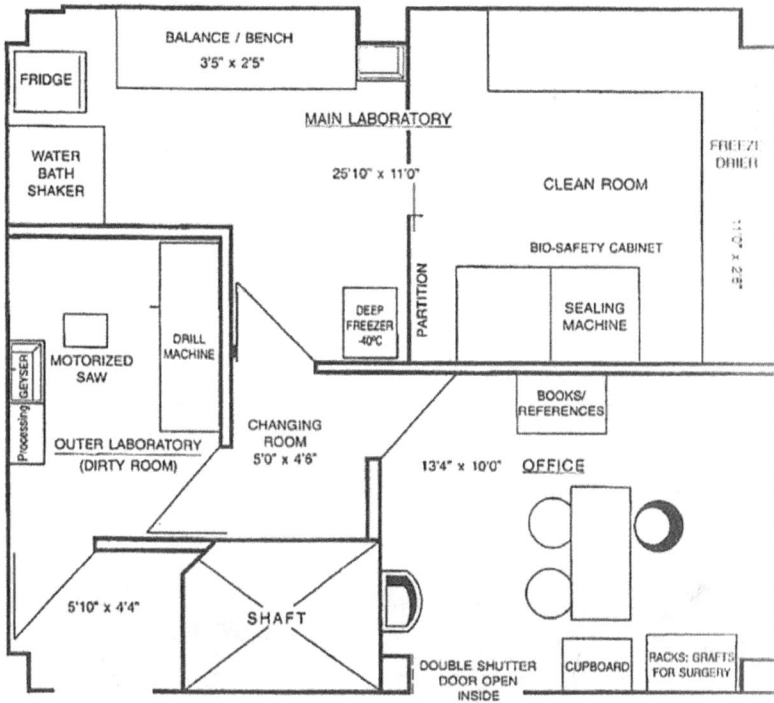

Fig. 5. Building plan of the Tata Memorial Hospital Tissue Bank in Mumbai, India.

Case Study: NUH Tissue Bank's Building Plan

The NUH Tissue Bank is located within the National University Hospital to facilitate the procurement of tissues and the application of sterilised bone graft products.

The building plan of the NUH Tissue Bank is shown in Fig. 6. As shown in the plan, the NUH Tissue Bank has met the minimum requirements of building design of a tissue bank, i.e. two separate rooms for processing and an area for documentation.

The NUH Tissue Bank consists of a documentation room, flanked by a Wet Processing Laboratory (Fig. 7) and a Dry Processing Laboratory

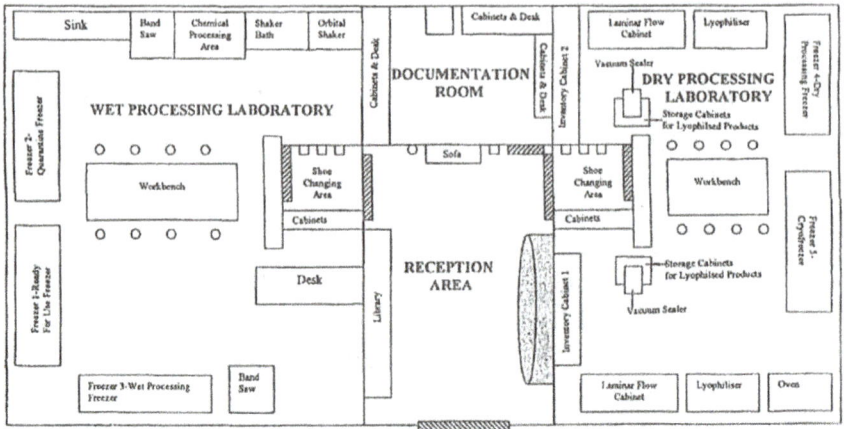

Fig. 6. Building plan of the NUH Tissue Bank in Singapore.

Fig. 7. Wet Processing Laboratory.

(Fig. 8), along with a central bay for reception/documentation/distribution (Fig. 9). Each laboratory has its own changing area to allow tissue bank technicians to change his/her attire and shoes before entering the laboratory.

Fig. 8. Dry Processing Laboratory.

Fig. 9. Reception/documentation/distribution area.

References

Nather A and Lee CCW (2007). Setting up a tissue bank. In Nather A, Yusof N, and Hilmy N (eds.), *Radiation in Tissue Banking*, World Scientific, Singapore, p. 67.

Nather A and Wang LH (2005). Setting up a tissue bank. In Nather A (ed.), *Bone Grafts and Bone Substitutes*, World Scientific, Singapore, p. 155.

Chapter 7

Equipments and Facilities in Tissue Banks

Poh Lin Chan* and Aziz Nather*

Introduction

In addition to well-trained tissue bank personnel, a tissue bank should also possess good facilities for it to operate smoothly and manufacture grafts efficiently. The bank should be equipped with the following basic facilities to support the tissue banking activities within it:

- Clean and adequate water supply
- Laminar flow provision
- Emergency electrical supply
- Biohazard disposal facility
- Equipment for tissue graft production

Clean and Adequate Water Supply

Water is an integral part of tissue manufacturing as it is both a solvent and cleaner. The water supply utilised by the bank should be both clean and adequate. In developing countries, water contamination is often a problem. If the water is contaminated by suspended particles and dirt, filtration

---

*NUH Tissue Bank, Department of Orthopaedic Surgery, Yong Loo Lin School of Medicine, National University of Singapore, Singapore.

should be carried out. Water may also be contaminated with microorganisms. In such cases, a water purification plant/facility may be required.

Laminar Flow Provision

A laminar flow room with filtered incoming air is ideal as it creates a pollution-free environment for the procurement and processing of sterile tissues. This could potentially reduce the dosage of radiation sterilisation required and hence, allow for greater tissue integrity preservation after radiation. Unfortunately, such facilities are costly and countries with limited financial resources are unable to afford these rooms. Alternatively, tissue banks could install more affordable laminar flow cabinets instead.

Emergency Electrical Supply

Equipments in the tissue bank are highly sensitive and usually run overnight. Hence arrangements must be made for a regular and stable electricity supply. In areas where termination of power is frequent, certain equipments such as electrical freezers and lyophilisers have to be backed up by an uninterruptible power supply (UPS). As the UPS only lasts for a limited period of time, an electrical generator has to be provided should the duration of power failure exceed the operating period of the UPS.

Tissue banks within hospitals possess the advantage of having red sockets installed in their banks. These red sockets are linked to the emergency backup generators within the hospital, and are capable of automatically providing emergency electrical supply in the event of an electrical failure. The NUH Tissue Bank is an example, with all five electrical freezers provided with red sockets (Nather, 2000). The power line of these red sockets in the NUH Tissue Bank is linked to the backup generator for the Operating Theatre Complex, allowing equipment in the bank to continue functioning despite a power failure.

If red sockets are not available, as in the case of most non-hospital tissue banks, all electrical freezers must be equipped with battery-operated CO_2 backup systems (Fig. 1). This backup system works by

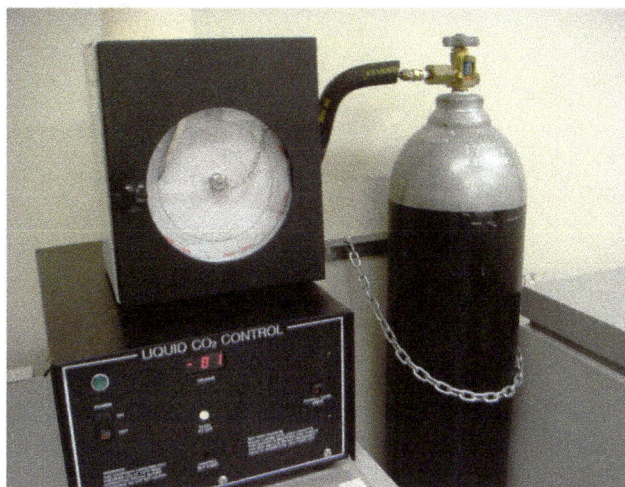

Fig. 1. Liquid CO_2 backup system for electrical freezers in the event of a power failure.

automatically infusing liquid CO_2 from a cylinder into the freezer when the cabinet temperature of the freezer rises to above $-65°C$ during an electrical failure, preventing the temperature from rising further (Nather and Wang, 2005).

Biohazard Disposal Facility

Facilities for the speedy and safe disposal of biohazard wastes must be available. In a hospital facility, this is easily provided, as biohazard wastes are collected in disposal bags and sent to the mortuary for disposal by incineration.

Equipment for Tissue Graft Production

Equipment are important assets to the tissue bank, as they facilitate the manufacturing of high quality tissue products. Equipment not only allow precise conditions (temperature, pressure, etc.) required during the manufacturing process to be attained, but also present a defined set of procedures that can be easily followed and replicated.

Fig. 2.   A −80°C electrical freezer with thermograph and liquid CO_2 backup system.

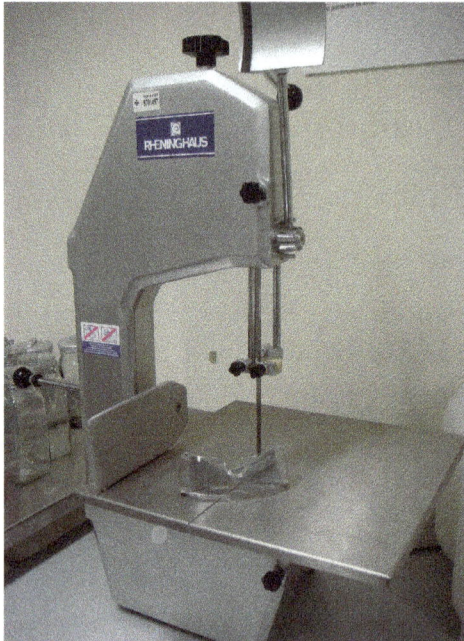

Fig. 3.   A stainless steel band saw.

Different processes are conducted in the Wet Processing Laboratory and the Dry Processing Laboratory and hence, require different sets of instruments. Furthermore, the bioburden in the Wet Processing Laboratory is higher and thus, instruments used within this room should not be used during processing in the Dry Processing Laboratory. The following sections describe the basic equipment found in a Bone Bank.

Equipment in a Wet Processing Laboratory

The equipment in this laboratory in a Bone Bank include:

1. *–80°C electrical freezer with thermograph and liquid CO_2 backup system (Fig. 2)
2. –160°C liquid nitrogen cryo freezer
3. Table for dissection
4. Wash basin with facility for jet flushing
5. *Stainless steel band saw (Fig. 3)
6. *Chemical processing instruments
7. *Shaker bath (Fig. 4)
8. *Orbital-wrist shaker (Fig. 5)
9. Autoclave machine (Fig. 6)

Equipment in a Dry Processing Laboratory

Equipment in this laboratory in a Bone Bank include:

1. *Lyophiliser (Fig. 7)
2. *Laminar airflow cabinet (Fig. 8)
3. *Vacuum sealer (Fig. 9)
4. *Electronic balance (Fig. 10)
5. *Oven

---

*Minimum equipment required.

Fig. 4. A shaker bath.

Fig. 5. Orbital-wrist shaker.

Fig. 6. Autoclave machine.

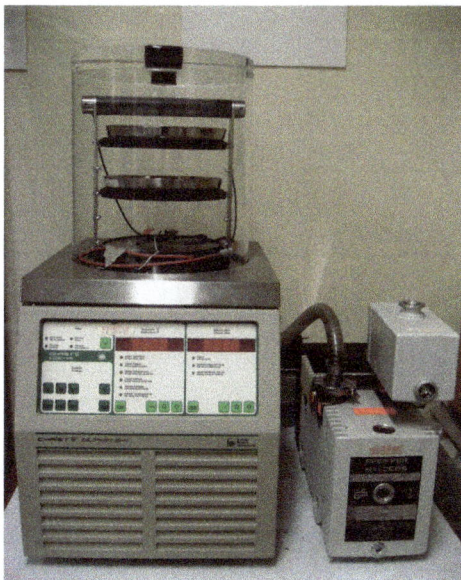

Fig. 7. A lyophiliser.

Fig. 8. A laminar airflow cabinet.

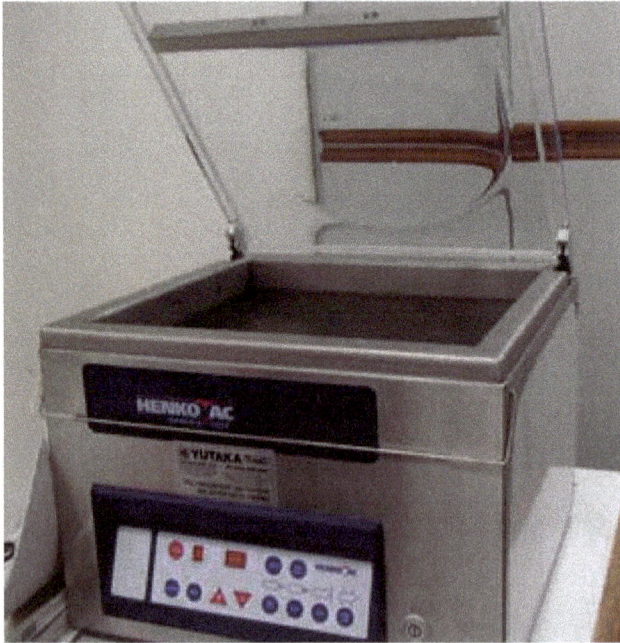

Fig. 9. A vacuum sealer.

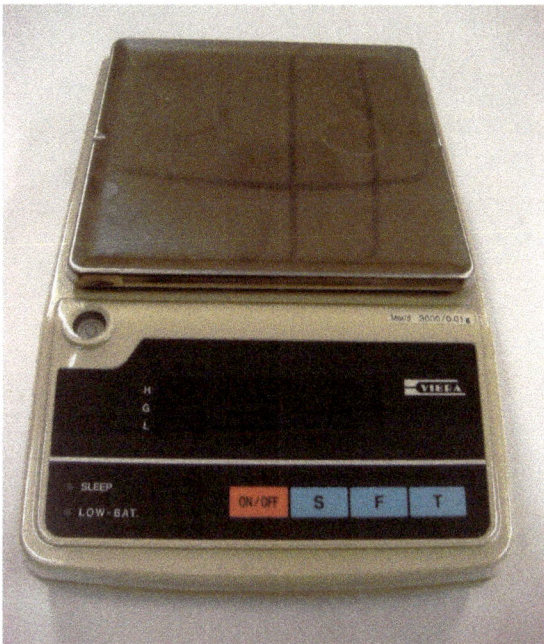

Fig. 10.   An electronic balance.

References

Nather A (2000). Organisation systems. In Phillips GO (ed.), *Radiation and Tissue Banking* World Scientific, Singapore, p. 237.

Nather A and Lee CCW (2007). Setting up a tissue bank. In Nather A, Yusof N, and Hilmy N (eds.), *Radiation in Tissue Banking*, World Scientific, Singapore, p. 67.

Nather A and Wang LH (2005). Setting up a tissue bank. In Nather A (ed.), *Bone Grafts and Bone Substitutes*, World Scientific, Singapore, p. 155.

Chapter 8

Setting Up a Tissue Bank

Aziz Nather* and Poh Lin Chan*

Introduction

Setting up a tissue bank is a serious business, requiring careful consideration of various factors, the availability of financial and physical assets, as well as commitment and resolution of the leaders setting up the bank. This chapter discusses the fundamental issues that are of concern to an individual interested in establishing a tissue bank.

Factors Affecting the Success of a Tissue Bank

In setting up a tissue bank, the following factors need to be considered in detail:

- Religion and culture of the population in the country
- Legal status of organ or tissue procurement and transplantation in the country
- Level and extent of government assistance, in particular, assistance from the Ministry of Health
- Support from the Institution involved (hospital, university or radiation institution)

*NUH Tissue Bank, Department of Orthopaedic Surgery, Yong Loo Lin School of Medicine, National University of Singapore, Singapore.

- Demand for allograft transplantation in the country
- Commitment of the personnel setting up the tissue bank

Religion and Culture

These two issues play a crucial role in determining the success of a tissue bank. Tissue banking is likely to succeed in a country where the predominant religion is Buddhism. This is due to the philosophy of Buddhism being in favour of organ and tissue donation (Nather, 2000a). In Sri Lanka, Thailand and Vietnam, where the population is predominantly Buddhists, donors are often forthcoming.

In contrast, countries with a predominantly Muslim population, such as Bangladesh, Pakistan, Malaysia and Indonesia, face a shortage of organ and tissue donors. Whilst Islam does not explicitly prohibit tissue donation, culturally, the Muslims prefer to return to God whole, burying everything alongside a deceased's body, including amputated limbs, foreskin from circumcision, and even amnion from delivery (Nather, 2000a).

Legal Status

The presence of a Tissue Transplantation Act has been a significant factor in promoting the development of tissue banks in several countries, such as Singapore, Hong Kong and India (Nather, 2000a). In Japan, however, tissue donation has not progressed significantly, even with the concept of brain death legislated in 1997. In contrast, the introduction of the brain death concept into law in Korea in 2000, has since allowed tissue banking to flourish and spurred the setting up of many tissue banks. On the other hand, there are countries, such as China and Thailand, where the absence of procurement and transplantation laws have not impeded the progress of tissue banking. Nevertheless, it is important to consider the possible effects of the presence or absence of such laws in one's country before embarking to set up a tissue banking programme.

Government Assistance

A tissue banking programme (usually in affiliation with kidney, liver and cornea transplantation programmes) is likely to succeed in countries where the government — particularly, the Ministry of Health — is keen to provide assistance to the programme. Strong government support has been responsible for the success of tissue banking in Singapore, Hong Kong and Malaysia (Nather, 2000a). In contrast, the development of tissue banking is laden with obstacles in countries where government priority is on other aspects of health.

Support from the Institution Involved

Equally important, if not more, the institution involved — hospital, university or radiation institution — must provide full support, including provision of manpower requirements, physical space for building the tissue bank and financial grants for running the tissue bank. In several instances, financial grants are obtained as research grants for conducting basic and clinical research on tissue allografts. As the bank begins to offer a clinical service, more grants gradually become available from institutions (or even the government). If the bank is not set up in a radiation facility, support must also be available in the beginning from a radiation institute to provide gamma irradiation of tissue grafts.

Demand for Allograft Transplantation

Before any tissue bank can progress, there must be a demand for tissue grafts. Despite the large number of donors available, the Model Human Tissue Bank in Sri Lanka could not grow due to the small number of grafts being utilised by surgeons in the country (Nather, 2000a). Several campaigns had been carried out to raise professional awareness on tissue banking and transplantation in the country but they have been unsuccessful thus far.

Commitment of the Personnel Setting Up the Tissue Bank

The single most important factor for the success of a tissue bank is the vision, mission and commitment of the director which must be equally shared with members of the team, who are responsible for setting up and running the bank. Setting up a tissue bank is a complex venture. Dynamism and devotion are required to face the numerous obstacles — from the institution, government, etc. — that need to be overcome. Without a committed team who is prepared to persevere and overcome these obstacles, the bank is unlikely to succeed (Nather and Wang, 2005).

Planning Required for Setting up a Tissue Bank

From the beginning of planning, it must be decided which organisation or institution will be responsible for supporting and financing the tissue bank. Approval must first be sought from the relevant organisation or institution to accept this responsibility before any plan to set up a tissue bank can proceed further.

In the Asia-Pacific Region, tissue banks have been developed by three different institutions or organisations:

1. Universities
2. Hospitals
3. Radiation institutions

In Singapore, the NUH Tissue Bank was established as a hospital venture because it was perceived to be financially viable as a cost centre of the National University Hospital (Nather and Wang, 2002).

Following this, careful planning is required on all aspects of tissue procurement and tissue transplantation, in order to set up a tissue bank which can function efficiently and productively.

Financial Considerations in Setting Up a Tissue Bank

Setting up a tissue bank is akin to starting a business venture — it requires investments. It is important to consider the financial aspects of the following factors when establishing a tissue bank:

- Manpower
- Building space and design
- Facilities and equipment

Manpower

The tissue bank should be run by a minimum of two full-time laboratory technologists for it to function efficiently. The technologists must be proficient and have received prior training in the following aspects of tissue banking:

- Donor selection
- Procurement techniques
- Processing techniques
- Documentation
- Distribution of tissue grafts

In the case of Singapore, NUH Tissue Bank staff is required to have at least a diploma in tissue banking from the National University of Singapore (Nather, 2000b).

Labour costs are high in Singapore; each laboratory technologist is paid a salary of ≈ S$30 000 a year. Thus, the annual manpower cost for running a tissue bank is at least S$60 000 a year.

Building Space and Design

Capital may be required to obtain premises for the setting up of the tissue bank. In the case of a tissue bank housed in its own building, funds are necessary for the construction of the tissue bank facility. For example, the purpose-built Model Human Tissue Bank in Sri Lanka required US$62 500 to secure a piece of land measuring 25 by 25 m² and US$66 667 for the construction of a 20 by 20 m² tissue bank building. In addition, renovation costs are also required to convert the space given into the various rooms and laboratories according to the adopted flow chart for tissue processing.

Facilities and Equipment

Expenditure is required for the setting up of certain facilities, such as the creation of a water purification facility, installation of power lines, as well

Table 1. Minimum equipment costs.

Equipment	SGD ($)	USD[a] ($)	MYR[a] ($)	IDR[a] ($)
−80°C electrical freezer with thermograph, liquid CO_2 backup	30 000	21 131	73 653	205 818 120
Stainless steel band saw	7000	4932	17 189	48 037 766
Shaker bath	3000	2114	7369	20 591 966
Orbital-wrist shaker	2000	1410	4915	13 733 784
Lyophiliser	30 000	21 131	73 653	205 818 120
Laminar airflow cabinet	10 000	7049	24 570	68 649 565
Vacuum sealer	6000	4228	14 746	41 201 352
Oven	10 000	7049	24 570	68 649 565
Electronic balance	2000	1410	4915	13 733 784
Total	93 000	65 553	228 518	638 440 958

[a]Rates as of September 2009.

as setting up a laminar flow facility. As for equipment costs, Table 1 serves as a guide to the estimated costs for the minimum equipment that must be acquired to set up a tissue bank (Nather and Lee, 2007). As indicated in the table a fund of about S$100 000 is necessary to purchase the minimum equipment required in a bank.

Financial Considerations in Running a Tissue Bank

After the tissue bank has been set up, the director has to submit an annual budget for approval by the funding institution. Only then can the funding institution provide the necessary funds to meet the annual costs incurred by the tissue bank. An example of an annual budget is shown in Table 2 (Nather and Lee, 2007).

Tissue banking is a business venture, requiring capital expenditure, maintenance costs, manpower costs, etc. On average, an allowance of about S$100 000 is provided by a grant from a hospital or a relevant institution to fund an annual running cost of approximately S$84 500 (Table 1). Unfortunately, no organisation is willing to fund such a large amount indefinitely. Instead, the supporting institution usually funds the running costs for a period of between two to five years, after which the tissue bank is expected to become self-sustaining.

Table 2.   Example of an annual budget.

	SGD ($)	USD[a] ($)	MYR[a] ($)	IDR[a] ($)
Manpower costs (salary for 2 technologists)	±60 000	±42 289	±147 378	±411 897 392
Equipment maintenance costs	±10 000	±7049	±24 570	±68 649 565
Electricity	±10 000	±7049	±24 570	±68 649 565
Water consumption	±1500	±1057	±3684	±10 297 434
Consumables	±10 000	±7049	24 570	68 649 565
Total	84 500	59 565	207 587	580 088 827

[a]Rates as of September 2009.

To achieve this, tissue processing costs must be charged to recipients utilising the tissue grafts provided by the tissue bank, on the condition that the recovery of such tissue processing costs is legal in the country of concern. To arrive at the actual cost to be charged for tissue processing, the tissue bank must do a detailed analysis of all the running expenditure in one year versus the number of tissue grafts processed annually, and then calculate the production costs for each type of graft produced, i.e.

Average cost per graft
= (annual running expenditure)
÷ (number of tissue grafts processed annually).

The recommended tissue processing cost charged is 20% higher than the cost of production in order to allow for cost recovery.

Conclusion

Setting up a tissue bank is a serious business requiring detailed planning and approval from various levels, including the department, the institution and the government. It must be approached in an organised and comprehensive manner as such to:

1. Secure space allocation from an institution.
2. Obtain approval from the institution to fund this venture.
3. Obtain approval from the key personnel in the institution to be part of the manpower organisation structure that has to be established.

4. Initiate plans for the building design of the tissue bank.
5. Recruit technologists to run the bank.
6. Purchase necessary equipment in stages (after funds have been procured from a research grant).

References

Nather A (2000a). Tissue banking in Asia Pacific region — Ethical, legal, religious, cultural and other regulatory aspects. *J ASEAN Orthop Assoc* **13**: 60–63.

Nather A (2000b). Diploma training for technologists in tissue banking. *Cell Tissue Banking* **1**: 41–44.

Nather A and Lee CCW (2007). Setting up a tissue bank. In Nather A, Yusof N, and Hilmy N (eds.), *Radiation in Tissue Banking*, World Scientific, Singapore, p. 67.

Nather A and Wang LH (2002). Bone banking in Singapore — fourteen years of experience. *J ASEAN Orthop Assoc* **15**: 20–29.

Nather A and Wang LH (2005). Setting up a tissue bank. In Nather A (ed.), *Bone Grafts and Bone Substitutes*, World Scientific, Singapore, p. 155.

Part III

Rules and Regulations

Chapter 9

Ethical, Religious, Legal and Cultural Issues for Tissue Banking

Aziz Nather*, Jia Ming Low* and Yi Lin Sim*

Introduction

Each country has its own set of values regarding ethics, religious beliefs, legal considerations, and cultural issues. These values will affect the development of tissue banking in the country. This chapter discusses the common issues encountered in tissue banking in the Asia-Pacific region.

Ethical Considerations

Donation of tissue is an act of humanity. It enables a person to alleviate the sufferings of fellow human beings. Tissue banks should not be allowed to sell tissues. Rather, they should provide tissue grafts on a non-commercial basis. There should not be any profit motive in the operation of tissue banks. Since costs are incurred during procurement, processing, and distribution, tissue banks may incur "processing costs", provided the law in the country (if any) makes provisions to allow for the retrieval of such costs. A common acceptable practice in institution banks is to work out the total costs of such procurement and processing — including man-power costs, equipment maintenance costs, cost of consumables, electricity and water consumption costs, etc. — and then charge the recipients using

*NUH Tissue Bank, Department of Orthopaedic Surgery, Yong Loo Lin School of Medicine, National University of Singapore, Singapore.

these tissues to pay for such costs (Nather, 2000). Some tissue banks in the Asia-Pacific region charge processing costs, e.g. Japan, Singapore, Malaysia, Sri Lanka, and India.

In countries in the Asia-Pacific region, there are very few commercial tissue banks operating in most countries. Korea is the only country in the Asia-Pacific region which has two or three commercial tissue banks. In the remaining countries, commercial banks are generally not favored. The view held by most countries in terms of developing commercial or private banks is to favor not-for-profit institutions, and most of them are against for-profit institutions if commercialisation is to take place at all.

In Europe, tissue banks comply with the Ethical Code of the European Association of Tissue Banks (EATB, 2003). In the United States, the American Association of Tissue Banks (AATB) provides standards to help ensure that tissue banks follow good ethical practice. However, no ethical code has been produced in the Asia-Pacific region. Nevertheless, tissue banks in this region comply with all the principles embodied in the Ethical Code of EATB. The guiding principle followed by all tissue bank operators is based on moral principles, human duty, and proper conduct, as enshrined in the Hippocratic Oath (which requires doctors and health workers to (*non nocere* or "not injure")).

Tissue banking helps to reduce a country's healthcare costs (Hachiya *et al.*, 1999). Allogeneic bone grafts are less expensive than custom-made prostheses or ceramics, which can be costly. Commercially produced bone allografts are also prohibitive in costs. However, bone allografts produced by non-commercial or institution tissue banks are often either provided free or at nominal costs ("processing costs only"), hence reducing the patient's medical costs.

Religious Considerations

Tissue donation is a sensitive issue that invokes important concerns regarding the dignity of the living and the dead, the concept of brain death, and the belief that tissue donation is the greatest gift one can bestow upon fellow human beings after one's death. The answers to these questions are inextricably tied to the dominant religious and cultural mindsets within each country. In this regard, the culture of an ethnic group is often inseparable from the religion followed by that group. Hence, religion plays a major role in promoting or retarding the development of tissue banking in each country. In the Asia-Pacific region, Buddhism, Islam, Christianity, and Hinduism are the major religions.

Buddhist View on Tissue Donation

Buddhism is in perfect agreement with tissue donation (Nather, 2000). Buddhism honours those people who donate their bodies and organs to advance medical science and to save lives. In Buddhist scriptures, there are stories where the donation of tissues has been referred to as acts of charity that earn merit. Buddhists are expected to meditate about the impermanence of life. The body will decay, just as a beautiful fragrant flower withers and decays. The concept of tissue donation is encouraged not only after death; even while living, tissue donation is considered to be a noble act.

In countries where Buddhism is the predominant religion, there is no shortage of tissue donors. These countries include Vietnam, Thailand, Sri Lanka, and Myanmar. The most successful public awareness programs on tissue donation have been achieved in Sri Lanka, Thailand, and Vietnam. The decision to set up a Model Human Tissue Bank in Sri Lanka by the International Atomic Energy Agency was greatly influenced by the world-renowned success of the Eye Donation Society — which, led by Dr Hudson Silva, achieved its target of procuring 40 000 eyes by May 1999 — coupled with the abundance of tissue donors in this predominantly Buddhist country (more than 90% of the population are Buddhists).

Buddhism is also one of the major religions in Korea (about 50%) and Singapore (about 42.5%). The success of the National University Hospital Tissue Bank in Singapore is largely due to the fact that the Buddhist community in Singapore strongly supports the tissue transplantation program. Most of the tissue donors in Singapore are Buddhists.

Islamic View on Tissue Donation

Muslims are by far the most controversial group for tissue donation. The Islamic states in the Asia-Pacific region include Malaysia, Bangladesh,

Pakistan, and Brunei. In addition, Islam is the predominant religion in Indonesia, a secular country that follows the five principles of *Pancasila*. There are about 200 million Muslims in China, another secular country. Islam is also an important religion in other secular countries such as Singapore (about 15%) and India (about 15%) (Nather, 2000).

The Koran respects life and values the needs of the living over those of the dead. In 1983, the Muslim Religious Council initially rejected organ donation, but it has now reversed its position. This means that organ donation and transplantation can be considered in circumstances when it would save a person's life. No mention is made about allowing transplantation to improve the quality of life of a recipient. As a result, Muslims are more likely to allow kidney donation and less likely to allow tissue donation, as the latter is perceived as merely improving the quality of (rather than saving) life. Nevertheless, while interpretations of the Koran vary according to different religious leaders, e.g. between the *ustazs* and the *ulamas*, tissue donation is not explicitly forbidden in the Koran (Nather, 2000).

Countries with a significant Muslim presence have their own muftis. A *mufti* is a religious official who is appointed by the government to deal with all Islamic matters in the country, including the issue of organ and tissue donation. *Fatwas* are religious rulings made by a fatwa committee as an official stand by the government on certain issues, e.g. the tissue donation and transplantation issue. The fatwa committee — chaired by the mufti — may include prominent religious leaders, doctors, lawyers, and members of the public.

A common misconception among Muslims is that organ and tissue donation is not permitted by the Islamic Law. However, fatwas concerning organ donation have been declared in several countries in the Asia-Pacific region, including Brunei, Malaysia, and Singapore. A fatwa was passed in Saudi Arabia in 1985 sanctioning both the live and cadaveric donation of organs. Likewise, in 1998, a fatwa was passed in the United Arab Emirates that sanctioned live and cadaveric organ donation, as well as organ donation from Muslims to non-Muslims, and also accepted the concept of brain death (El-Shahat, 1999).

A milestone event for *fatwas* specific to tissue donation occurred on September 4, 1995, when the first fatwa on bone, skin, and amnion was introduced by the Malaysian Islamic Centre. This was followed on June 29, 1997, by a fatwa on bone, skin, and amnion in Indonesia, sanctioning tissue procurement from deceased donors. This was a great leap forward for Indonesia, as the previous law — the Indonesia 1992 Health Regulation — allowed tissue procurement only from living donors. Unfortunately, efforts by the author since 1995 to seek a Similar Fatwa for Muslims in Singapore from the *Majlis Ugama Islam Singapura* (Religious Council of Islamic Singapore) have not succeeded so far, although a fatwa for cornea was passed in 1999 — the first fatwa for tissues in Singapore.

However, fatwas are not legally binding, and so the decision to donate remains very much the prerogative of the individual and his/her family. Hence, the introduction of favourable fatwas is only the first step in promoting public acceptance of tissue donation among Muslims.

Another important consideration for Muslims is that they must bury the body as soon as possible after death. Therefore, procedures like tissue procurement, which may delay the burial, are not taken very kindly.

Culturally, Muslims accept that God created them whole and they prefer to return to Him whole. It is a common practice among many Muslims to bury amputated limbs, foreskins from circumcision, and amnions from delivery. Not all Muslims follow this practice as it is a cultural practice and not a religious requirement.

Therefore, it does not come as a surprise that there is a big shortage of bone donors in countries where Islam is the predominant religion, including Pakistan, Bangladesh, Malaysia, and Indonesia. The lack of donors also slows down the development of tissue banking in these countries. They have been successful in procuring amnion, but not bones and ligaments, from deceased donors.

In Islam, the *waris* issue is also important. When a person dies, the *waris* or "next of kin" plays a key role as to what happens to the body of the deceased. Even if the donor has consented, the next of kin must also consent if tissue donation is to be allowed. For Muslims, therefore, consent is required not only from the patient, but also from his/her next of kin or *waris*.

Nevertheless, the demand for tissues in these countries is great and steadily increasing. Indonesia and Malaysia have resorted to the production of bovine bone xenografts for bone transplantation until such time that public awareness programs can produce better results (Nather, 2000). Specific fatwas for bone have been passed in both countries, but they have not produced significant changes in the Muslim population's attitude towards tissue donation. More public education is needed to change entrenched cultural practices and beliefs, along with the passing of fatwas, before more Muslims will come forward and pledge to be tissue donors.

Christian View on Tissue Donation

Christianity is the predominant religion in the Philippines and Australia (Nather, 2000). It is also one of the major religions, though not the predominant one, in Korea (about 26.3%) and Singapore (about 14.6%). Tissue donation is considered to be consistent with the ecclesiastical Christian dogma of loving one's neighbour as oneself, as it is thought to be an act of genuine altruism — of giving something up at little or no cost to the donor to save the lives of others. This was reiterated by the late Pope John Paul II while attending the Congress of the Society for Organ Sharing on June 20, 1991, when he mentioned the words of Jesus Christ as narrated by the apostle Luke: "'Give, and it will be given to you; good measure, passed down, shaken together, running over, will be put into your lap' (Luke 6:38). We shall receive our supreme reward from God according to the genuine and effective love we have shown to your neighbour." These words are in full support of organ and tissue donation and transplantation. Christian communities in Europe and the USA support tissue transplantation. For instance, on the weekend of November 13–15, 1998, churches and synagogues across the US encouraged their faithful to sign donor cards (*Japan Times*, 1998).

Despite the strong Christian presence in the Philippines and Korea, however, other factors (including cultural factors) have led to a shortage of donors in these countries. The cultural concept is again that God created them whole and they would like to return to Him whole, not physically altered by the act of tissue donation.

Hindu View on Tissue Donation

Hinduism is the predominant religion of India, a secular country. It is also an important religion in Sri Lanka (about 7.1%), Singapore (about 4%), and Malaysia (6.3%).

Hinduism is parallel to Buddhism in many ways, both religions practice cremation of the body, which is in fact an act of destruction of the body, in front of and with full knowledge of the relatives. Hindus are not prohibited by religious law from donating their organs and it has no objection to tissue donation and transplantation. Donation is an individual decision, but "dharma" ("good duty") suggests that doing good deeds for others is desirable. (London Health Sciences Centre, 2005) Therefore, resistance to the concept of tissue donation is not expected (Nather, 2000).

Legal Considerations

There is no universal law governing tissue procurement and tissue transplantation for the various countries in the Asia-Pacific region. If regulatory laws are present in some of these countries, they are based on similar human transplantation acts practiced in Europe and the USA. These acts cover a wide range of issues, which include the definition of brain death, the definition of tissues and organs, the issue of consent for organ donation (either by the donor or next of kin), and the prohibition of trade in human organs.

Two different legal frameworks are seen to be operating in the Asia-Pacific region: the "opting-in" system based on informed consent, and the "opting-out" system based on presumed consent. For example, while Malaysia follows the opting-in system for kidneys and corneas, Singapore follows the opting-out system under the 1987 Human Organ Transplant Act for kidneys only; in 2004, this opting-out act was extended to include corneas. It should be noted that in almost all countries, these laws are specifically designed for organ transplantation. Tissue procurement can be carried out only by following such laws for organs.

Singapore

It is stated that tissue procurement follows the Medical (Therapy, Education and Research) Act of 1973, whereby "any person of sound mind and eighteen years of age or above may give all or any part of his body for education ...transplantation.... The gift takes effect upon death" (Nather, 2000).

In 1987, the Human Organ Transplant Act (HOTA) was passed. It is based on presumed consent, and "makes provision for the removal of kidneys from the bodies of persons who are citizens or permanent residents, who have died from accidents, for transplantation purposes only. Muslims and persons over 60 years old are exempted from provisions of this Act." The seven criteria for brain death are also described in this Act. An amendment to the Act in 2004 extended the list of tissues procured to include the heart, liver, and cornea, and extended the donor pool to non-accidental causes of death, among other changes. With effect from 2008, Muslims are also included in the HOTA act.

Australia

The donation of human tissue is regulated by the legislation in each of the eight states and territories under substantially uniform acts (known as the Human Tissue Act in some states, and as the Transplantation and Anatomy Act in others), which were passed in the late 1970s and early 1980s. Most provisions require consent from the donors or from the families of brain-dead, heart-beating donors, with the exception of tissues removed at autopsies that can be used for transplant, therapeutic, educational, and research purposes without further reference to the next of kin. Nonetheless, the tissue banking sector in Australia has, for the most part, sought to include consultation with next of kin in their protocol for practical and ethical purposes (Ireland and McKelvie, 2003).

India

The procurement of tissues for transplantation is governed by the Transplantation of Human Organs Act, which was enacted in 1994.

However, as healthcare comes under the purview of state governments, this Act is applicable only when a state adopts it. Fourteen states have yet to approve of it (Gajiwala, 2003).

A human organ as defined by this Act is any part of the human body consisting of a structured arrangement of tissues, which cannot be replicated by the body if wholly removed. There is no specific mention of particular organs. Under the Act, therapeutic purposes are defined as the systemic treatment of any disease or the measures to improve health according to any particular method or modality. The Act recognises brain stem death, which needs to be certified by four qualified medical practitioners approved by the state. The removal of organs is subject to prior written consent from the deceased or from the person who is lawfully in possession of the dead body, with the exception of the removal of organs from bodies sent for postmortem examination. In the latter case, the Act authorises the removal of organs for therapeutic purposes, provided there were no known objections from the deceased person.

While regulations specific to tissue banking have yet to be developed, the Transplantation of Human Organ Rules was issued by the Government of India in 1995 to combat the illegal trading of human organs. It has since been adopted by the state of Maharashtra (Gajiwala, 2003).

Malaysia

The transplantation of cadaveric tissues in Malaysia is governed by the Human Tissues Act 1974. The Act enables the removal of tissues from cadavers for therapeutic, medical education, and research purposes under two conditions: at the express request of the donor, which may be given at any time either in writing or orally stated during the deceased's last illness in the presence of two witnesses; or in the absence of objection from the deceased and with the consent of the next of kin (Kassim, 2005).

In this act, the word "tissue" is not defined. Likewise, "the person lawfully in possession of the body" is not defined, nor is there an articulation of a hierarchy of relatives deemed to be the next of kin. More significantly, the current Act does not provide an exact definition of death. Presently, the Act only requires two fully registered medical practitioners to confirm (upon personal examination of the body) that life is extinct. There is no inclusion of brain death in this Act as a method of determining death, although brain-dead donors are a source of organs in cadaveric organ transplantation. With regard to fatwas specific for tissue donation, the first fatwa on bone, skin, and amnion was introduced by the Malaysian Islamic Centre on September 4, 1995 (Nather, 2000).

Sri Lanka

Sri Lanka follows The Human Tissue Transplantation Act No. 48 of 1987, which requires consent from the donor or next of kin.

The Philippines

Tissue donation in the Philippines follows The Republic Act 7170, 1991, which authorises the legacy or donation of all or part of the human body after death for specified purposes (Nather, 2000).

Vietnam

Tissue procurement is provided for by the Civil Code, Article 32, Chapter 2, where consent is needed from the donor or next of kin; and by The People's Health Protection Code, Chapter 4, which provides for tissue transplantation (Nather, 2000).

Indonesia

In the Asia-Pacific region, Indonesia is unique in that its legislation for tissue procurement is incomplete. The Indonesian 1992 Health Regulation provides for the procurement of tissues from living donors, but not from deceased donors. A fatwa for bone, skin, and amnion was introduced by the Religious Council on June 29, 1997, permitting tissue procurement from cadaveric donors (Nather, 2000).

Japan

In Japan, a new law concerning human organ transplants was passed in 17 October 1997 (K. Ota, 1999). According to the Law, organs can be removed only if the donor has expressed his/her intention with respect to the definition of death and organ donation in a written document beforehand, and if the family has already signed the donor card and also agreed with organ removal (the family has the authority to veto an individual's organ donation by refusing to sign the donor card). Brain death is defined under this Law as "an irreversible cessation of all functions of the entire brain including brain stem". Under the Law, individuals are able to choose a definition of death, either brain death or traditional cardiac death, according to their own personal views on human death.

As of December 2005, only 33 brain-dead cases have been used for organ transplantation (Bagheri, 2005). The first of these cases was performed February 28, 1999, when a liver, a heart, and two kidneys were legally procured from a brain-dead donor in Kochi and air-flown for transplantation to four recipients in four other Japanese cities (Reuters/AFP, 1999). It was hoped that this landmark event would change the legal environment to favor tissue transplantation in Japan (Nather, 2000). Unfortunately, this has not happened.

Korea

The Organ Transplantation Law was passed in Korea in 2000. Under the Law, brain death is defined as the irreversible cessation of the whole brain function, and has to be diagnosed by two specialist doctors and the patients' physician as well as approved by a brain-death determination committee. Donor consent is required for organ removal, but the family has strong veto power towards organ transplantation. The Korean Network for Organ Sharing is a centralised authority for organ procurement. Since the Law was enacted, the number of brain death diagnoses and donations has decreased. Before the Law was enforced, 162 cases were diagnosed as brain dead; as of 2003, only 43 cases were diagnosed (Bagheri, 2005).

Bangladesh

The Tissue Donation and Transplantation Act was passed in April 1999, permitting donation from live and cadaveric donors (Nather, 2000).

Countries in the Asia-Pacific Region with No Laws on Tissue Donation

In contrast, several countries in the Asia-Pacific region do not have any law concerning tissue procurement and transplantation. Such countries include Thailand, China, and Myanmar. In China, although there is no human transplantation act, transplantation is nevertheless, practiced in accordance with the principle of written consent for donation prior to the patient's death or from the next of kin. In Myanmar, apart from the Eye Donation law, no law for tissue donation exists (Shroff, 2002). In Thailand, no law exists. However, the Medical Council is responsible in regulating human organ transplantation. They decide the criteria from time to time, and determine whether any punitive action should be taken against the doctors.

Cultural Issues Pertaining to Organ Donation and Transplantation

A common principle followed by all religions in the world is that saving lives overrides all objections. What holds people back from organ donations and transplantation are cultural reservations such as ignorance of the process of organ donation, fear of mutilation, and fear that organs will be sold or used only by the rich. Hence, certain cultural issues have to be resolved before organ donation and transplantation can progress in any country.

Muslims culturally accept that God created them whole and they prefer to return to Him whole. Islam does not explicitly forbid donation. However, the cultural practice among many Muslims is to bury amputated limbs, and even foreskin from circumcision and amnion from delivery. There is thus a big shortage of bone donors in Muslim countries. Likewise, Christianity promotes organ and tissue donation. However, there is a shortage of donors in Christian-dominant countries like the Philippines, due to the cultural attitudes of the people who also prefer to return to God whole.

To overcome cultural barriers, there is a need for more public awareness programmes on the need for and benefits of tissue donation and tissue transplantation in the community. For example, it has been found that in cases where the wishes of the deceased are not known, only 50% of their relatives would consent to tissue donation. Yet, it is shown that by encouraging people to discuss about organ donation and tissue transplants and make their wishes known to their relatives, the number of people consenting to organ donations and tissue transplants could increase by 93–94% (Shroff, 2002). Hence, there is a need to address the cultural and religious issues that may hinder tissue donation. In addition, there is also a need for professional awareness among doctors to encourage more surgeons to carry out more tissue transplantations.

References

American Association of Tissue Banks *Online* (2007). About AATB. http://www.aatb.org/.

Bagheri A (2005). Organ transplantation laws in Asian countries: a comparative study. *Transplant Proc* **37**: 4159–4162.

El-Shahat YIM (1999). Islamic viewpoint of organ transplantation. *Transplant Proc* **31**: 3271–3274.

European Association of Tissue Banks *Online* (2003). Common Standards for Tissues and Cells Banking. EATB. http://www.eatb.de/.

Gajiwala AL (2003). Setting up a tissue bank in India: the Tata Memorial Hospital experience. *Cell Tissue Bank* **4**: 193–201.

Hachiya Y, Sakai T, Narita Y, Izawa H, and Yoshizawa K (1999). Status of bone banks in Japan. *Transplant Proc* **31**: 2032–2035.

Human Tissue Act *online* (2006). HTA. http://www.hta.gov.uk/about_hta.cfm.

Ireland L and McKelvie H (2003). Tissue banking in Australia. *Cell Tissue Bank* **4**: 151–156.

Ota K (1999). Proceedings of the XVIIth World Congress of the Transplantation Society. *Transplantation Proceedings.* **31**(1): 205–209.

Kassim PN (2005). Organ transplantation in Malaysia: a need for a comprehensive legal regime. *Med Law* **24**: 173–189.

Laws of Malaysia (1974). Act 130: Human Tissues Act, 1974. Ketua Pengarah Percetakan.

Nather A (2000). Tissue banking in Asia Pacific region — ethical, legal, religious, cultural and other regulatory aspects. *J ASEAN Orthop Assoz* **13**: 60–63.

Reuters/AFP (1999). Media frenzy shocks heart-donors family. Breaking of taboo. *The Straits Times*, March.

Shroff Sunil. *Online* (2002) Cadaver Organ Donation. http://www.medindia.net/slides/ppt/Cadaver-Organ-Donation.ppt.

Chapter 10

Ensuring Safety — Donor Evaluation and Screening

Aziz Nather* and Shushan Zheng*

Many hospitals have household refrigeration or freezers containing small quantities of allografts removed in the course of routine surgical procedures that are used for subsequent transplant needs...

This practice is not only inadequate to ensure reliable preservation but it is potentially hazardous because unexpected pathogenic factors or diseases may be transferred with the graft.

Friedlaender 1976

Introduction

Tissue banking started out as a cottage industry in the 1950s, with orthopaedic surgeons collecting femoral heads from their own patients and storing them in hospital freezers. As and when needed, surgeons would retrieve the stored femoral heads for use (Tomford, 1994). However, this practice was highly unsafe as there were no measures in place to prevent the transmission of disease or infection through tissue transplantation.

*NUH Tissue Bank, Department of Orthopaedic Surgery, Yong Loo Lin School of Medicine, National University of Singapore, Singapore

With time, guidelines for ensuring safety for allografts have evolved, and protocols detailing standards for tissue banking have been published. The safety measures put in place today are without a doubt superior to those that exist half a century ago. It is therefore unfortunate that in spite of these advancements, some tissue banks are still following the cottage industry practice of the old days, and may hence continue to endanger the lives of recipients.

Allograft transplantation has the potential to transmit diseases, including HIV-1, HIV-2, hepatitis B, hepatitis C, syphilis and other bacterial infections such as *Clostridium sordellii* infection and *Streptococcus pyogenes* infection. Cases of disease transmission have occurred in the past, with the first case of HIV-1 transmission in bones reported in 1984 (Centers for Disease Control, 1988). In the next section, we shall explore some of these cases of infection.

Femoral Head Banking NUH Tissue Bank Experience

(Adapted from Nather A and David V, *Orthopedics*
30: 308–312, 2007.)

Introduction

A study was undertaken by a group of orthopaedists at the National University Hospital, Singapore, between the period of September 1989 and December 1984. The study population included 273 patients aged 60 and above, who have undergone hemiarthroplasty for fracture of neck of femur (Garden's Grade III and IV), and from whom femoral heads may be procured. The NUH team screened each patient by taking a detailed donor medical history, performing a thorough clinical examination, sending samples for laboratory tests and having a chart review.

(Continued)

(*Continued*)

Results

Positive Laboratory Tests

Screening Test	No. of Patients
Positive culture	6
Hepatitis B	4
Syphilis	3
Total	13

Microorganism	No. of Patients
*S. epidermis**	3
S. aureus	2
Bacteriodes fragilis	1
Total	6

*on enrichment

Medical screening by detailed medical history had excluded 103 patients out of 273, i.e. 37.7% of femoral heads were excluded. Of the remaining 170 femoral heads, 13 were rejected after laboratory test result shown positive, i.e. 7.6%.

Conclusion

The study results showed that it is important to take detailed medical history — it excluded 37.7% femoral heads. The percentage of femoral heads rejected due to laboratory tests was small — only 7.6%. Still, 116 femoral heads out of 273 (42.5%) were unsafe to use.

Hence, it is unsafe to use femoral heads from living donors without any prior screening procedure and screening tests.

Transmission of HIV Type 1 Virus from a
Seronegative Organ and Tissue Donor

(Adapted from Simonds RJ *et al.*, *New Engl J Med*
326: 726–732, 1992.)

Introduction

Since 1985, donors of blood, organ, tissues and semen have been routinely screened for HIV-1 antibody. In October 1985, a 22-year-old HIV-1 seronegative man died from a gunshot in the head. Four solid organs and 54 tissues from the man were procured by Lifenet Transplant Services.

Findings

Seven recipients became infected with HIV-1, of which 4 were recipients of organs and 3 were recipients of unprocessed frozen bone. In contrast, the rest of the recipients — 2 receiving corneas, 3 receiving lyophilised soft tissue, 25 receiving ethanol-treated bone, 3 receiving irradiated dura mater and 1 receiving marrow-evacuated frozen bone — tested negative for HIV-1.

Conclusion

All tissue grafts processed in one way or another did not transmit HIV-1. Hence, all tissue grafts produced should be processed and frozen bones should be gamma-irradiated.

Risk of Transmission of HIV

It has been found that the risk of HIV transmission through blood transfusion is 1 in 250 000 while the risk through deep-frozen allograft is 1 in 1 667 600 (Buck *et al.*, 1989). The risk of HIV transmission through lyophilised allograft is virtually zero.

Transmission of Hepatitis C by Tissue Transplantation

(Adapted from Conrad EU *et al.*, *J Bone Joint Surg*
77A: 214–224, 1995.)

Introduction

The introduction of a second-generation immunoassay for antibodies to hepatitis C virus (HCV 2.0) provided the opportunity to determine if HCV can be transmitted through tissue transplantation. Banked sera from tissue donors that had previously been found to be non-reactive to the first-generation HCV antibody assay (HCV 1.0) and non-reactive to antibodies of hepatitis-B core antigen were retested with HCV 2.0. After the sera from two donors were found to be reactive, the transplant records of recipients of tissues from these donors were reviewed, and hospitals or the surgeons in charge were contacted.

DONOR 1, Woman, died of suicide attempt in 1989
- 23 grafts were procured
- 6 frozen bones, 12 freeze-dried bones, 3 soft tissue grafts and 2 corneas
- Except for corneas, all other grafts were irradiated
- 1 recipient was reactive: irradiated freeze-dried bone for spinal fusion (history of multiple risk factors, including sexual partner with hepatitis C)
- Hepatitis C attributed to recipient himself rather than transmission from Donor 1

DONOR 2, Man, died of multiple injuries in 1991
- 14 grafts were procured
- 4 frozen bones, 8 cryopreserved soft tissues and 2 corneas
- 9 of 12 grafts were transplanted in 9 recipients
- 4 recipients were reactive

(Continued)

(Continued)

The results of the present report demonstrate that HCV can be transmitted by bone, ligament, and tendon allografts. They also support the need for testing of all tissue donors for antibodies to HCV before the tissue is released for transplantation. The results also suggest that 17 kiloGrays of gamma irradiation may inactivate HCV in tissue.

Hepatitis C Virus Transmission from an Antibody-Negative Organ and Tissue Donor — United States, 2000–2002

(Adapted from Cieslak PR *et al.*, *Morb Mortal Wkly Rep*
52: 273–276, 2002.)

Centers for Disease Control (CDC) Report

In June 2002, a case of acute hepatitis C in recipient of a patellar tendon allograft (processed aseptically) was reported in Oregon. The donor had shown no risk nor signs of hepatitis C virus (HCV) infection during donor screening. Neither were antibodies to HCV detected during serological testing (anti-HCV assay). However, further investigation by CDC using nucleic acid testing (HCV RNA assay) on the donor's stored serum confirmed HCV infection.

In total, 40 grafts had been transplanted from the same cadaveric donor, of which 6 were organs, 32 were tissues and 2 were corneas. The recipients were located in 16 states in the USA and in two foreign countries. All tissues had been treated with surface chemicals or antimicrobial treatment. All bone grafts underwent gamma irradiation.

Findings

Further investigation by CDC revealed at least 8 other organ and tissue recipients infected with Hepatitis C. However, there were

(Continued)

(Continued)

no HCV transmission in recipients of skin ($n = 2$) or irradiated bone ($n = 16$).

Conclusion

At the time of death, the donor was probably in the 8–10 week window period (Fig. 1), hence the serological test results were negative.

The transmissions could have been prevented if donors had been tested for HCV RNA. This shows that more sensitive HCV assays with a shorter window period are important for preventing transmission of HCV infections from organs and tissues.

HCV transmission had occurred despite surface chemicals or antimicrobial treatment. This suggests that tissue processing methods (e.g. gamma irradiation) would affect the likelihood of HCV transmission and other viruses from infected donors.

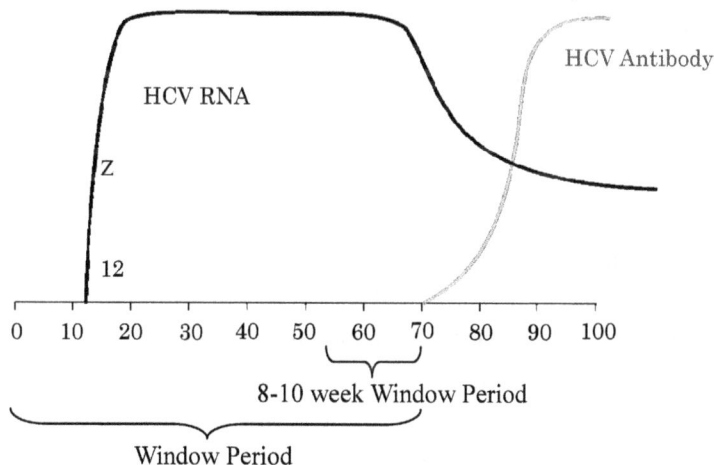

Fig. 1. Graph demonstrating the window period of serological tests for hepatitis C virus (HCV). The window period is defined as the time between infection with HCV (Day 0) and the development of a detectable HCV-antibody response (Day 70). The donor had no detectable antibodies to HCV during initial serological testing as he had been in the 8–10 week window period. This led to a false negative result.

Update: Allograft-Associated Bacterial Infections — United States, 2002

(Adapted from Archibald LK *et al.*, *Morb Mortal Wkly Rep* **51**: 207–210, 2002).

Centers for Disease Control (CDC) Report

A 23-year-old man in Minnesota underwent reconstructive knee surgery in November 2001. He died 3 days after receiving the fresh femoral condyle bone-cartilage allograft. His pre-mortem blood culture revealed the presence of *Clostridium sordellii*.

A second recipient also underwent knee reconstructive surgery in Illinois using fresh femoral condyle and frozen meniscus graft from the same donor. A day later he developed fever. Eight days after surgery, he was admitted to a local hospital for septic arthritis. Cultures for anaerobic bacteria, including *C. sordellii*, were not obtained. The patient has since recovered.

Ten tissues from same donor, including the two cases above, were transplanted into 9 patients in 8 states. Three resulted in infection. CDC tested 19 non-implanted tissues from same donor and identified *C. sordellii* in two tissues (fresh femoral condyle and frozen meniscus). This result confirmed that the allografts were the source of infection.

It was found that the tissues were cultured after suspension of allograft in antibiotic/antifungal solution. Furthermore, all tissues were processed aseptically but did not undergo terminal sterilisation, e.g. gamma irradiation.

Findings of Further Investigation

This incident led the CDC to launch investigations into additional cases of infections associated with musculoskeletal grafts.

(Continued)

(Continued)

As of March 2002, 13 cases of Clostridium infections identified, 12 of which were caused by *C. septicum*, 1 by *C. sordellii*. The allografts implicated included:

- 8 frozen tendons used for anterior cruciate ligament (ACL) reconstruction,
- 2 fresh femoral condyles,
- 2 frozen bone, and
- 1 frozen meniscus.

All allografts associated with infections were processed aseptically but did not undergo terminal sterilisation. In 11 out of 13 cases, evidence confirmed that the allograft was the source of infection. In fact, 11 out of 13 Clostridium infections were traced to one tissue processor, later identified by the FDA as CryoLife, Inc. at Kennesaw, Georgia. All tissues processed by the tissue processor after October 3, 2001, except heart valves, were recalled ("CDC says Clostridium infections were caused by musculoskeletal tissue allografts", *Transplant News*, 2004).

CDC Recommendations

The investigations led to several recommendations by the CDC, two of which are stated here:

1. Tissues should be processed using a method, e.g. gamma radiation, that can kill bacterial spores. Otherwise, aseptically processed tissues cannot be considered sterile.
2. Allograft tissues should be cultured before suspension in antimicrobial solutions. If Clostridium or other bowel flora is present, the tissue should not be used.

Invasive *Streptococcus pyogenes* After Allograft Implantation — Colorado, 2003

(Adapted from Bos J *et al.*, *Morb Mortal Wkly Rep*
52: 1173–1176, 2003.)

Centers for Disease Control (CDC) Report

In September 2003, a 17-year-old male recipient underwent anterior cruciate ligament repair with a hemipatellar tendon allograft in Colorado. Six days after the procedure, he was admitted to a local hospital with pain and erythema at the incision site, fever of 102°F (39°C), and chills. The allograft tissue was removed, and the patient underwent surgical exploration and fasciotomy of the affected thigh. Cultures of his blood, wound aspirate, and explanted tissue grew Group A Streptococcus (GAS).

The recipient's surgeon alerted the Food and Drug Administration (FDA) about the possible allograft infection and this prompted an investigation. By then, tendon allografts from the same cadaveric donor had been implanted in 5 other patients. However, as of 1 December, 2003, no adverse outcome was reported.

It was revealed that the preprocessing cultures of the tissue had yielded Group A Streptococcus. However, upon processing of allografts using aseptic technique and antimicrobial solution, the postprocessing cultures yielded negative results. No end sterilisation procedure had been used.

Investigations by the CDC, FDA, and the Colorado Department of Public Health and Environment confirmed the allograft as the source of infection.

CDC Recommendations

As in the case of *C. sordellii* infection, the antimicrobial solution used failed to eradicate bacteria; postprocessing culture also failed to

(Continued)

(*Continued*)

detect bacteria. Hence, given the apparent ability of the organism to endure tissue processing with antimicrobial treatment, the presence of Group A Streptococcus in donor tissue should prompt the rejection of tissue unless a sterilising procedure, e.g. gamma irradiation, is used.

General Standards for Tissue Banking

In order to ensure the safety of tissue grafts across the board, tissue bank associations have published their own set of standards on tissue banking to be followed by tissue banks accredited by the association. Some examples of standards include that of the American Association for Tissue Banks (AATB), the European Association for Tissue Banks (EATB), the Asia-Pacific Association of Surgical Tissue Banks (APASTB) and the Australasian Tissue Banking Forum (ATBF).

In the Asia-Pacific region, countries follow the APASTB General Standards (Fig. 2), which is an adaptation of the IAEA General Standards Template.

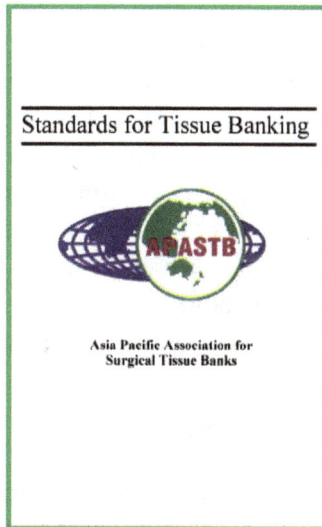

Standards for Tissue Banking

APASTB

Asia Pacific Association for
Surgical Tissue Banks

Fig. 2. The APASTB General Standards.

In addition, each tissue bank must follow a procedure manual and a quality manual and be audited and licensed by their country's Ministry of Health.

Donor Selection Procedure

In accordance with the APASTB General Standards, under the section *B 1.100 General,*

> The suitability of a specific donor for tissue allograft donation is based upon medical and behavioural history, medical records review, physical examination, cadaveric donor autopsy findings (if an autopsy is performed) and laboratory tests.

Donor History Review

Donor's Medical Records

Data pertaining to the donor's age, medical history, medications taken, family medical history, sexual history and exposure to toxic materials are recorded. In the case of deceased donors, the probable cause of death must also be obtained.

These data should be obtained from the living donor or from the next of kin for cadaveric donors. The interview should be carried out by suitably trained personnel. A qualified physician shall approve the donor evaluation process. At the time of the interview, the risk of transmission of these infections via tissue donation and the need to avoid transplantation of tissues from individuals who have engaged in high-risk behaviour should be explained to the living donor or the next of kin for cadaveric donors.

A suitable questionnaire shall be used for reviewing of the donor's medical history (see Fig. 3). If any adverse information is found in the Donor Exclusion Criteria, the donation is refused in the interest of the recipient.

General Criteria for Potential Donors for Bone

In general, the ideal donor is a young, healthy adult under 55 years old. Cadaveric bones are harvested within 12 hours of death if body is not kept

⊙LIFENET	Donor Medical & Social History Questionnaire	DONOR ID _____

Person Interviewed: _____ Relationship to Deceased: _____

Do you feel that you knew the deceased well enough to answer questions regarding medical/social history? ☐ Yes ☐ No

Person Conducting Interview _____
and Completing Form: Print Name   Title

                                       Signature   Date of Interview

Donor Name/I.D. #: _____ UNOS I.D. # (if Organ Donor) _____
All potential donors must be screened according to the USPHS current criteria for exclusion of high risk donors.

1. Did the deceased have any history of heart disease, high blood pressure, or chest pain? Poor cirulation especially in the legs? Take any drugs for heart or B/P problems? If so, what?	☐ Yes ☐ No ☐ Other
2. Did the deceased suffer from any type of liver disease? Any history of yellow jaundice? Been told they had any type of Hepatitis? Any close contact with persons diagnosed with viral Hepatitis in the past 12 months?	☐ Yes ☐ No ☐ Other
3. Did the deceased suffer from any type of neurologic or brain disease such as Alzheimer's seizures, periods of confusion or recent memory loss, history of brain tumor?	☐ Yes ☐ No ☐ Other
4. Did the deceased have any kidney related disease? Kidney stones? Frequent infections? Ever been treated with kidney dialysis?	☐ Yes ☐ No ☐ Other
5. Did the deceased have a history of diabetes? Required oral medication or insulin injections?	☐ Yes ☐ No ☐ Other
6. Did the deceased have any history of digestive or intestinal problems? Ever have bloody stools or intestinal surgery?	☐ Yes ☐ No ☐ Other
7. Did the deceased have any history of arthritis or joint decease? History of broken bones? Any complaints of stiff or sore joints?	☐ Yes ☐ No ☐ Other
8. Did the deceased have any history of asthma, emphysema, or any lung disease? Ever have a positive skin test for Tuberculosis? Ever treated for TB?	☐ Yes ☐ No ☐ Other
9. Has the deceased been seen by a physician, or hospitalized in the past two years? What physician, hospital, psychiatric, or long term care facility?	☐ Yes ☐ No ☐ Other
10. Has the deceased ever had cancer or received radiation therapy or drugs for cancer?	☐ Yes ☐ No ☐ Other

Fig. 3. Questionnaire for reviewing donor's medical history.

⊕LIFENET Donor Medical & Social History Questionnaire　DONOR ID _____	
11. Please name any surgical procedures the deceased has had in the past.	☐ Yes ☐ No ☐ Other
12. Has the deceased experienced any periods of explained or unexplained weight loss?	☐ Yes ☐ No ☐ Other
13. Did the deceased ever use non-prescribed drugs or other substances, i.e. cocaine, marijuana, steroids, inhalants?	☐ Yes ☐ No ☐ Other
14. Has the deceased ever received blood transfusions or blood products prior to this admission?	☐ Yes ☐ No ☐ Other
15. Was the deceased ever refused as a blood donor or told not to donate? Why?	☐ Yes ☐ No ☐ Other
16. Did the deceased ever receive an organ or tissue transplant, ie, bone, cornea, skin, heart, kidney?	☐ Yes ☐ No ☐ Other
17. In the past 12 months did the deceased have a tattoo, ear/body piercing, acupuncture, or accidental needle stick?	☐ Yes ☐ No ☐ Other
18. Was the deceased vaccinated for Hepatitis B?	☐ Yes ☐ No ☐ Other
19. In the past 4 weeks was the deceased vaccinated for any reason?	☐　　☐　　☐
20. was the deceased ever given human growth hormone?	☐ Yes ☐ No ☐ Other
21. What medications, if any, did the deceased take on a regular basis?	☐ Yes ☐ No ☐ Other
22. Did the deceased use tobacco products? Cigarettes? Packs per day? For how long? Other tobacco products?	☐ Yes ☐ No ☐ Other
23. Did the deceased drink alcohol? How much? What type?	☐ Yes ☐ No ☐ Other
24. Has the deceased ever been exposed to a toxic substance, i.e. lead, pesticides?	☐ Yes ☐ No ☐ Other
25. Has the deceased recently exhibited flu-like symptoms such as cough, colds, swollen lymph nodes, nausea, vomiting, persistent diarrhea, or fever>100°F.? Currently taking antibiotics?	☐ Yes ☐ No ☐ Other

Fig. 3.　(*Continued*)

LifeNet Donor Medical & Social History Questionnaire	DONOR ID _____

* USPHS CURRENT CRITERIA	
*26. Male Donors: Has the deceased had sex with another male in the Past 5 years?	☐ Yes ☐ No ☐ Other
*27. In the past 5 years has the deceased used a needle to inject drugs into their veins, muscle, or under their skin for nonmedical use?	☐ Yes ☐ No ☐ Other
*28. Has the deceased received human-derived clotting factor concentrates for hemophilia or related clotting disorders?	☐ Yes ☐ No ☐ Other
*29 Has the deceased engaged in sex in exchange for money or drugs in the past 5 years?	☐ Yes ☐ No ☐ Other
*30. Was the deceased exposed to known or suspected viral Hepatitis or HIV-infected blood through accidental needlestick or through contact with an open wound, non-intact skin, or mucous membrane in the past 12 months?	☐ Yes ☐ No ☐ Other
*31. Was the deceased an inmate of a correctional system or jail, or released from a correctional system or jail in the past 12 months?	☐ Yes ☐ No ☐ Other
*32. Has the deceased had sex in the past 12 months with any person known or suspected to have viral Hepatitis or HIV infection, or any person described in above questions #26-31?	☐ Yes ☐ No ☐ Other
PEDIATRIC DONORS	
*33. A. Was the child born to a mother with, or at risk for HIV infection, or who responded "yes" to question 26-32?	☐ Yes ☐ No ☐ Other
B. If "yes", was the child breast fed in the past 12 months?	☐ Yes ☐ No ☐ Other

Are there other individuals that may provide additional information regarding these medical and social history question: YES _____ NO _____

IF YES, PLEASE PROVIDE NAME AND PHONE NUMBER:

_____

_____

ADDITIONAL COMMENTS: (please refer to question numbers where applicable)

_____

_____

_____

Fig. 3. (*Continued*)

Table 1.   Age criteria for various types of musculoskeletal tissue.

Musculoskeletal		
Structural support	Women	< 50 years old
Structural support	Men	< 55 years old
Closed epiphysis		> 15 years old
Morsellised tissue		No age criteria

refrigerated and within 24 hours of death if body is kept refrigerated at +4°C in special mortuaries.

Age Prerequisite

Age plays an important role in the procurement of musculoskeletal structures. In general, suitable female donors should be 18–50 years old while male donors should be 18–55 years old. Further criteria for the various types of tissue are presented in Table 1.

Donor Exclusion Criteria

- Persons with haemophilia or related clotting disorders
- Acute or chronic infection or sepsis
- Malignancy
- Autoimmune disease, e.g. rheumatoid arthritis
- Significant history of connective tissue diseases (e.g. systemic lupus erythematosus, rheumatoid arthritis)
- Prolonged corticosteroid therapy
- Inflammatory disease
- Debilitating, degenerative, neurological disease
- Suspicion of central degenerative neurological diseases including Alzhemier's disease, Creutzfeldt-Jakob disease or familial history of Creutzfeldt-Jakob Disease, multiple sclerosis
- Hepatitis, syphilis, slow virus infection, AIDS, ARC (AIDS-related complex)
- History of chronic viral hepatitis
- History, clinical evidence, suspicion, or laboratory evidence of HIV

- High risk factors for AIDS and/or hepatitis
- Acute viral hepatitis or jaundice of unknown aetiology
- Intravenous drug abuse
- History of chronic haemodialysis
- Septicaemia and systemic viral disease or mycosis or tuberculosis at the time of death
- Exposure to toxic substances
- Serious illness of unknown aetiology
- Treatment with human pituitary-derived growth hormone
- Unknown cause of death
- More than 20% total surface area burnt
- Evidence of dementia
- Bone disease (osteoporosis, osteopenia, steroid-induced osteonecrosis)

General Criteria for Potential Donors for Skin

In general, skin is procured from deceased donors under the age of 55 years old. If the skin graft procured is meant for use without secondary sterilisation, it should be procured within 4 hours (at room temperature) or 12 hours (with refrigeration) post-mortem. If the skin graft is to be freeze-dried and sterilised, it should be procured within 24 hours post-mortem.

Furthermore, the potential skin donor must have an Estimated Body Surface Area (EBSA) of at least 1.75 m² (EBSA = Height$^{0.725}$ × Weight$^{0.425}$ × 0.007184 in m²; Dubois and Dubois, 1916) for sufficient skin tissues to be harvested.

Donor Exclusion Criteria

(i) Infection

    a. Septicaemia

    b. Untreated syphilis

    c. Leprosy

    d. Generalised acute or chronic infections of the skin

    e. Tuberculosis

(ii) Active or viral diseases

   a. HIV infection
   b. Viral hepatitis
   c. Slow virus infection
   d. Systemic mycosis
   e. Highly infectious and potentially fatal viral diseases such as SARS

(iii) Malignant diseases, except primary basal carcinoma of the skin, primary brain tumours and any treated and healed carcinoma *in situ*

(iv) Risk of Creutzfeldt-Jacob disease transmission, i.e. donors who receive human growth hormone or who have a family history of dementia and/or neurodegenerative disease

(v) Death of unknown cause or diseases of unknown aetiology

(vi) High risk category for HIV infections or hepatitis

(vii) Multi-transfusion 48 hours prior to death and does not meet haemodilution requirement

(viii) Structurally damaged skin due to

   a. Autoimmune dermatosis
   b. Extensive dermatitis
   c. Collagen disease
   d. Acute/healed burns in area of harvesting

Due to the March 2003 Severe Respiratory Syndrome (SARS) outbreak in Singapore, additional exclusion criteria have been implemented (Chua *et al.*, 2004):

   (i) Febrile patients (temperature >37°c) anytime in the last 10 days where SARS cannot be excluded

   (ii) Patients with pneumonia and lymphopenia

   (iii) Immunocompromised patients, i.e. ESRF, chronic steroid therapy, cancer patients on chemotherapy

   (iv) Patients with recent history of travel to SARS affected areas for emergency surgery even if afebrile

   (v) Patients with history of contact with SARS patients on home quarantine

(vi) All suspect or probable SARS patients requiring emergency operations but were not fit for transfer to the SARS designated hospital

(vii) Patients from hotspots or isolation wards

Further Exclusion Criteria for Donors of Bone and/or Skin

For the prevention of the transmission of HIV, hepatitis B, hepatitis C and Creutzfeldt-Jacob disease through tissue transplantation, a further set of behavioural exclusion criteria has been designed. These criteria are applicable to both donors of bone and donors of skin.

(i) Persons who have been exposed in the preceding 12-month period to known or suspected HIV, hepatitis B virus and/or hepatitis C virus infected blood through percutaneous inoculation or through contact with an open wound, non-intact skin or mucous membrane.

(ii) Persons who report ever having non-medically administered/ prescribed intravenous, intramuscular or subcutaneous injections of addictive drugs.

(iii) Men who have had sex, even once, with another man.

(iv) Persons who have engaged in sex in exchange for money or casual sex with a person other than their regular sexual partner during the preceding 12 months.

(v) Persons who have had or have been treated for syphilis, gonorrhoea, or any other sexually transmitted diseases in the preceding 12-month period or who have had a reactive screening test for syphilis in the absence of a confirmatory test in the preceding 12-month period.

(vi) Persons with haemophilia or related clotting disorders, or who have received human-derived clotting factors.

(vii) Persons who have had sex in the preceding 12-month period with individuals from categories (ii–vi) above, or with any person who has or is suspected of having HIV, hepatitis B virus, or hepatitis C virus, or in whom the next of kin has reported high risk behaviour.

Physical Examination

A physical examination should be performed and documented before accepting a donor or retrieving tissue(s) from a cadaveric donor. This examination confirms the patient's age, nutritional status and general well-being (before death, in the case of deceased donors).

A qualified physician shall perform the physical examination. All details are recorded and distinctive marks are identified on a suitable body map (Fig. 4). Special attention is given to the presence of infected wounds, tattoos, decubitus ulcers and abscesses.

In addition, for the prevention of HIV, hepatitis B, hepatitis C and Creutzfeldt-Jacob disease, the physician is to look out for the presence of the following:

  (i) Blue or purple spots on the skin or mucous membranes typical of Kaposi's sarcoma
 (ii) Unexplained generalised lymphadenopathy
(iii) Unexplained fever >38.6°C for more than 10 days
 (iv) Needle tracks or other signs of parenteral drug abuse
  (v) Oral thrush
 (vi) Signs of sexually transmitted diseases such as genital ulcerative disease, herpes simplex, syphilis or chancroid
(vii) For male donor, physical evidence of anal intercourse, including perianal condyloma
(viii) Unexplained jaundice, hepatomegaly, ascites or stigmata of chronic liver disease

Laboratory Tests

To screen donors thoroughly, it is imperative to conduct serological as well as microbiological examinations. A qualified physician shall obtain the blood sample from the patients. In the case of living donors, consent must be taken using a Consent Form before the tissue bank can send the blood sample to the laboratory for serological testing. Tests shall be performed by a qualified, and if applicable, licensed laboratory and according to Good Laboratory Practice (GLP).

⑨LIFENET Tissue Donor
Information Form

Tissue: Donor I.D. # _____

Procurement Observations

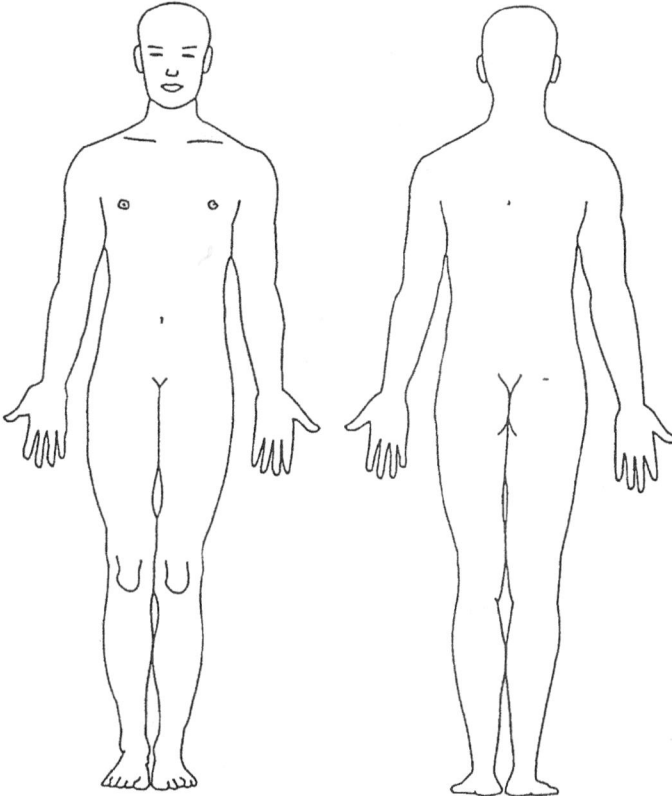

Body Identified by: ☐ Toe Tag   ☐ Wrist Band   ☐ Other _____

Name: _____ Donor I.D.: _____

UNOS I.D.: _____ Decedent's Height: _____ Weight: _____

Examined By: _____ Date: _____

Comments:

Fig. 4.   A suitable body map to use for physical examination.

Storing of Donor Serum

In accordance with the APASTB General Standards, under the section *B 1.516 Donor Serum Archive*:

> A sample of donor serum shall be securely sealed and stored frozen in a proper manner until 5 years after the expiration date of the tissue or according to applicable Inter-governmental, National, Regional and Local Law or Regulation.

Minimum Testing Requirements

The extent and number of laboratory tests done varies among countries and among donors. Even so, a set of minimum requirements exist. The minimum testing accepted universally includes[1]:

1. Syphilis   — VDRL/RPR (if positive, TPHA)
2. Hepatitis B — HBsAg
3. Hepatitis C — Anti-HCV (using 2nd generation test)
4. AIDS   — Anti-HIV$_1$, Anti-HIV$_2$
5. Microbiological (culture sensitivity) testing for anaerobes, aerobes and fungi

From this baseline of tests required, tissue bank associations around the world may make compulsory additional tests based on the endemicity of diseases (Figs. 5 and 6) and the level of screening technology available locally.

Table 2 demonstrates the difference in minimum testing requirements between the Asia-Pacific Association of Surgical Tissue Banks (APASTB), American Association of Tissue Banks (AATB), Australasian Tissue Banking Forum (ATBF) and European Association of Tissue Banks (EATB).

---

[1] Refer to the end of the chapter for "A Brief Guide to Screening Assays".

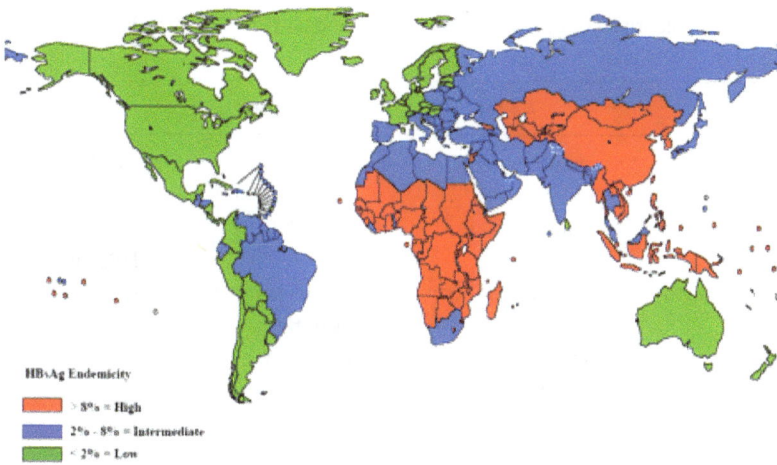

Fig. 5. World prevalence of hepatitis B based on figures obtained in 2004. There is a high prevalence of hepatitis B in the Asia-Pacific region.[2]

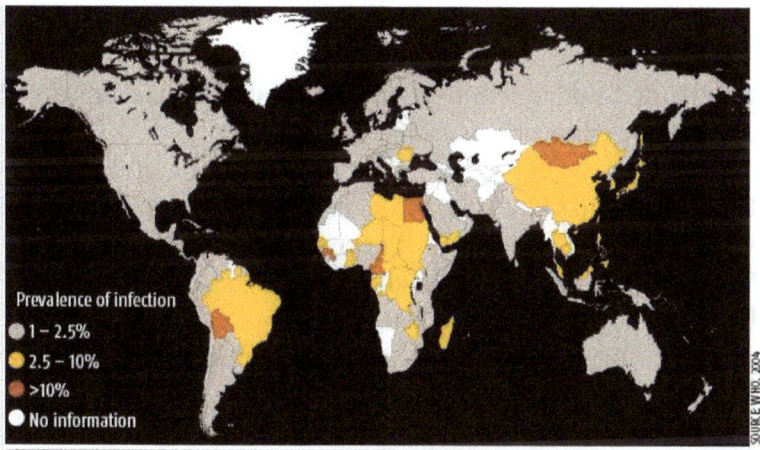

Fig. 6. World prevalence of hepatitis C based on figures obtained in 2003. There is a high prevalence of hepatitis C in Southeast Asia and China.[3]

---

[2] Reproduced from "Hepatitis B Surface Antigen Assays: Operational Characteristics (Phase I): Report 2" by the World Health Organisation, 2004.

[3] Reproduced from "Cheap Drug Dodges Big Pharma Patents" by *New Scientist*, 2007.

Table 2. Comparison of minimum testing requirement between APASTB, AATB, ATBF and EATB.

APASTB	AATB	ATBF	EATB
anti-HIV$_1$ & anti-HIV$_2$	anti-HIV$_1$ & anti-HIV$_2$	anti-HIV$_1$ & anti-HIV$_2$	anti-HIV$_1$ & anti-HIV$_2$
anti-HCV	anti-HCV	anti-HCV	anti-HCV
HBsAg	HBsAg	HBsAg	HBsAg
syphilis antibody	syphilis antibody	syphilis antibody	syphilis antibody
	anti-HBc	anti-HBc	anti-HBc
	anti-HTLV-I & anti-HTLV-II	anti-HTLV-I & anti-HTLV-II	anti-HTLV-I[†]
	NAT for HIV$_1$ and HCV*		

*For cadavers only.
[†]For endemic countries only.

Optional Tests

Other tests can be done selectively on donors of bone and/or skin when indicated. These optional tests include:

1. Cytomegalovirus (CMV), toxoplasmosis antibodies, and Ebstein-Barr virus (EBV) tests for immunosuppressed patients
2. Test for hepatitis B core antibody
3. Histopathological examination
4. Antigen testing for HIV, using antibody or nucleic acid testing (NAT) for HIV
5. Test for human T-lymphotropic virus I+II in endemic regions
6. Alanine aminotransferase (ALT) for living donors

Repeat Testing for Living Donors

For donations from living donors, tests for hepatitis and HIV should be repeated after three months. Meanwhile, donated tissues are held in quarantine and are released only when repeat examinations show negative results.

Safety Net Under Multi-Organ and Tissue Procurement System

For donations from deceased donors under the Multi-Organ and Tissue Procurement System, the tissues are subjected to an optional quarantine period of about three months. During those three months, recipients of organs from the deceased donor are monitored for any possible transmission of disease or infection.

A Brief Guide to Screening Assays

Current Tests for HIV-1 Screening

- Anti-HIV$_1$ – ELISA
- Confirmatory Test – Western Blot

For the initial screening of HIV, anti-HIV$_1$ is used. It uses enzyme-linked immunosorbent assay (ELISA) to test for antibodies to HIV-1 using plated HIV antigens. If ELISA is positive, then Western Blot is used as confirmatory test. Western Blot is a technique to test for HIV antibodies by using HIV proteins attached on a membrane.

However, anti-HIV$_1$ is limited by its window period, which is defined as the time from infection to appearance of detectable antibodies (Fig. 7). The window period of HIV using current anti-HIV$_1$ test is 22 days, though it is expected to drop with the introduction of new generation tests which are more sensitive in their detection of HIV antibodies. HIV-infected donors who donate during the window period do not have sufficient HIV antibodies to test positive in the antibody test. This leads to false negatives.

Supplemental Test for HIV-1 Screening

p24 Antigen Assay

p24 antigen assay is a supplemental test to anti-HIV$_1$. It uses ELISA to detect the capsid antigen, p24, a structural protein that makes up most of the HIV viral core particle (Fig. 8), by testing the blood sample against antibodies specific to p24 antigen.

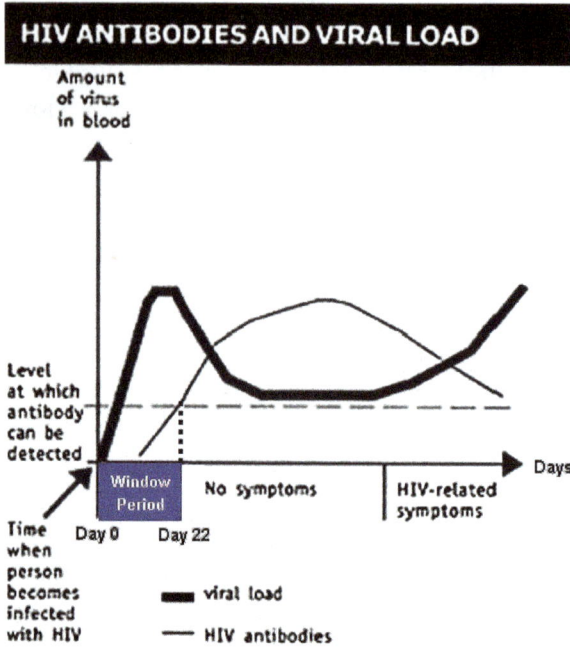

Fig. 7. Graph demonstrating the window period of HIV.

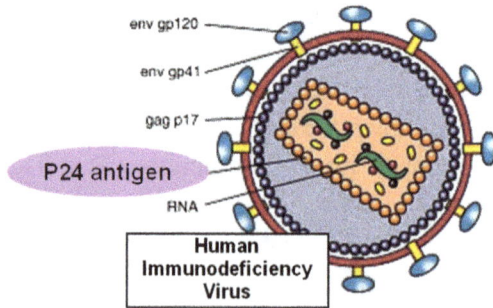

Fig. 8. Human immunodeficiency virus (HIV).

The advantage of p24 antigen assay is that the p24 antigen is detected earlier than antibodies. According to a Food and Drug Administration (FDA) Talk Paper in 2001 (Fig. 9),

Average window period with antibody tests is 22 days. [p24] Antigen testing cuts the window period to approximately 16 days.

This reduction in window period helps to reduce the number of false negatives caused by window period donors.

In the US and UK, p24 has already been supplanted by nucleic acid testing (NAT) as the preferred test to confirm HIV-1 infection as NAT is more effective for reducing the window period (U.S. Food and Drug Administration, 2004; Laperche, 2005). NAT will be explained in a later section.

Average window period with antibody tests is 22 days... NAT (Nucleic Acid Testing) further reduces this period to 12 days for blood donors.

U.S. Food and Drug Administration, 2001

New Generation Screening Tests for HIV

As assays for HIV improve in sensitivity, the window period is reduced. Third-generation assays employ more capture antigens to test for HIV

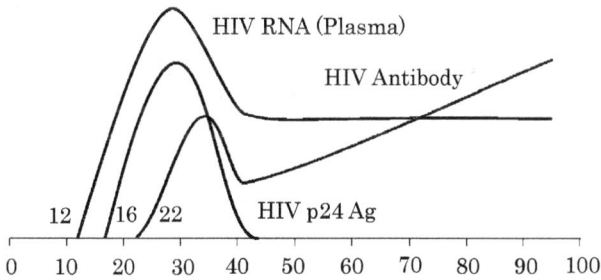

Fig. 9. Graph demonstrating the appearance of HIV markers during early infection. With Day 0 as the start of infection, detectable HIV antibody response appears on Day 22, detectable level of p24 antigen appears on Day 16 and detectable level of HIV RNA appears on Day 12. This puts the window period of anti-HCV, p24 antigen assay and NAT for HCV at 22 days, 16 days and 12 days respectively.[4]

---

[4] Graph reproduced from the presentation "The biology of window period infections: implications for donor screening and inventory hold" by Dr. Steven Kleinman during the FDA BPAC meeting on September 21, 2001.

antibodies (Louie *et al.*, 2006) while fourth-generation assays combine ELISA with p24 antigen assay to improve sensitivity (Weber *et al.*, 1998).

Current Tests for HIV-2 Screening

Anti-HIV$_2$

For the screening of HIV-2, HIV-1/HIV-2 ELISA is used to simultaneously test for antibodies against HIV-1 and HIV-2.

Current Tests for Hepatitis B Screening

Hepatitis B Surface Antigen (HBsAg)

HBsAg is a very sensitive assay that uses ELISA to detect hepatitis B surface antigen (Fig. 10) in serum. HBsAg develops between six weeks and six months following infection, but prior to the onset of symptoms.

One of the limitations of the test is its inability to detect mutant HBsAg, which occur due to mutations, e.g. amino acid changes at residues 143, 144, and 145 of the HBsAg. This leads to false negatives in donors who are infected with hepatitis B but escape detection because their mutant HBsAg are not detected. Fortunately, more sensitive HBsAg assays are being developed. The improvement in design of these new assays results in the ability to detect mutant HBsAg and in improved sensitivity.

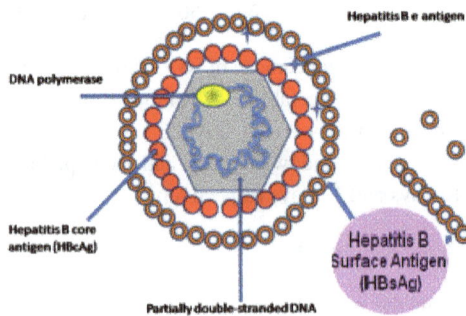

Fig. 10.   Hepatitis B virus.

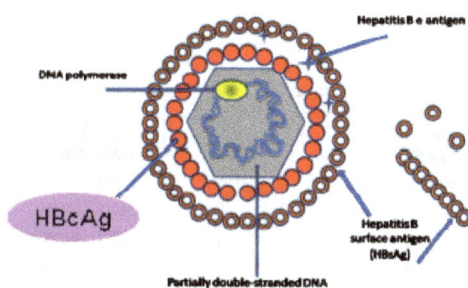

Fig. 11. Hepatitis B virus.

Total Hepatitis B Core Antibody (Anti-HBc)

Anti-HBc is a supplemental test to HBsAg. It uses ELISA to detect total hepatitis B core antibody (Fig. 11), which are antibodies that appear as a response to hepatitis B core antigen (HBcAg).

Current Tests for Hepatitis C Screening

- Anti-HCV
- Confirmatory test — recombinant immunoblot assay (RIBA)

Anti-HCV uses ELISA to test for antibodies against the hepatitis C virus. Antibodies to hepatitis C virus are produced by the immune system after more than two months of infection (Møller and Krarup, 2003).

Fourth-generation ELISA tests for anti-HCV are already available. The new assays include a greater number of HCV-encoded antigens to allow more specific antibody detection and to improve sensitivity.

If the results of anti-HCV are positive, recombinant immunoblot assay (RIBA) is used as a confirmatory test. RIBA is a qualitative ELISA test in which four recombinant HCV-encoded antigens are immobilised as individual bands on test strips. The strips are incubated with the patient's serum. Any HCV antibodies present in the serum would then bind to the corresponding antigen bands on the strip, and show up as bands on the strip (Fig. 12).

However, RIBA has its limitations. RIBA results fall into three categories — negative (showing no bands), indeterminate (showing one band)

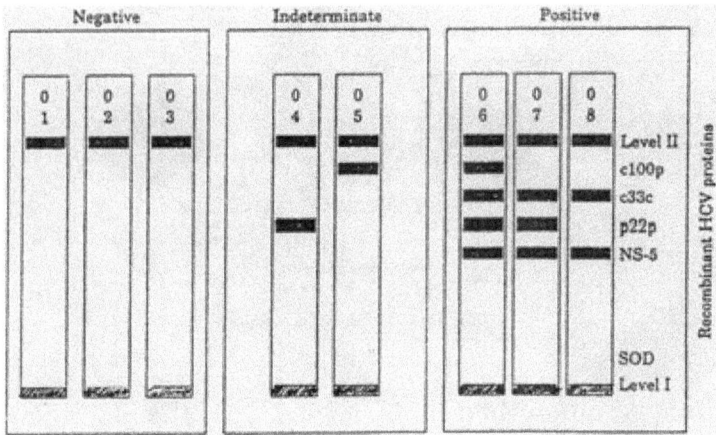

Fig. 12. Test strip for recombinant immunoblot assay (RIBA).

or positive (showing two or more bands). Results for RIBA are often inde-terminate. This undermines their overall effectiveness in confirming a hepatitis C infection.

New generation tests of RIBA have been introduced. These new gen-eration tests include different combinations of capture molecules in different quantities. For example, the third generation RIBA, RIBA 3.0, includes a combination of recombinant antigens and synthetic peptides. Improvements in the design have led to greater specificity and sensitivity (Martin *et al.*, 1997).

RIBA has already been supplanted by NAT as the preferred test to confirm HCV infection (Scott and David, 2007).

New Trend in HIV-1 and Hepatitis C Screening — Nucleic Acid Testing

In recent years, Nucleic Acid Testing has been introduced as an alterna-tive to the current antibody-based test for HIV-1 and hepatitis C Virus. It is used to detect HIV-1 RNA and HCV RNA using polymerase chain reaction (PCR), real-time PCR in particular, or transcription-mediation amplification (TMA). The advantage of NAT lies in the reduction of win-dow period for both HIV-1 (22 to 12 days) and HCV (70 to 12 days) (U.S. Food and Drug Administration, 2001) (Fig. 13).

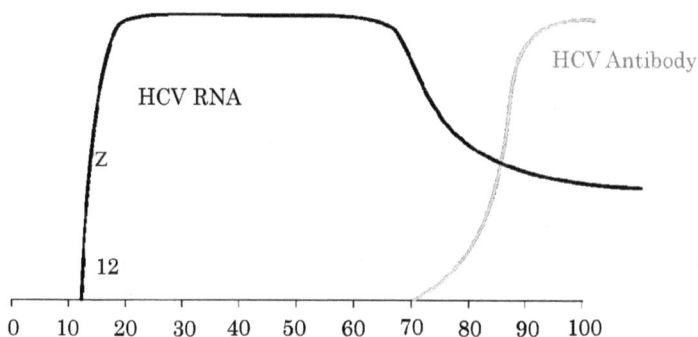

Fig. 13. Graph demonstrating the appearance of hepatitis C virus markers during early infection. With Day 0 as the start of infection, detectable antibodies to HCV appear on Day 70, while detectable level of HCV RNA appears on Day 12. This puts the window period of anti-HCV and NAT for HCV at 70 days and 12 days respectively.[5]

The trend of using NAT for donor screening started among blood donation centres, and has since been caught on by tissue banks in the US. In August 13, 2004, the American Association of Tissue Banks (AATB) announced that "all donors retrieved by AATB-accredited tissue banks for processing, storage or distribution must be tested for HIV-1 and HCV using a licensed NAT for cadaveric specimens" (Warren, 2004).

However, like all tests before it, NAT has its own limitations. The logistics needed for its implementation is an issue. The cost-effectiveness of NAT is also debatable. According to AuBuchon *et al.* (1997), "HIV NAT prevents 8 more cases of transfusion-associated HIV infection annually, at an additional cost of $96 million per year in the US". The exact figure for cost-effectiveness varies from country to country, and it is thus up to the various countries or associations to decide if NAT should be implemented.

Alternative Tests for Hepatitis C Screening

- HCV core antigen assay
- HCV combination test

---

[5] Graph reproduced from the presentation "The biology of window period infections: implications for donor screening and inventory hold" by Dr. Steven Kleinman during the FDA BPAC meeting on September 21, 2001.

For countries that do not implement NAT for hepatitis C screening, improved assays which allow detection of viral antigens alone or together with antibodies may be a viable alternative (Laperche, 2005). For example, hepatitis C can be screened using hepatitis C core antigen assay, an ELISA test that tests for hepatitis C core antigen. The sensitivity of the antigen test is comparable to NAT, and it would likely be less expensive and simpler to perform than most of the NAT on the market (Dawson, 2007). Another feasible alternative to NAT is the HCV combination test, which combines the detection of HCV antigen and HCV antibodies (Ansaldi *et al.*, 2006).

Current Tests for Syphilis Screening

Syphilis is caused by *Treponema pallidum* (Fig. 14), a Gram negative spirochaete bacterium. In response to the bacterial infection, the immune system produces a broad range of antibodies, including non-specific and specific treponemal antibodies.

Rapid Plasma Reagin (RPR)

For the intial screening of syphilis, RPR is used to detect non-specific treponemal antibodies (reagin) by using cardiolipin antigens. When the test

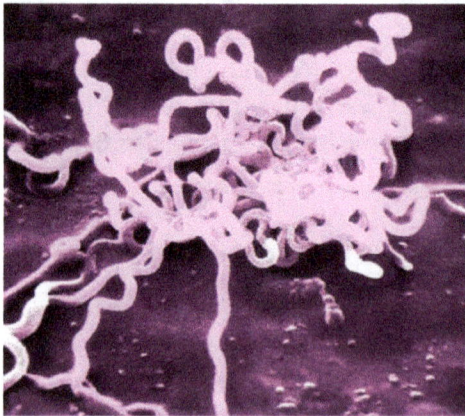

Fig. 14.   Treponema pallidum.

serum is added to a suspension containing carbon particles and cardiolipin antigen, any antibodies present would bind to the antigen and clump together, forming a lumpy mass (flocculate) that is made visible to the eye by the carbon particles. In contrast, no flocculation is seen in a test serum without non-specific treponemal antibodies.

If RPR is positive, TPHA is carried out as confirmatory test.

Venereal Diseases Research Laboratory (VDRL)

VDRL is an alternative test for the initial screening of syphilis. It runs on the same principle as RPR, but it is an older test. Like RPR, it is a flocculation test that uses cardiolipin-antigen to detect non-specific treponemal antibodies.

If VDRL is positive, TPHA is carried out as confirmatory test.

Confirmatory test — Treponema pallidum Haemagglutination assay (TPHA)

TPHA detects specific treponemal antibodies by using treponemal antigens attached to gelatin particles to test the patient's serum. When the test serum is added to the mixture of gelatin and antigens, any antibodies in the serum would cause agglutination and the formation of a mat across the bottom of the testing well. In contrast, no agglutination is seen in a test serum without specific treponemal antibodies.

Microbiological Testing

In accordance with the APASTB General Standards, under the section *B 1.610 Bacteriological Testing Methods*,

> Representative samples of each retrieved tissue have to be cultured, if the tissues are to be aseptically processed without terminal sterilisation. Samples shall be taken prior to exposure of the tissue to antibiotic containing solution.

Microbiological testing should be carried out on small donor tissue samples or swabs. The culture medium should allow the detection of aerobes, anaerobes and fungi. The results of the culture are to be documented in the donor record.

In addition, blood culture from cadaveric donor may be useful in evaluating the state of the cadaver and interpreting the cultures performed on the grafts themselves.

References

Ansaldi F, Bruzzone B, Testino G, Bassetti M, Gasparini R, Crovari P, and Icardi G (2006). Combination hepatitis C virus antigen and antibody immunoassay as a new tool for early diagnosis of infection. *J Viral Hepat* **13**: 5–10.

Archibald LK, Jernigan DB, and Kainer MA (2002). Update: Allograft-Associated Bacterial Infections — United States, 2002. *Morb Mortal Wkly Rep* **51**: 207–210.

AuBuchon JP, Birkmeyer JD, and Busch MP (1997). Safety of the blood supply in the United States: opportunities and controversies. *Ann Intern Med* **12**: 904–909.

Bos J, Crutcher JM, Gershman K, Cote' T, Greenwald MA, Polder J, Srinivasan A, Arduino M, Jernigan DB, Beall B, Elliott JA, Facklam RR, Schuchat A, *et al.* (2003). Invasive *Streptococcus pyogenes* after allograft implantation — Colorado. *Morb Mortal Wkly Rep* **52**: 1173–1176.

Buck BE, Malinin TI, and Brown MD (1989). Bone transplantation and human immunodeficiency virus. An estimate of risk of Acquired Immunodeficiency Syndrome (AIDS). *Clin Orthop* **240**: 129–136.

Chua A, Song C, Chai A, Chan L, and Tan KC (2004). The impact of skin banking and the use of its cadaveric skin allografts for severe burn victims in Singapore. *Burns* **30**(7): 696–700.

Cieslak PR, Thomas AR, Kohn MA, Chai F, Nainan OV, Williams IT, Bell BP, Tugwell BD, and Patel PR (2002). Hepatitis C virus transmission from an antibody-negative organ and tissue donor — United States. *Morb Mortal Wkly Rep* **52**: 273–276.

CDC says clostridium infections were caused by musculoskeletal tissue allografts, more oversight needed (2004, July 30). *Transplant News.* Retrieved from http://findarticles.com/p/articles/mi_m0YUG/is_14_14/ai_n17208423.

Centers for Disease Control (1988). Transmission of HIV through bone transplantation: case report and public health recommendations. *Morb Mortal Wkly Rep* 37: 597–599.

Conrad EU, Gretch DR, Obermeyer KR, Moogk MS, Sayers M, Wilson JJ, and Strong DM (1995). Transmission of hepatitis C by tissue transplantation. *J Bone Joint Surg* **77A**: 214–224.

Dawson GJ (2007). HCV core antigen and combination (antigen/antibody) assays for the detection of early seroconversion. *J Med Virol* **79**: 54–58.

Dubois D and Dubois EF (1916). A formula to estimate the approximate surface area if height and weight be known. *Arch Intern Med* **17**: 863–871.

Laperche S (2005). Blood safety and nucleic acid testing in Europe. *Euro Surveill*. Retrieved from http://www.eurosurrveillance.org/ViewArticle.aspx?Articled=516.

Louie B, Pandori MW, Wong E, Klausner JD, and Liska S (2006). Use of an acute seroconversion panel to evaluate a third-generation enzyme-linked immunoassay for detection of human immunodeficiency virus-specific antibodies relative to multiple other assays. *J Clin Microbiol* **44**(5): 1856–1858.

Martin P, Fabrizi F, Quan S, Brezina M, Kaufman E, Sra K, DiNello R, Polito A, and Gitnick G (1997). Automated RIBA hepatitis C virus (HCV) strip immunoblot assay for reproducible HCV diagnosis. *J Clin Microbiol* **36**(2): 387–390.

Møller JM and Krarup HB (2003). Diagnosis of acute hepatitis C: anti-HCV or HCV-RNA? *Scand J Gastroentero* **38**(5): 556–558.

Nather A and David V (2007). Femoral head banking NUH Tissue Bank experience. *Orthopedics* **30**: 308–312.

Simonds RJ, Holmberg SD, Hurwitz RL, Coleman TR, Bottenfield S, Conley LJ, Kohlenberg SH, Castro KG, Dahan BA, Schable CA, *et al.* (1992). Transmission of HIV Type 1 virus from a seronegative organ and tissue donor. *New Engl J Med* **326**: 726–732.

Scott JD and David RG (2007). Molecular diagnostics of hepatitis C virus infection: A systemic review. *J Am Med Assn* **297**: 724–732.

Tomford WW (1994). A perspective in bone banking in the United States. In Urist MR, O'Connor BT, and Burwell RG (eds.), *Bone Grafts, Derivatives and Substitute*, Butterworth-Heinemann, London, pp. 193–195.

U.S. Food and Drug Administration (2002). FDA approves first nucleic acid test (NAT) system to screen whole blood donors for infections with human immunodeficiency virus (HIV) and hepatitis C virus (HCV). *U.S. Food and Drug Administration.* Retrieved from http://www.fda.gov/default.htm.

U.S. Food and Drug Administration (2004). Use of nucleic acid tests on pooled and individual samples from donations of whole blood and blood components (including source plasma and source leukocytes) to adequately and appropriately reduce the risk of transmission of HIV-1 and HCV. *U.S. Food and Drug Administration.* Retrieved from http://www.fda.gov/cber/guide-lines.htm.

Warren J (2004, Sept 30). Beating FDA to the punch, AATB requires accredited tissue banks to implement NAT testing for HIV-1 and HCV. *Transplant News.* Retrieved from http://findarticles.com/p/articles/mi_m0YUG/is_18_14/ai_n17208495.

Weber B, Fall EMB, Berger A, and Doerr HW (1998). Reduction of diagnostic window by new fourth-generation human immunodeficiecy virus screening assays. *J Clin Microbiol* **36**(8): 2235–2239.

Part IV

Basic Sciences

Chapter 11

Anatomy of Musculoskeletal System

Aziz Nather*, Zameer Aziz* and Cui Lian Chong*

Tissue bank operators are required to assist the orthopaedic surgeons in the procurement of tissues from consenting donors. To do this they need to be trained in sterile technique. It is also important that they have basic knowledge of the anatomy of the lower limb, upper limb, the pelvis, the skin, and the amnion. In the case of bone procurement, it is essential that the technician is trained to identify several landmarks in each bone which will guide them to decide if the bone specimen procured is from the left side or the right side.

Lower Limb

Bones of the Lower Limb

The bones of the lower limb include the:

- Femur
- Patella
- Tibia
- Fibula
- Tarsal/metatarsal/phalanges of the foot (Nather, 2001)

---

*NUH Tissue Bank, Department of Orthopaedic Surgery, Yong Loo Lin School of Medicine, National University of Singapore, Singapore.

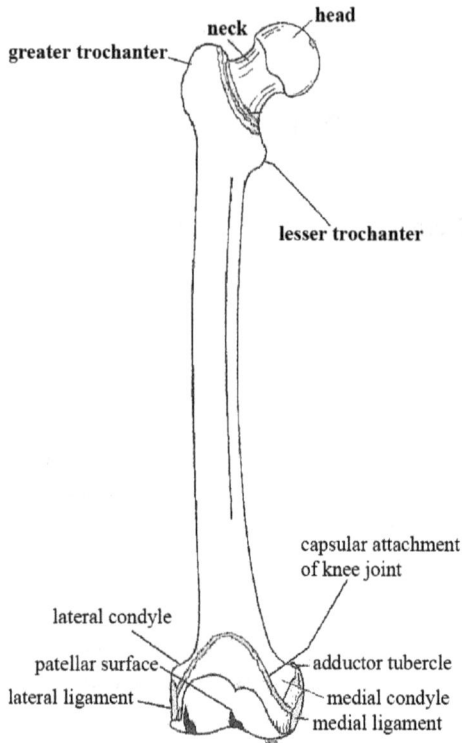

Fig. 1. Femur.

Femur

The femur or thigh bone is the longest bone in the human body. It consists of a shaft and two extremities. The upper end of the bone comprises a head, a neck and a lesser trochanter (Fig. 1). The head forms slightly more than a sphere and is directed upwards, medially and slightly anteriorly. The greater trochanter is a large quadrangular eminence located laterally at the junction of the neck with the shaft. The lesser trochanter is a smaller eminence projecting medially and posteriorly from the neck-shaft junction. The lower end of the femur is widely expanded into two prominent masses: the medial and the lateral condyles. Anteriorly, the articular surfaces of both condyles are joined together to form a grooved articular surface, the patellar surface for articulation with the patella (knee cap).

Patella

The patella (knee cap) is the largest sesamoid bone situated in front of the knee joint in the quadriceps femoris tendon. It is flat and triangular in shape, with the apex pointing inferiorly. Its undersurface is divided by a ridge into a medial facet and a lateral facet.

Tibia

The tibia is the larger of the two bones on the medial side of the leg. It is the second largest bone in the body next to the femur. It consists of a shaft and two ends. The upper end is expanded, consisting of two prominent masses — the medial and lateral condyles. The medial condyle is larger than the lateral condyle. The tibial tuberiosity is placed at the upper end of the anterior border of the shaft, the upper part giving attachment to the ligamentum patella. The lower end of the tibia is smaller than the upper end and projects downwards on the medial side beyond the rest of the bone to form the medial malleolus (Fig. 2).

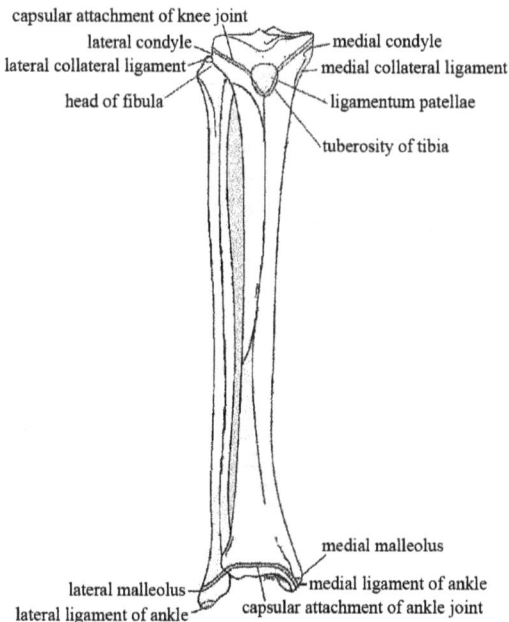

Fig. 2.　Tibia.

Fibula

The fibula is the lateral bone of the leg. It is very slender compared to the tibia. Not much weight transmission occurs through this bone. It consists of the head, an upper end, the shaft and a lower end (Fig. 2).

Foot

The skeleton of the foot consists of three segments: the tarsal bones, the metatarsal bones and the phalanges or bones of the toes (Fig. 3). The tarsus comprises seven bones, making up the posterior half of the foot. The proximal row consists of the talus and the calcaneum. The talus lies above

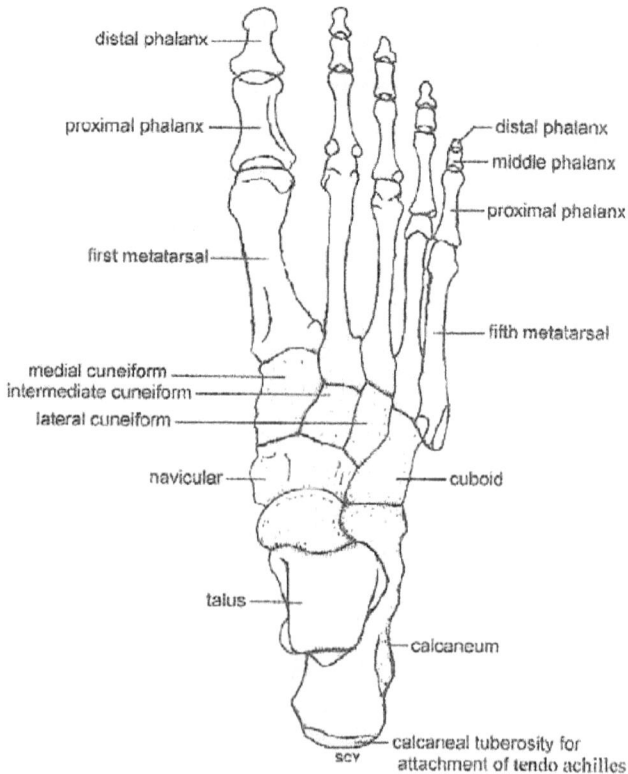

distal phalanx

proximal phalanx

distal phalanx

middle phalanx

proximal phalanx

first metatarsal

fifth metatarsal

medial cuneiform
intermediate cuneiform
lateral cuneiform

navicular

cuboid

talus

calcaneum

scv

calcaneal tuberosity for
attachment of tendo achilles

Fig. 3.   Foot.

the calcaneum. The distal row contains four bones: the medial cuneiform, the intermediate cuneiform, the lateral cuneiform and the cuboid. On the medial side, the navicular bone is interposed between the talus and the medial three bones of the distal row. The calcaneum projects backwards beyond the bones of the leg to form the calcaneal tuberiosity, which gives insertion to the tendo Achilles. The forefoot consists of five metatarsal bones and phalanges. The big toe had only two phalanges — proximal and distal phalanges. The other four toes has three phalanges each — proximal, middle and distal phalanges.

Muscles of the Lower Limb

The muscles of the lower limb can be classified into four groups:
  (i)   Muscles of the iliac region
 (ii)   Muscles of the thigh
(iii)   Muscles of the leg
(iv)   Muscles of the foot

Muscles of the Thigh

The anterior femoral muscles include the tensor fascia lata, the sartorius and the quadriceps femoris, which are the extensors of the leg. The quadriceps femoris consist of the vastus lateralis (the largest part), the vastus medialis, intermedius and the rectus femoris (Fig. 4).

The muscles of the gluteal region included the gluteus maximus, the gluteus medius and the gluteus minimus. The posterior femoral muscles consist of the hamstrings, the biceps femoris, the semitendinosus and the semimembranosus (Fig. 5).

Muscles of the Leg

They are divided into three groups: anterior, lateral and posterior. The anterior crural muscles include the tibialis anterior, the extensor hallucis longus, the extensor digitorum longus and the peroneus tertius (Fig. 6).

These muscles are dorsiflexors of the foot. The tibialis anterior becomes tendinous in the lower third of the leg and passes down the

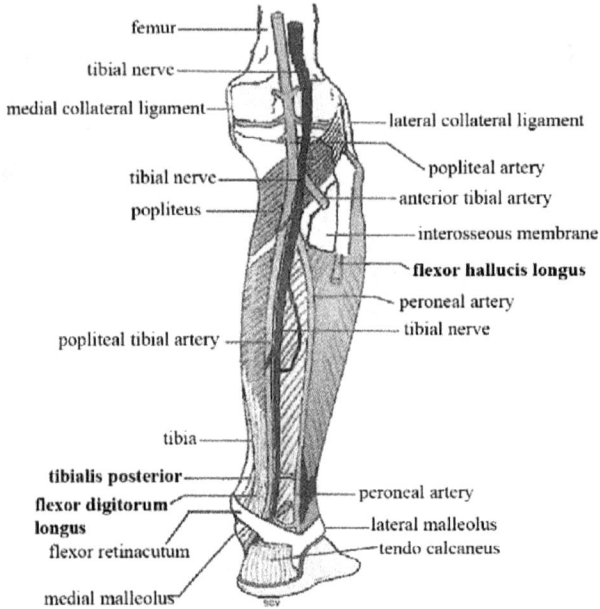

Fig. 4. Muscles of thigh — the anterior aspect.

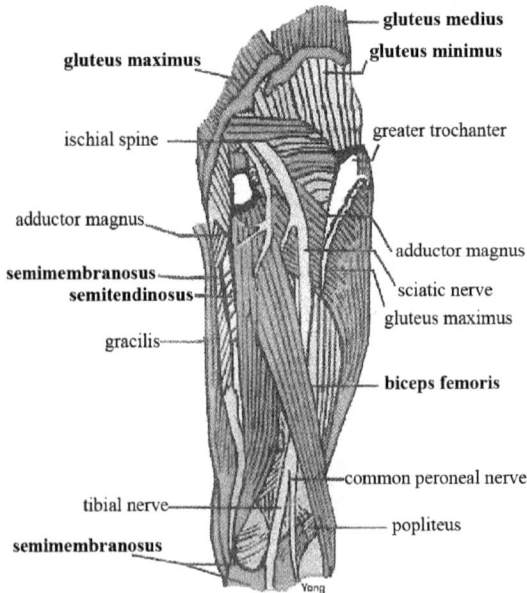

Fig. 5. Muscles of thigh — the posterior aspect.

medial side of the foot to be inserted into the undersurface of the medial cuneiform bone and adjoining part of the base of the first metatarsal bone. The lateral crural muscles consist of the peroneus longus and peroneus brevis and they evert the foot (Fig. 6). The posterior crural muscles can be subdivided into two groups — superficial and deep. The superficial group forms the muscle mass of the calf of the leg. The superficial group consists of the gastrocnemius, soleus and the plantaris (Fig. 7). The

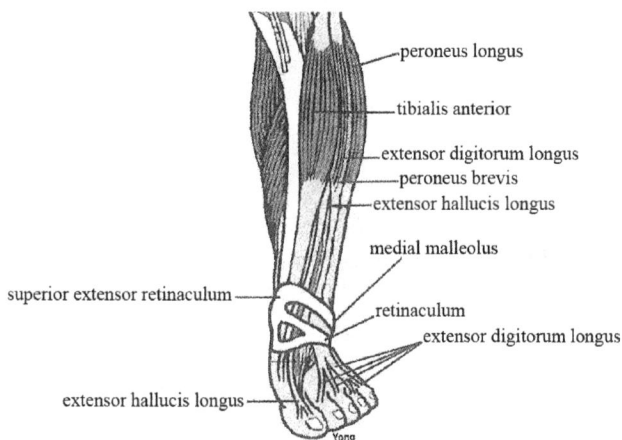

Fig. 6. Muscles of leg — the anterior aspect.

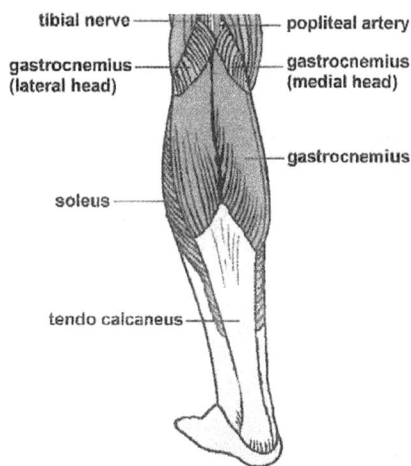

Fig. 7. Muscles of leg — the posterior aspect (superficial group).

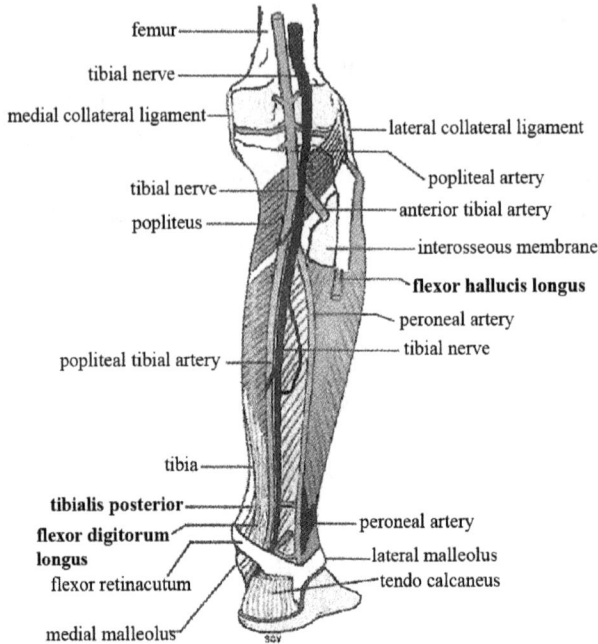

Fig. 8. Muscles of leg — the posterior aspect (deep group).

gastrocnemius and the soleus form a muscle mass called the triceps surae, which forms the tendo calcaneus (Fig. 7), the thickest and strongest tendon in the body. This tendo Achilles expands and is inserted into the middle and posterior surface of the calcaneum.

The deep group consists of the popliteus, flexor hallucis longus, flexor digitorum longus and the tibialis posterior (Fig. 8). The tibialis posterior becomes tendinous in the lower quarter of the leg, passes in a groove behind the medial malleolus and inserts into the tuberiosity of the navicular bone. It is the principal invertor of the foot.

Joints of Lower Limb

Hip Joint

There is a synovial ball-and-socket joint formed between the hemispherical head of the femur and the cup-shaped acetabulum of the hip bone (Fig 9).

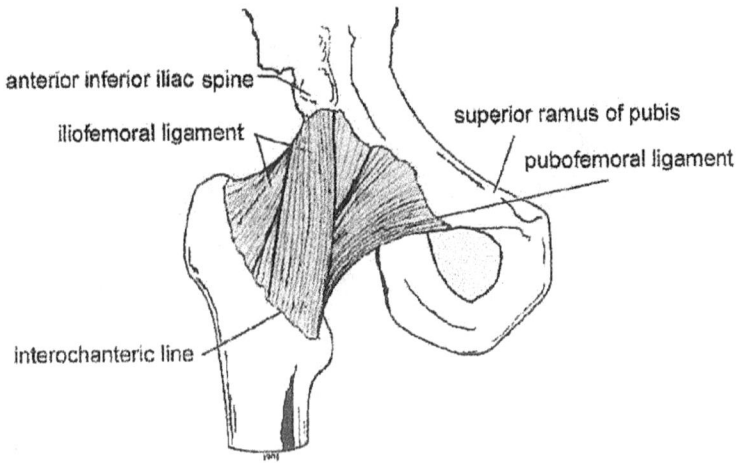

Fig. 9. Hip joint.

Knee Joint

The knee joint is the largest and most complicated joint in the body. It consists of the articulation between the medial and lateral condyles of the femur and the corresponding condyles of the tibia and a gliding joint, between the patella and the patellar surface of the femur. The articular surfaces of the femur, tibia and patella are covered with hyaline cartilage. The tibio-femoral joint is a synovial joint of the hinge type whilst the patella-femoral joint is a synovial joint of gliding variety. The suprapatellar bursa is present anteriorly beneath the quadriceps tendon (Fig. 10a). The ligamentum patellae is attached above the lower border of the patella and below the tuberiosity of the tibia. The lateral collateral ligament is attached above the lateral condyle of the femur and below the head of the fibula. The medial collateral ligament is attached above the medial condyle of the femur and below the medial surface of the tibial shaft. Both ligaments are extracapsular.

The cruciate ligaments are two very strong intracapsular ligaments crossing each other within the joint cavity (Fig. 10b). The anterior cruciate ligament prevents posterior displacement of the femur on the tibia. The meniscus or semi-lunar cartilages within the joint act as cushions between

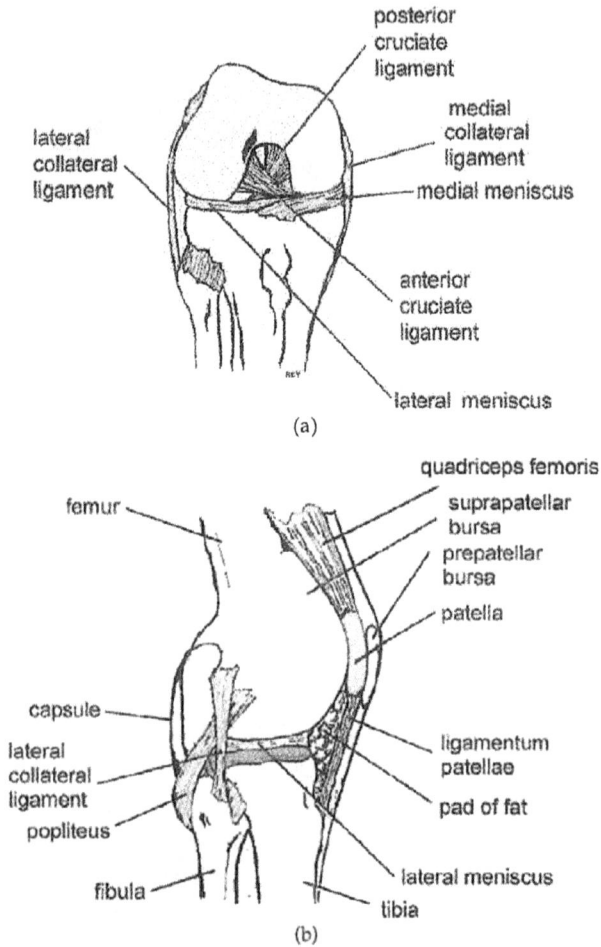

posterior
cruciate
ligament

medial
collateral
ligament

lateral
collateral
ligament

medial meniscus

anterior
cruciate
ligament

lateral meniscus

(a)

quadriceps femoris

femur

suprapatellar
bursa

prepatellar
bursa

patella

capsule

lateral
collateral
ligament

popliteus

ligamentum
patellae

pad of fat

fibula

lateral meniscus

tibia

(b)

Fig. 10.   Knee joint.

the femur and the tibia. The medial meniscus is nearly semi-circular in shape whilst the lateral meniscus is nearly circular in shape.

Pelvis

The bony pelvis consists of four bones: the innominate bones forming the lateral and anterior wall and the sacrum and coccyx, which are the

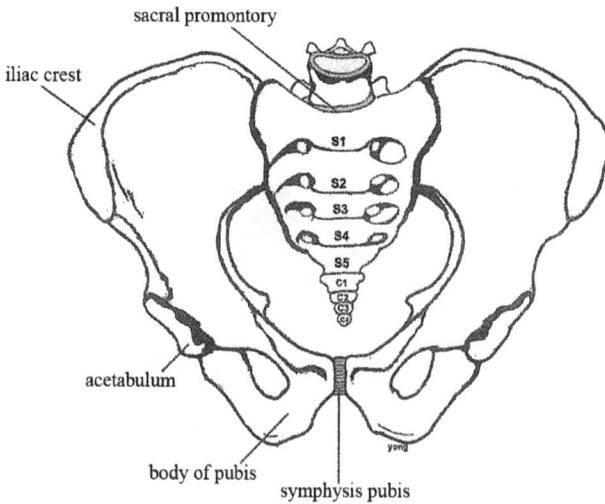

Fig.11. Bones of pelvis.

continuation of the vertebral column forming the posterior wall (Fig. 11). The iliac crest is a prominent ridge running between the anterior superior iliac spine and the posterior iliac spine (Nather, 2001).

Upper Limb

The upper limb of man is built for prehension. The hand is a grasping mechanism, with four fingers flexing against an opposed thumb (Chia, 2001).

Bones of the Upper Limb

Bones in the upper limb include:

- Clavicle (Fig. 12)
- Humerus (Fig. 13)
- Radius
- Ulna
- Bones in the wrist and hand

Clavicle

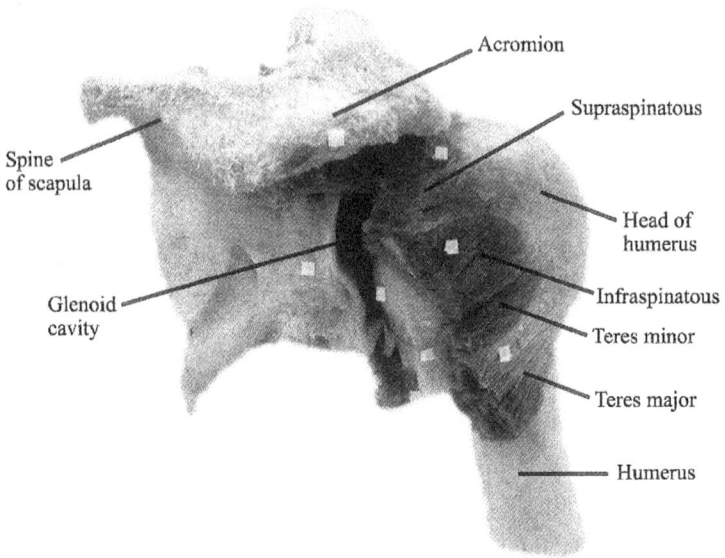

Fig. 12. Clavicle.

Humerus

The humerus bone (Fig. 13) consists of:

- Head with articular surface
- Anatomical neck
- Lesser tuberiosity (anterior and medial in anatomical position)
- Intertubercular (bicipital) groove
- Greater tuberiosity (posterior and lateral in position)
- Surgical neck
- Medial epicondyle
- Lateral epicondyle
- Trochlea (medial articulation, lower end humerus)
- Capitelleum (lateral articulation, lower end humerus)

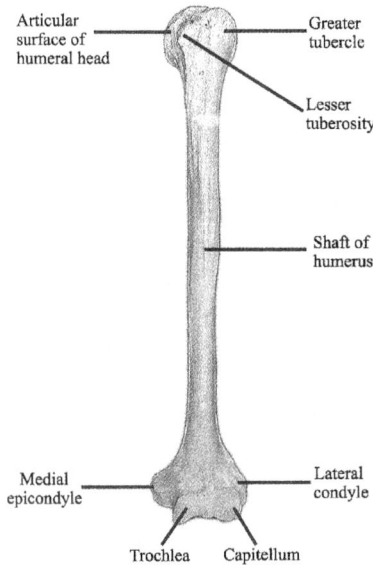

Fig. 13. Humerus — anterior view.

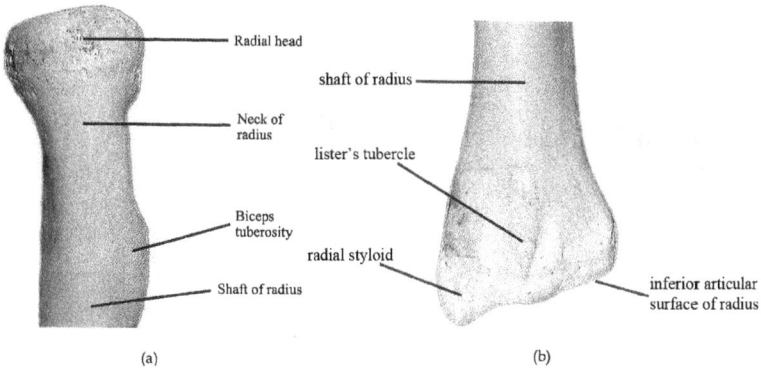

Fig. 14. (a) Proximal end of radius. (b) Distal end of radius (dorsal view).

Radius

The radius consists of:

- Proximal end: head, neck and radial (biceps) tuberiosity (Fig. 14a)
- Distal end: inferior articular surface, styloid process (radial styloid), dorsal radial tubercle (lister's tubercle) (Fig 14b)

Ulna

The ulna consists of:

- Proximal end: olecranon and coronoid processes; trochlear notch (Fig. 15a)
- Distal end: head, styloid process (Fig. 15b)

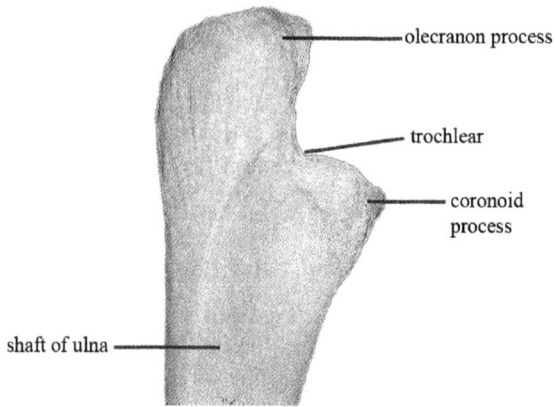

(a)

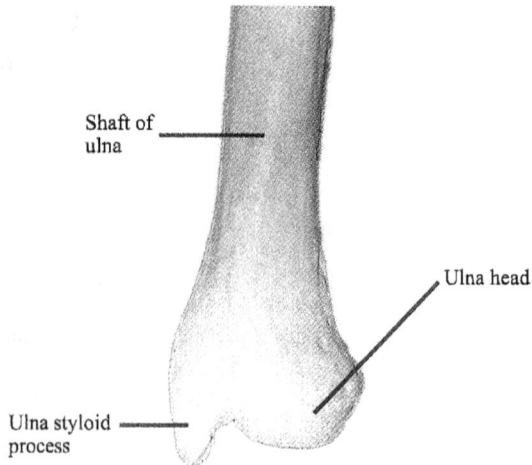

(b)

Fig. 15. (a) Proximal end of ulna. (b) Distal left ulna.

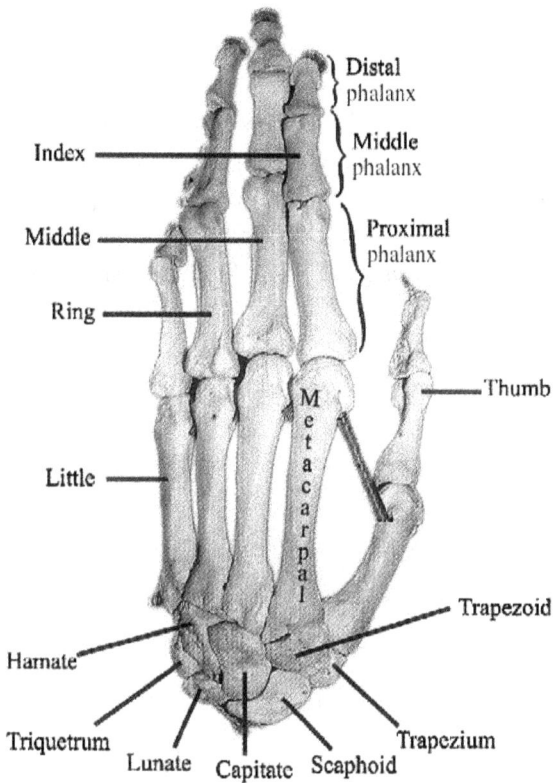

Fig. 16. Dorsal view of left hand bones.

Bones of the Wrist and Hand

The wrist and hand (Fig. 16) consists of:

- Eight carpal bones (divided into four proximal and four distal)
  - Four proximal: scaphoid, lunate, triquetrum (these three are articulate with the radius at the wrist joint), pisiform
  - Four distal: trapezium, trapezoid, capitate, hamate
- Five metacarpal bones
- 14 phalanges: three on each finger, two for the thumb

References

Chia J (2001). Anatomy of the upper limb. In Nather A (ed.), *The Scientific Basis of Tissue Transplantation*, World Scientific, Singapore, p. 3–9.

Nather A (2001). Anatomy of the pelvis. In Nather A (ed.), *The Scientific Basis of Tissue Transplantation*, World Scientific, Singapore, p. 51–52.

Nather A (2001). Anatomy of the lower limb. In Nather A (ed.), *The Scientific Basis of Tissue Transplantation*, World Scientific, Singapore, p. 25–41.

Chapter 12

Biology of Healing of Bone Allografts

Aziz Nather* and Yan Yi Han*

Introduction

Bridging a large bone defect is one of the most challenging problems in orthopaedic and maxillofacial surgeries. Common causes of a large bone defect are tumour resection and gap non-unions due to trauma. Such a defect can be reconstructed using a massive bone allograft (Mankin *et al.*, 1996).

Alternatives to the use of a massive bone allograft include vascularised bone transplant (Weiland and Daniel, 1979), non-vascularised bone autograft (Enneking *et al.*, 1980), prosthesis (Sim *et al.*, 1987; Natarajan, 1998; Schindler *et al.*, 1997) and ceramic (Yamamuro, 1990).

An allograft refers to the tissue transplanted between genetically non-identical members of the same species. In contrast, an autograft refers to the tissue removed from one portion of the skeleton and transferred to another part of the body of the same individual.

The disadvantage of using vascularised bone transplant is that the highly specialised technical expertise required is not always available. Furthermore, the demanding surgical technique requires prolonged operating time, and therefore incurs higher cost. Such expertise is often not available in the developing countries. Besides, the bone transplanted, the fibula (Pho, 1981) is not always large enough to match the bone defect in

---

*NUH Tissue Bank, Department of Orthopaedic Surgery, Yong Loo Lin School of Medicine, National University of Singapore, Singapore.

a bigger bone, such as the femur, to provide for immediate biomechanical stability and allow weight bearing.

With non-vascularised bone autograft, the size and amount of bone graft available are limited. In addition, it is often associated with considerable donor site morbidity (Montgomery *et al.*, 1990) — in particular, donor site pain and donor site infection.

With modular prosthesis, cost is the main limiting factor. The price of hip and shoulder modular prostheses ranges from US$10 000 to US$15 000 on the average. The cost of elbow modular prosthesis is more affordable at about US$7000. With custom-made prosthesis (Sim *et al.*, 1987), the same prohibitive cost remains. In addition, there is added cost of transportation and the need to cope with a delay of one to two weeks to allow for the fabrication of individually tailored prostheses. With the use of prosthesis, complications include long-term loosening of prostheses secured by cement and onset of "polyethylene disease" (Harris, 1994). "Polyethylene disease" arises through wear and tear of the prosthesis with the surrounding tissue.

Similarly, ceramic incurs a high cost of approximately US$10 000 to US$15 000. It is not available locally, although it is commercially available in Japan (Yamamuro, 1990). Whilst ceramic is a simpler option, osteointegration and host–graft union of the ceramic with the host bone are poor, as compared to an autograft or allograft.

On the other hand, allograft presents a very favourable option, provided there is a good tissue bank locally to provide bone allografts of high quality for use. There is no limitation to the number and size of the bone grafts that could be provided. Unlike prosthesis, which requires customisation to fit the defect (Schindler *et al.*, 1997), an allograft can be tailored intraoperatively to match the size and extent of the bone resection. In addition, the local cost of using an allograft is lower than that of a prosthesis or ceramic. In Singapore, the cost of a whole femur allograft is only about US$800. This makes an allograft about 13 to 19 times cheaper than a prosthesis. Besides, the use of an allograft would obviate the long-term problems of loosening of the prosthesis requiring revision surgery. For this reason, biological reconstruction is the best option especially for children. In addition, bone allograft is the only source of osteochondral transplant. A shortage of autografts and the attendant morbidity present with procuring an autograft are also reasons why allografts are being used more frequently.

It is not surprising, therefore, that in the Asia-Pacific region, where cost considerations are very important, bone allograft has become the option of choice. There is an increasingly large demand for bone allografts in the region. This in turn is responsible for the increasing number of new tissue banks that have been set up in the region within the past decade.

Biological Reconstruction Using Bone Allografts

The choice of an allograft with suitable properties is crucial to the success of an allograft transplantation. The types of bone allografts transplanted include:

- cortical allograft,
- cancellous allograft,
- cortico-cancellous allograft, and
- osteoarticular allograft.

Cortical bones are biomechanically strong and provide good structural support. They are ideal for massive bone defect reconstruction requiring load-bearing. Cancellous bones, on the other hand, are spongy and biomechanically weak. Cancellous bone grafts are used for the filling of bone cavities or for packing purposes only.

Some of the methods by which bone allografts are processed include:

- deep freezing,
- cryopreservation, and
- freeze-drying.

Deep-frozen cortical bone allograft must be used where biomechanical demand is high and weight-bearing function is required. Such cases include the intercalary reconstruction of bone or bone reconstruction where joint function is sacrificed (resection arthrodesis) in the lower limb. Freeze-dried cortical bone allograft is too weak and should not be used in such situations. Deep-frozen or freeze-dried cortical allograft can be used for spinal reconstruction following corpectomy. In the case of bone reconstruction where joint function is to be preserved, cryopreserved, osteoarticular allograft is the only allograft that can be used.

Gamma irradiation is an additional sterilisation process, which further sterilises the bones procured. In addition, it destroys the immunogenicity of the allograft (Dziedzic-Goclawska *et al.*, 1991; Fideler *et al.*, 1994) and inactivates the hepatitis C virus (Conrad *et al.*, 1995). Where articular cartilage transplantation is not required, deep-frozen bones can be used either without irradiation — "sterile-procured" and relying totally on the sterility of the procurement technique for safety — or with gamma irradiation to a dose of 25 kGy where the sterility of the graft is ensured. For these reasons, deep-frozen and gamma-irradiated bones are recommended for use by the authors.

The best method for reconstructing a bone defect is by using bone, because bone will unite with bone tissue itself. For successful biological reconstruction using bone allograft, one must first understand the biology of healing of massive deep-frozen cortical allograft (Nather, 1990).

Biological Healing of Bone Transplant

There are two important issues in the biological healing of both autogenic and allogenic bone transplants:

1. *Union of host–graft junctions* — fracture healing of host–graft junctions by the formation of osteoid callus. This is very important since non-union of the host–graft junction can lead to graft resorption.
2. *Graft incorporation* — healing of the graft itself by resorption activity, new bone formation and "callus encasement".

Biological Healing of Bone Autograft

In the biology of healing of an autograft, the following processes occur:

- osteoconduction,
- osteoinduction, and
- osteogenesis.

The process of osteoconduction refers to the ingrowth from the recipient (host) bed into the graft of capillaries, perivascular tissue and osteoprogenitor cells. The grafts act as inert scaffolds for the ingrowth of the host tissue. In other words, if a bamboo were to be used as a scaffold, osteoconduction would be akin to the ingrowth of host tissue and cells into the bamboo.

Osteoinduction is the mechanism in which new bone is formed by the active recruitment of host pluripotent cells that differentiate into osteoblasts. It is produced by the diffusion of osteogenic bone matrix proteins, referred to as bone morphogenic proteins (BMPs) from the demineralised bone matrix.

Osteogenesis is the formation of new bone from surviving cells within a bone graft — namely cells from the inner cambium layer of periosteum that survive autogenous transplantation. It does not occur with allograft transplantation, as allograft is processed without its periosteum.

Cortical autograft, specifically, undergoes "callus encasement", bone resorption, revascularisation and new bone formation in the biological healing of the graft. The resorption–apposition process occurs in an irregular fashion and takes a long time.

"Callus encasement" refers to callus formation around the graft after transplantation. The "callus" is an irregular mass of cells, which serves to protect the graft. The next step in graft healing is bone resorption, where lysis and assimilation of the graft into host tissue take place. Revascularisation is a process in which the bone repaired with the graft regains its blood supply. Finally, new bone formation occurs when new bone is formed following graft transplantation. New bone can only be formed after graft resorption happens to make room for it.

Resorption and apposition, as termed by Axhausen, is also known as the process of "creeping substitution". This refers to the initial resorption of the grafted bone, followed by the replacement of bone and bone cells from the host (Nather *et al.*, 1990a).

In order to understand the biological healing of allografts, it is important to understand how a non-vascularised cortical autograft heals.

Biological Healing of Bone Allograft

In order to use a deep-frozen cortical allograft successfully for biological reconstruction, it is important to understand its biological healing process.

Any form of cortical bone healing takes place through the process of "creeping substitution". Burchardt (1983) showed that this process takes place at the host–graft interphase.

Non-Vascularised Cortical Bone Autograft Transplantation

This refers to a cortical bone autograft, which is transplanted without its blood supply.

Research Study

Nather *et al.* (1990a) showed that in the tibia of adult cats, union of a large cortical autograft (two-thirds of the diaphysis) — a 4-cm segment — occurred by eight weeks (Fig. 1). Revascularisation, resorption activity, new bone formation and "callus encasement" took place readily (Fig. 2), with increased new bone formation occurring with time. These observations were quantitated by the bone resorption index, new bone formation index and "callus encasement" index.

Bone resorption reached a peak of 13% at 12 weeks and then dipped to 6.5% at 16 weeks (Fig. 3). New bone formation steadily increased

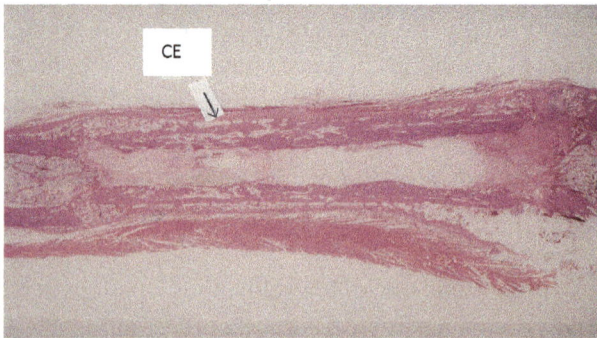

Fig. 1. Histological section of an autograft specimen (D_6) at eight weeks in a cat with osteoid callus at both host–autograft junctions. Several resorption cavities could be seen in both cortices of the autograft. "Callus encasement" could be seen enveloping the whole segment (CE).

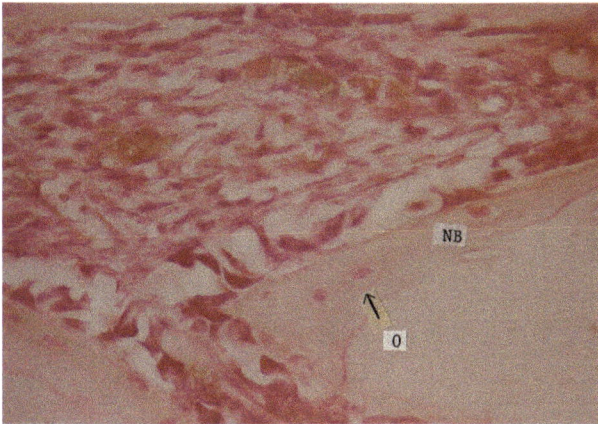

Fig. 2. Higher power magnification of a resorption cavity (40×) in the same autograft specimen (D_6) showing new bone formation lining the cavity. Osteoblasts could be seen in the new bone being laid down.

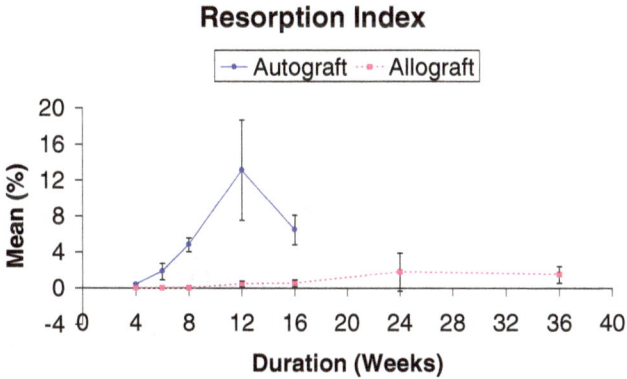

Fig. 3. Resorption index of autografts and allografts in adult cats.

with time, reaching 4.3% at 16 weeks (Fig. 4). Callus encasement appeared from four weeks onwards and reached a value of 8% at 16 weeks (Fig. 5).

Revascularisation was first seen at two weeks (Fig. 6). By six weeks, the entire segment became revascularised (Nather *et al.*, 1990b).

Cortical New Bone Formation Index

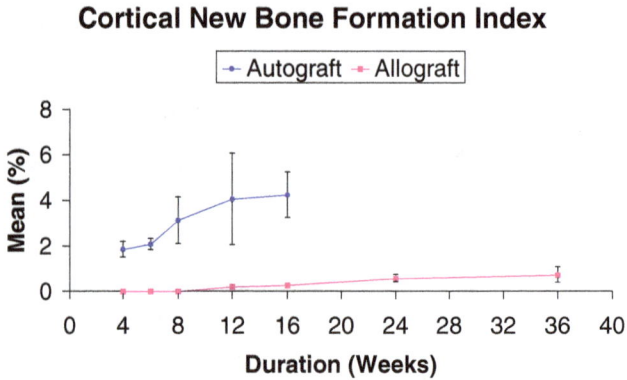

Fig. 4. Cortical new bone formation index of autografts and allografts in adult cats.

'Callus Encasement' Index

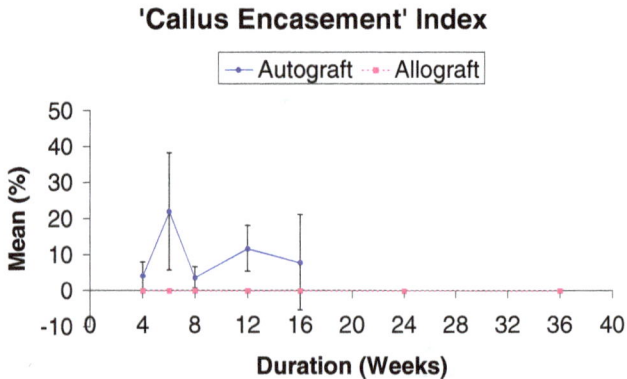

Fig. 5. Callus encasement index of autografts and allografts in adult cats.

The revascularisation index gradually increased till six weeks with a value of 0.14 mm/sq mm. It reached a plateau at 8, 12 and 16 weeks (Fig. 7).

Excision of the periosteum did not produce any difference to the healing of the autografts, as compared to controls. Obliteration of the medullary canal also produced no differences. However, when the muscle bed was isolated from the autograft (Fig. 8), the reparative processes were

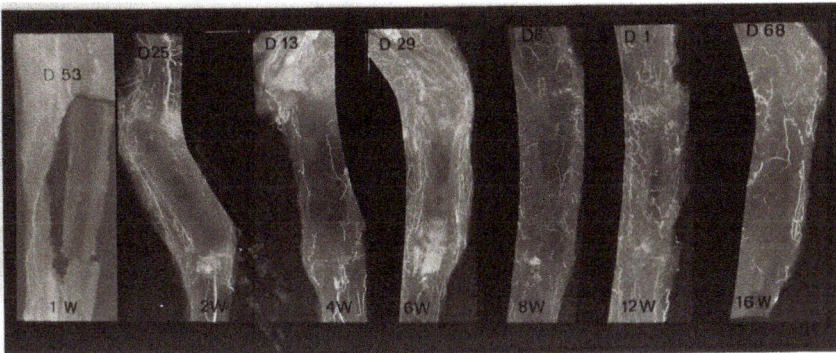

Fig. 6. Microangiograms of non-vascularised autograft specimens. Revascularisation started at two weeks. By six weeks, the entire autograft specimen had become vascularised.

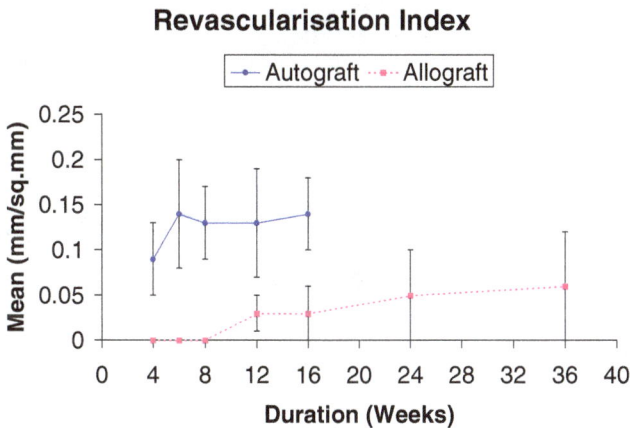

Fig. 7. Revascularisation index of autografts and allografts in adult cats.

markedly impaired (Nather *et al.*, 1990a). This showed that the muscle bed is most vital for the healing of the non-vascularised and large cortical autograft segment.

Soil–Seed Theory

Based on these observations, Nather proposed the "soil–seed theory": when the soil is good, any seed planted in it will grow. The soil is the

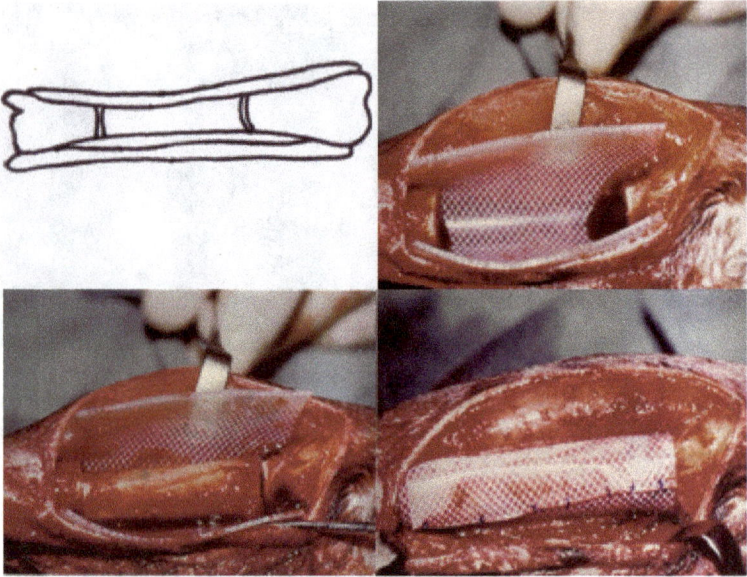

Fig. 8. Silastic sheath isolating the large cortical autograft from the surrounding muscle bed.

vascular muscle bed, whose muscles provide the vessels, which revascularise the bone graft. The seed could be a vascularised autograft (the best seed), a non-vascularised autograft (a good seed) or an allograft (a poor seed).

Vascularised Autograft Transplantation

With the transplantation of a vascularised autograft, the process of resorption–apposition does not occur. The bone transplanted with its blood supply re-anastomosed maintains total viability of its living cells — the osteocytes (Ostrup and Fedrickson, 1974), making it the "best seed".

Deep-Frozen Cortical Allograft Transplantation

Methodology of Research Study

Using the same experimental model as before, Nather (1990) reported the healing of deep-frozen cortical allograft. The tibia of adult cats was

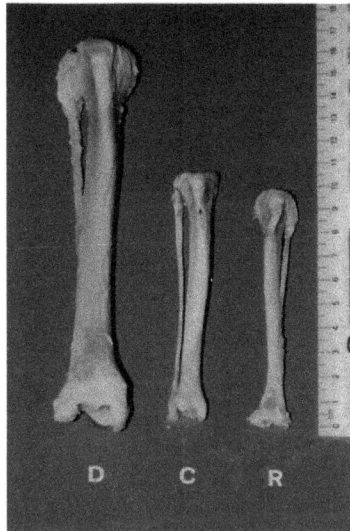

Fig. 9. The feline tibia (C) is fully weight-bearing as in man, compared to the fibula in the dog (D) or in the rabbit (R), which is not fully weight-bearing.

chosen as the experimental model because, in comparison to dogs and rabbits, the feline tibia and fibula most resembles that of man. Furthermore, the canine fibula is less ideal than the feline tibial model (Nather, 1990a), since the latter is fully weight-bearing, while the former is not (Fig. 9).

In this experiment, a total of 24 cats were employed. Four cats were studied for each observation period at 1, 2, 3, 4, 6 and 9 months from the start of the experiment. The parameters studied were again that of resorption activity, new bone formation activity and "callus encasement" activity. Newly included parameters were those of fracture union and revascularisation.

First, the allografts in adult cats were procured under sterile conditions, with the periosteum stripped off. They were stored using a sterile double-jar technique (Fig. 10) in an electrical freezer at −80°C for at least a month before use. Processing by deep-freezing kills all the living cells within an allograft in two weeks. This includes cells in the marrow (the most immunogenic part of the graft), thus leaving the allograft with greatly reduced immunogenicity.

Fig. 10. Sterile double-glass jar.

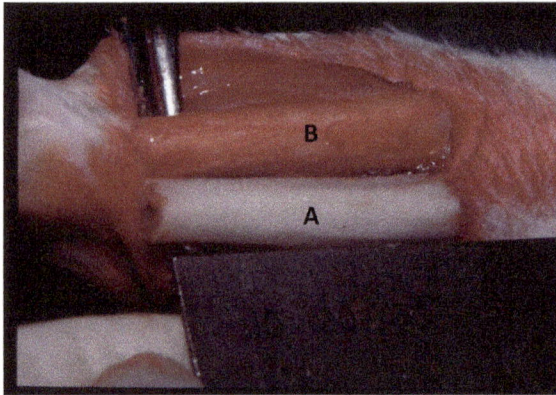

Fig. 11. Allograft devoid of periosteum (A) used to reconstruct the defect resulting from the excision of a large bone segment (B) in the recipient tibia.

Next, the deep-frozen allografts were transplanted into the recipient cats. Two-thirds of the tibial diaphysis in each cat was excised to create a large bone defect (4 cm), which the allograft was used to reconstruct (Fig. 11). Internal fixation was performed using an intramedullary rod that was 2.3 mm in diameter (Fig. 12).

Microangiography was then performed using barium sulphate perfusion, following which the leg specimen was retrieved. A central 2-mm

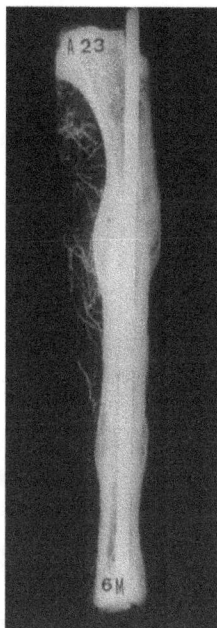

Fig. 12. Radiograph of allograft specimen at six months (A23) showing intramedullary rod fixation. Good callus formation could be seen at both host–allograft junctions.

thick longitudinal slice was cut from the decalcified specimen and subjected to soft X-rays, according to the technique described by Nather *et al.* (1990b). Histology was also performed on the same longitudinal 2-mm decalcified strip, which was embedded in paraffin wax and cut for staining.

Results of Research Study

Callus formed at the host–allograft junctions could be seen microscopically from about four weeks onwards. Histologically, the callus was mainly cartilaginous in nature. It was only from 12 weeks onwards that osseous callus was observed in all specimens (Fig. 13).

Allograft concurrently achieved osseous union after 12 weeks at host–allograft junctions (Nather, 1990). Solid union could hence only be seen macroscopically at both host–allograft junctions from 12 weeks

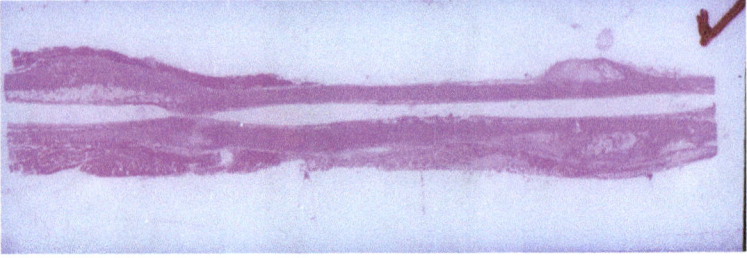

Fig. 13. Histological section of allograft specimen (A14) at 12 weeks, showing bridging of both host–allograft junctions by osteoid callus. Both cortices of allograft showed few and tiny resorption cavities only. No "callus encasement" could be seen around the allograft.

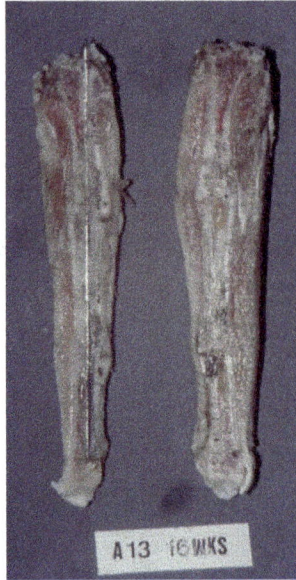

Fig. 14. Gross appearance of allograft specimen (A13) at 16 weeks. Intramedullary rod used for fixation readily seen on left side of the picture. Solid callus visible at both host–allograft junctions.

onwards (Nather, 1990), as shown in Figs. 14 and 15. In comparison, osseous union took place at host–autograft junctions after only eight weeks (Nather *et al.*, 1990a).

Unlike in the autograft, where "callus encasement" occurred covering the autograft, "callus encasement" did not occur to envelop the allograft.

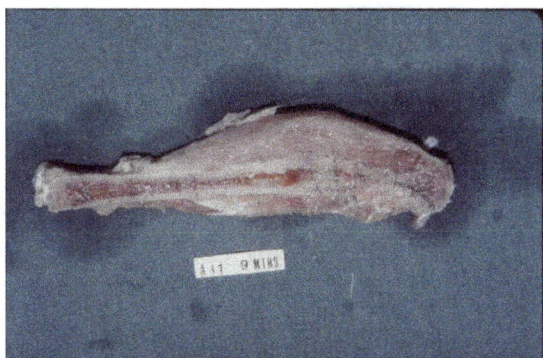

Fig. 15. Gross appearance of allograft specimen (A11) at nine months. Solid union seen at both host–allograft junctions.

"Callus encasement" index was 0% for all allograft specimens (Fig. 6) in the central 2-cm portion of the allograft. "Callus encasement" was confined only to 2 to 6 mm of the host–allograft junctions and did not extend to cover the central portion of the allograft.

The resorption activity occurring in allograft was markedly lower than that occurring in autograft (statistically significant). The resorption index measured only 0.48% at 12 weeks and was less than 2% even at nine months (Fig. 3). Microscopically, only a few very small resorption cavities appeared in the peripheral part of the cortex on the periosteal side at 12 and 16 weeks, up to a period of six and nine months (Fig. 16). No periosteum was reformed. No resorption cavities were seen in the inner cortex adjacent to the endosteal surface.

New bone formation was limited in allograft, which was significantly lower than that in autograft. No new bone formation was observed at 4, 6 and 8 weeks. Minimal new bone formation was seen at 12 and 16 weeks, up to a period of six to nine months (Nather, 1990). The cortical bone formation index was only 0.21% at 12 weeks and reached 0.74% at nine months (Fig. 5).

Revascularisation in allograft was proven to be significantly lower than that in autograft as well. For allograft, the revascularisation index remained very low at 0.06 mm/sq mm even at nine months (Fig. 8). Microangiographically, whilst some vessels could be seen to penetrate the

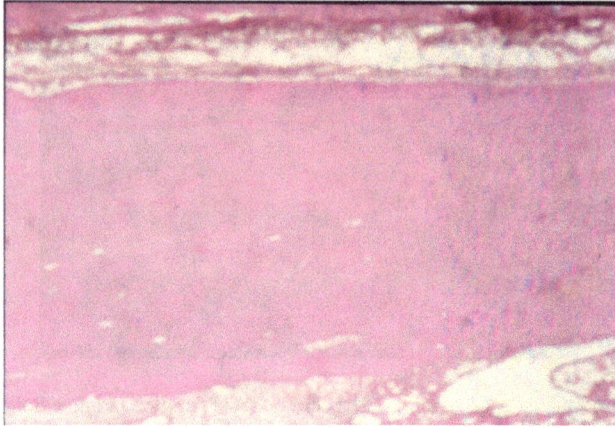

Fig. 16. Higher magnification (10×) of histological section of allograft (A15) at nine months. Very few and very small resorption cavities were visible in the cortex.

allograft in the region of the host–allograft junctions, very few vessels could be seen to penetrate the cortical allograft in the central 2-cm portion even at nine months (Figs. 17 and 18).

In contrast to the active repair processes occurring with non-vascularised autograft (Nather *et al.*, 1990a), very little reparative activity occurred in cortical allograft (Nather, 1990). Less than 2% of the allograft showed resorption and new bone formation. No callus encasement was seen with allograft. Large deep-frozen cortical allograft remained biologically inert for a long period of time in the adult cat.

Clinical Significance

The fact that resorption–apposition does not occur significantly in deep-frozen allograft means that allograft does not undergo much weakening with time after transplantation. In contrast, non-vascularised autograft initially become weak due to active resorption–apposition, until more new bone formation occurs (Enneking *et al.*, 1975).

An allograft therefore acts as a "spacer" which remains relatively inert, although union at the host–allograft junctions occurred without much difficulty. This union was delayed as compared to that at host–autograft junctions.

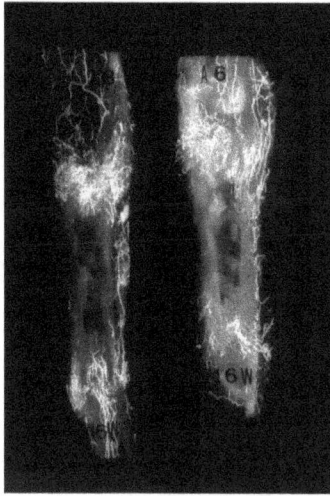

Fig. 17. Microangiogram of allograft specimen (A6) at 16 weeks. Only some vessels could be seen to enter the allograft in the region of host–allograft junctions. In the central portion of the allograft, no vessel was seen to enter the bone.

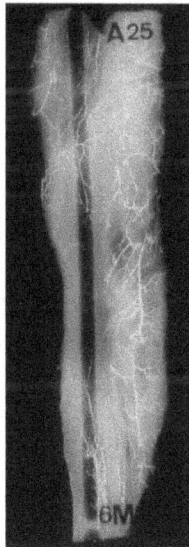

Fig. 18. Microangiogram of allograft specimen (A25) at six months. Hardly any vessels were seen to penetrate the cortex of the allograft. Only some vessels were seen at the host–allograft junctions.

Conclusions

An allograft is biologically inert. It exhibits minimal or no osteoinduction and osteogenesis, due to the lack of viable surface cells. At best, it acts as an "osteoconductor" or "spacer", providing a scaffold for host cells to repopulate their physical structure. Whilst the extent of osteoconduction that occurs with cortical allograft is generally small and very limited, it occurs more readily and to a larger extent in cancellous or cortico-cancellous allograft.

References

Burchardt H (1983). The biology of bone graft repair. *Clin Orthop Relat Res* **174**: 28–42.

Conrad EU *et al.* (1995). Transmission of hepatitis C by tissue transplantation. *J Bone Joint Surg Am* **77**: 214–224.

Dziedzic-Goclawska A, Ostrowski K, Stachowickz W, Michalik J, and Grzesik W (1991). Effect of radiation sterilization on osteoinductive properties and the rate of remodeling of bone implants preserved by lyophilization and deep-freezing. *Clin Orthop Relat Res* **272**: 30–37.

Enneking WF, Burchardt H, Puhl JJ, and Piotrowski G (1975). Physical and biological aspects of repair in dog cortical-bone transplants. *J Bone Joint Surg Am* **57**: 237–252.

Enneking WF, Eady JL, and Burchardt H (1980). Autogenous cortical bone grafts in the reconstruction of segmental skeletal defects. *J Bone Joint Surg Am* **62**: 1039–1058.

Fideler BM, Vangsness CT, Moore T, Li Z, and Rasheed S (1994). Effects of gamma irradiation on human immunodeficiency virus. *J Bone Joint Surg Am* **76**: 1032–1035.

Harris WH (1994). Osteolysis and particle disease in hip replacement. *Acta Orthop Scand* **65**(1): 113–123.

Mankin HJ, Gebhardt MC, Jennings LC, Springfield DS, and Tomford WW (1996). Long-term results of allograft replacement in the management of bone tumors. *Clin Orthop Relat Res* **324**: 86–97.

Montgomery DM, Aronson DD, Lee CL, and LaMont RL (1990). Posterior spinal fusion: allograft versus autograft bone. *J Spinal Disord* **3**: 370–375.

Natarajan MV (1998). Challenges and achievements in orthopaedic oncology. In *Proceedings of the 43rd Annual Conference of Indian Orthopaedic Association*, Jabalpur, India.

Nather A (1990). Healing of large diaphyseal allograft transplants. An experimental study. In Chao EYS (ed.), *Proceedings of the International Society for Fracture Repair*, Second Meeting, Mayo Clinic, p. 90.

Nather A, Balasubramaniam P, and Bose K (1990a). Healing of non-vascularised diaphyseal bone transplants. An experimental study. *J Bone Joint Surg Br* **72**: 830–834.

Nather A, Balasubramaniam P, and Bose K (1990b). Bone morphometry of revascularization a large avascular segment of bone. A microangiographic study. In Takahashi HE (ed.), *Bone Morphometry*, Nishimura, Smith-Gordon, Tokyo, London, pp. 92–95.

Ostrup LT and Fredrickson JM (1974). Distant transfer of a free, living bone graft by microvascular anastomoses. An experimental study. *Plast Reconstr Surg* **54**: 274–285.

Pho RWH (1981). Malignant giant-cell tumor of the distal end of the radius treated by a free vascularized fibular transplant. *J Bone Joint Surg Am* **63**: 877–884.

Schindler OS, Cannon SR, Briggs TW, and Blunn GW (1997). Stanmore custom-made extendible distal femoral replacements. Clinical experience in children with primary malignant bone tumour. *J Bone Joint Surg Br* **79**: 927–937.

Sim FH, Beauchamp CP, and Chao EY (1987). Reconstruction of musculoskeletal defects about the knee for tumor. *Clin Orthop Relat Res* **221**: 188–201.

Weiland AJ and Daniel RK (1979). Microvascular anastomoses for bone grafts in the treatment of massive defects in bone. *J Bone Joint Surg Am* **61**: 98–104.

Yamamuro T (1990). Replacement of the vertebrae with bioactive glass-ceramic prostheses. In Nather A (ed.), *Proceedings of the 13th Singapore Orthopaedic Association Meeting in Conjunction with the 2nd Asia Pacific Association of Surgical Tissue Banking Meeting*, Singapore, p. 60.

Chapter 13

Biomechanics of Healing of Bone Allografts

Aziz Nather*, Si Qi Koh* and Wen Hui Teng*

Introduction

Bone is a living tissue which is able to adapt and respond to its environment. It also has the ability to alter and repair itself. Its dynamic nature is demonstrated by the study of biomechanics, which examines the forces acting upon and within its structure.

From a biomechanical standpoint, the successful transplantation of large bone allografts depends on the following factors:

- type of physiological loads the transplant is subjected to,
- type of bone allograft used and
- type of processing used to produce the allograft.

Biomechanical Demand of Reconstruction

Bone is subjected to the following loading conditions:

- compression,
- tension and
- torsion.

* NUH Tissue Bank, Department of Orthopaedic Surgery, Yong Loo Lin School of Medicine, National University of Singapore, Singapore.

Fig. 1.   Bone subjected to compression.

Compression

Compression occurs when the graft is subjected to a force (Fig. 1) which results in the reduction or shortening of length.

Tension

Tension occurs when a bone is subjected to bending forces (Fig. 2).

Torsion

Torsion occurs when an object is twisted about its axis (Fig. 3).

When such a force is applied to a bone, the initial response of the bone is to resist the applied force. The tissue begins to deform when the internal resistance against the applied force is overcome. In the initial phase, the deformation is not permanent and the bone is able to revert to its original shape with the removal of the force (Fig. 4). This deformation is known as elastic deformation. As the force increases, the extent of the deformation will increase. A critical point known as the yield point will be reached when the bone starts to show signs of permanent damage such as tears and cracks. The deformation has now become plastic, as the bone is unable to revert to its original shape.

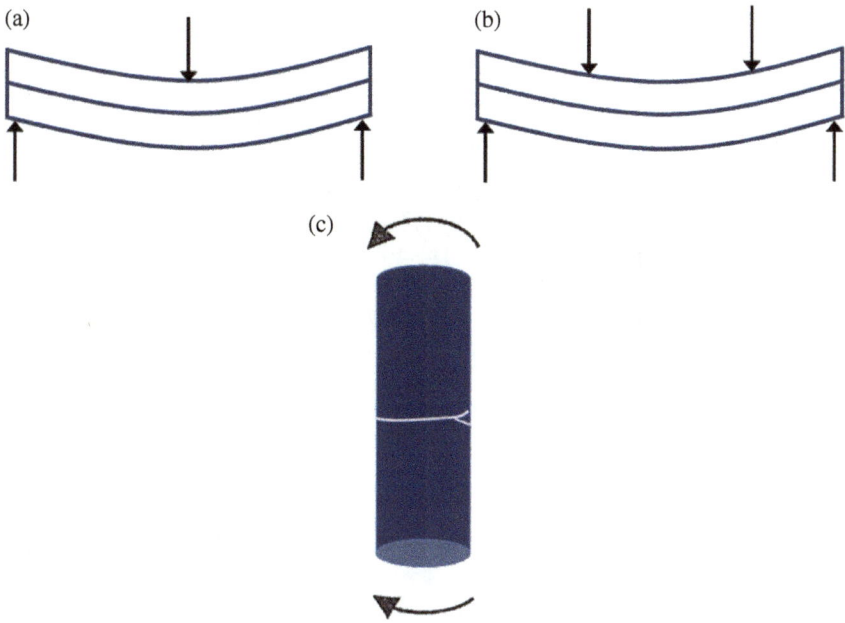

Fig. 2. (a) Bone subjected to three-point bending; (b) bone subjected to four-point bending, and (c) transverse fracture with small butterfly fragment.

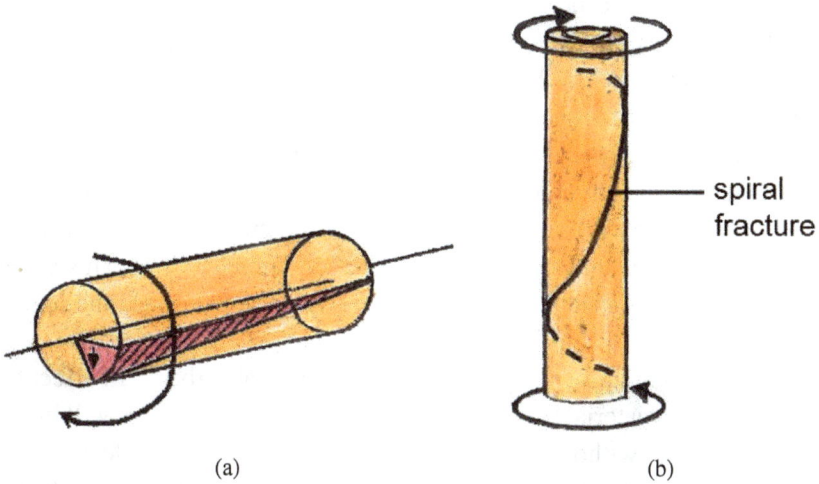

Fig. 3. (a) Bone subjected to torsion and (b) spiral fracture.

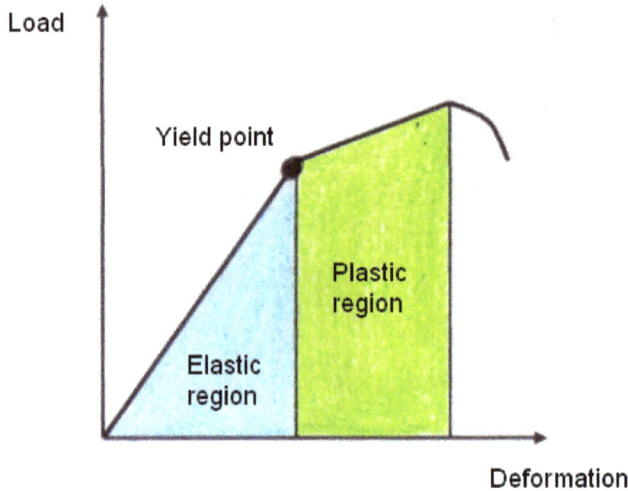

Fig. 4. Graph showing the structural behaviour of material when subjected to basic loading conditions.

Such deformation is reflected in the stretching of a rubber band. The rubber band will lengthen with increasing force applied and if it reverts to its original length when the force is removed, the rubber band is said to have undergone elastic deformation. If the rubber band is continually lengthened, tears will occur and the rubber band will soon snap in response to this stretching. When this occurs, the rubber band is said to have undergone plastic deformation as it will not return to its original shape.

Due to their dynamic nature, biological tissues such as large bone allografts will absorb and store a certain amount of energy when loaded. This stored energy is known as the elastic strain energy and can be represented by the area under the load deformation curve (Fig. 4). This stored energy allows the material to recover its original length once the load is removed.

The different amounts of elastic strain energy absorbed will affect the strength of the material (Fig. 5). A brittle material such as glass will break, fracture or tear without absorbing much energy. This is reflected by a steep curve with a very small area under the curve. A compliant or ductile material such as copper can undergo a greater degree of deformation

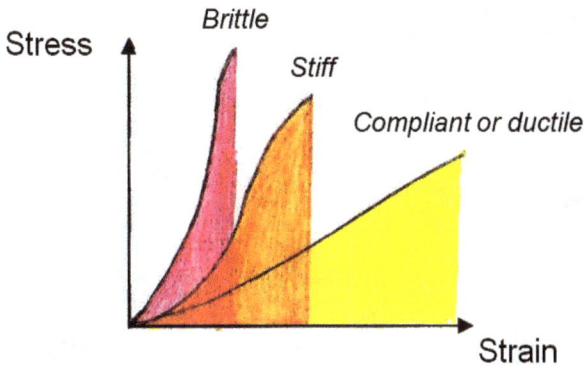

Fig. 5. Graph comparing the stress–strain curves of "brittle", "stiff" and "compliant or ductile" materials.

before failure and this relationship is represented by a curve with a gentler gradient and larger area under the curve.

In addition, it must be noted that unlike inert materials, biological tissues such as bones display viscoelastic behaviour. This means that if the load is applied onto the tissue at a faster rate, the resistance offered by the tissue will be greater. Hence, the tissue will experience greater stress if the force is applied at a faster rate than if it were to be applied onto the tissue at a slower rate.

Type of Allografts Used

For massive bone allograft reconstruction in a weight-bearing limb, e.g. lower limb, which demands large physiological loads for early ambulation, the strongest bone allograft must be used to ensure success. This is usually deep-frozen cortical allograft from deceased donors coupled with the strongest internal fixation, preferably a specially designed intramedullary nail with interlocking screws for added rotational stability (Fig. 6). Lyophilised cortical long bones from deceased donors should not be used. Despite using deep-frozen cortical allograft, Mankin *et al.* (1983) found serious fracture of the allografts occurred in 15 out of 91 patients, at an incidence of 16.5%.

In situations where bone allografts are used to pack bone cavities, the biomechanical demand on the transplantation is much less. Weight bearing

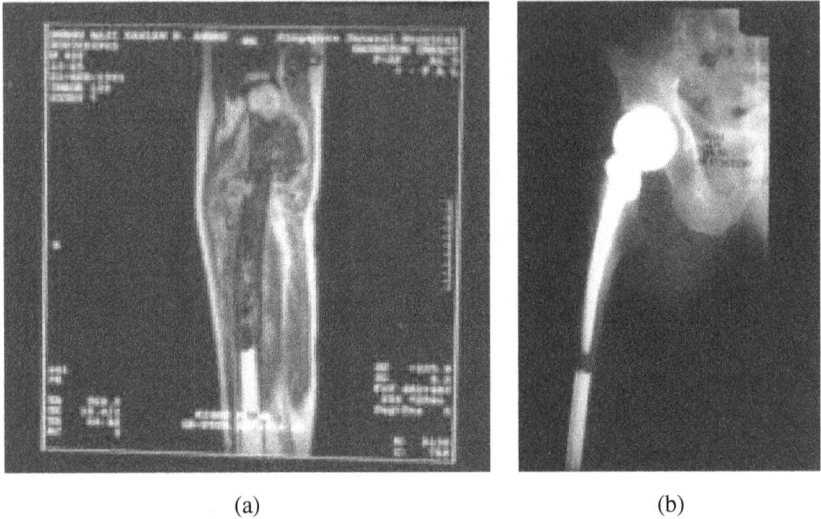

(a)   (b)

Fig. 6. (a) MRI showing chondrosarcoma involving proximal two-thirds of femur. (b) Whole deep-frozen femur allograft used for reconstruction with Moore's hemiarthroplasty proximally and interlocking nail distally.

is usually not needed for three months, though early mobilisation is required. No rigid internal fixation is necessary. For these purposes, morsellised bone allografts could be used either deep frozen or lyophilised.

On the other hand, in anterior spinal reconstruction following corpectomy for tumours and trauma of the spine, both deep-frozen and lyophilised cylindrical cortical allografts can be used (Fig. 7). In the author's series (Nather, 1999) of 40 massive anterior spinal reconstructions using 25 deep-frozen and 9 lyophilised cortical allografts, no graft failure was observed.

Effect of Different Methods of Processing on Biomechanical Strength of Bone Allograft

The three common methods employed for processing are:

1. deep-freezing,
2. radiation sterilisation and
3. freeze-drying.

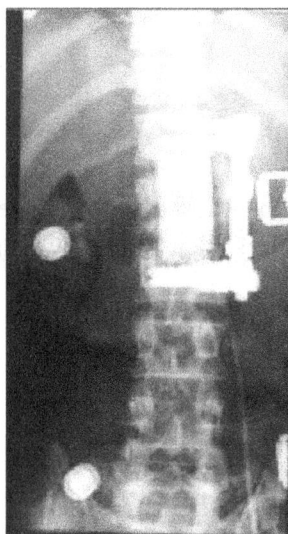

Fig. 7. Femoral cortical ring deep-frozen allograft used for reconstructing colonic metastasis to L1 lumbar vertebra with Kaneda instrumentation following anterior corpectomy.

Deep-Freezing

Deep-freezing involves the processing of allograft bones in electrical freezers at −70°C. Sedlin (1965) showed that bones frozen at −20°C do not undergo any change in physical properties. Komender (1976) showed that bones deep frozen at −78°C showed no change in bending, compression and torsion strength. In addition, several authors demonstrated that the torsional strength of bone remained unchanged at temperature ranges from −20°C to −196°C (Komender, 1976; Pelker *et al.*, 1983, 1984). Hamer *et al.* (1996) showed no change in bending strength when bones were frozen at −70°C.

Radiation Sterilisation

Several studies showed that the compression strength of bone allografts is not altered by radiation doses less than 30 KiloGrays (KGy) (Triantafyllou *et al.*, 1975; Komender, 1976; Bright and Burchardt, 1983). Komender

(1976) reported that 90% of torsion strength is maintained up to 30 KGy. In contrast, when the irradiation dosage was increased to 60 KGy, the specimens showed a reduction in bending, compression and torsion strength. The torsion strength was decreased to 65% by irradiation at a dose of 30 KGy and freeze-drying. Triantafyllou *et al.* (1975) showed that the bending strength of bone was markedly reduced to 10–30% of controls by a combination of lyophilisation and radiation sterilisation at a dose of only 33 KGy.

In contrast, Hamer *et al.* (1996) showed that the bending strength of bone was reduced to 64% of controls after irradiation with 28 KGy and the reduction in strength was also dose dependent. However, it is pertinent to note that the latest study by Zhang *et al.* (1994) showed that there was no statistical significant difference between irradiated and non-irradiated groups for both deep-frozen allografts and freeze-dried tricortical iliac crest allografts at a radiation dosage of 20–25 KGy. Hence, the authors recommended using 25 KGy for secondary sterilisation of human iliac crest wedges.

Freeze-Drying

Of the three methods, freeze-drying causes the greatest reduction in the strength of bones. Freeze-drying has been reported to cause a small increase of about 20% in the compression strength of the rehydrated bone (Pelker *et al.*, 1984). In contrast, Komender (1976) showed that lyophilisation increased the compression strength but rehydration of lyophilised bone restored the compression strength to normal. Triantafyllou *et al.* (1975) showed that lyophilisation decreased the bending strength of 55–90% of controls. Pelker *et al.* (1984) reported the appearance of longitudinal microscopic cracks in the bone when the freeze-dried specimens were rehydrated. This could explain the reduction in strength with lyophilisation. On the other hand, Wolfinbarger *et al.* (1994) showed that there was no significant change of compression strength with deep-frozen, freeze-dried and rehydrated freeze-dried tricortical iliac crest wedges used for spinal fusion surgery.

Nonetheless, whilst freeze-drying has been shown to generally weaken the bone, lyophilised grafts have been used successfully for clinical transplantation (Schneider and Bright, 1976; Spence *et al.*, 1976; Delloye, 1999).

Biomechanical Strength of Deep-Frozen and Non-Irradiated Cortical Allografts

Nather and Goh (2000) studied the biomechanical strength of deep-frozen and non-irradiated cortical bone allografts using the tibial diaphysis of the adult cat as the experimental model and compared it with the strength of vascularised and non-vascularised autografts (Fig. 8) (Nather *et al.*, 1990b).

This study (Fig. 9) showed that deep-frozen and non-irradiated cortical grafts did not achieve 100% torsional strength. The maximum torque strength at nine months was only about 60% of the control value. In contrast, non-vascularised autografts attained 100% strength by 12 weeks. With vascularised autografts, 100% strength was achieved by eight weeks (Nather *et al.*, 1990b).

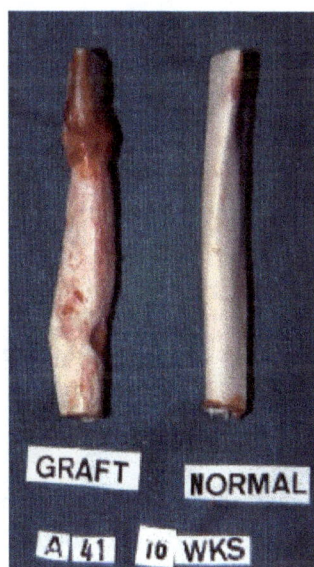

Fig. 8. At 16 weeks, the allograft was thicker than the unoperated tibia on the right. Callus is seen at both host–graft junctions.

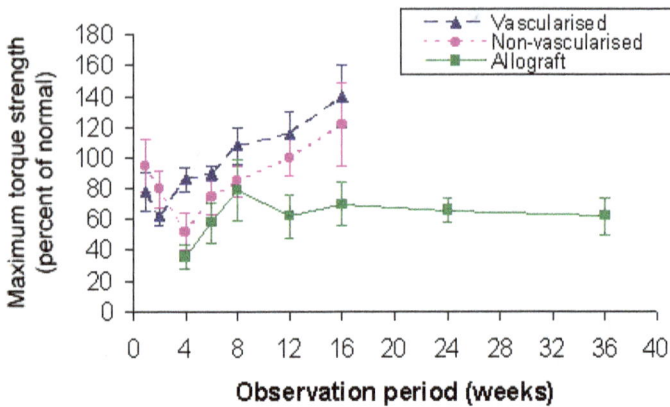

Fig. 9.   Graph showing maximum torque strength of deep-frozen cortical allografts, non-vascularised autografts and vascularised autografts.

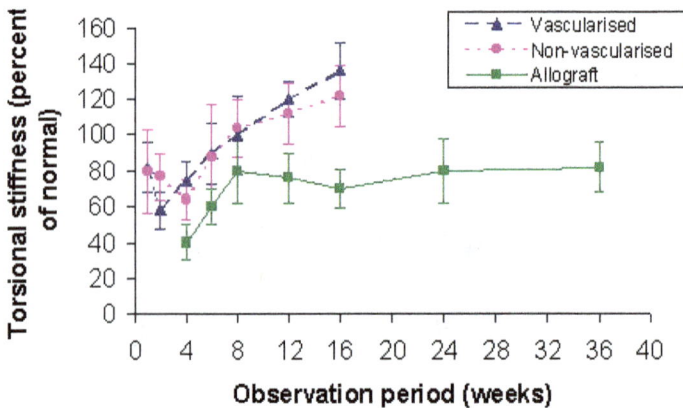

Fig. 10.   Graph showing torsional stiffness of deep-frozen cortical allografts, non-vascularised autografts and vascularised autografts.

Figure 10 showed that the allografts reached higher values for torsional stiffness — 80% at six and nine months.

Clinically, it must be recognised that deep-frozen cortical allografts are significantly weaker than autografts. Therefore, for massive recon-struction of the extremities, where load bearing is required to compensate the intrinsic weakness of the allografts, strong and rigid internal fixation must be employed. Specially designed intramedullary nails must be used for massive reconstruction of non-vascularised autografts (Enneking and Shirley, 1977). Nailing is preferred to plating for internal fixation of the lower limb to allow for immediate weight bearing and to reduce the rate of graft facture. Recently, to give added strength and rotational stability to the reconstruction, interlocking nails have been used.

Biomechanical Strength of Freeze-Dried and Gamma-Irradiated Cortical Allografts

Nather *et al.* (2004) further studied the biomechanical strength of freeze-dried and gamma-irradiated cortical allografts using the same tibial allograft model in the adult cat.

The study showed that the maximum torque of freeze-dried and gamma-irradiated cortical allografts was significantly weaker than deep-frozen cortical allografts. At 24 weeks, the maximum torque was only 12% of the normal strength compared to 64% for deep-frozen allografts (Fig. 11).

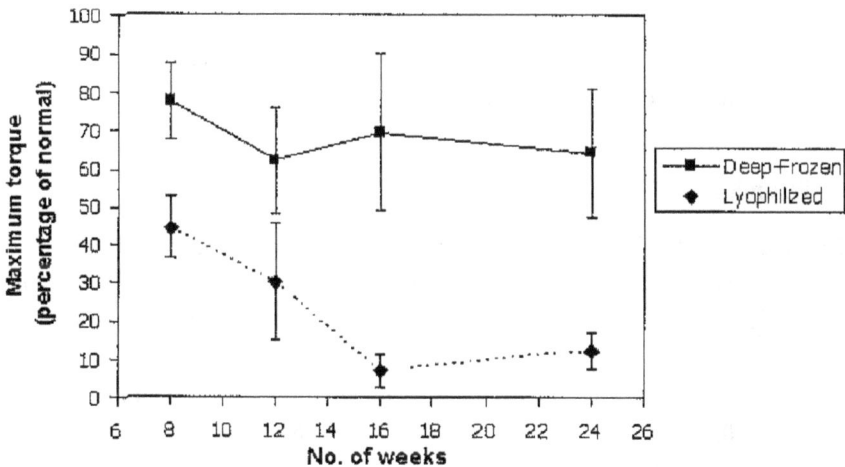

Fig. 11. Graph showing maximum torque of deep-frozen versus freeze-dried cortical allografts.

Nather *et al.* (2004) concluded that lyophilised and gamma-irradiated cortical bone allografts — which only possess one-fifth of the strength of large deep-frozen cortical bone allografts — are not suitable for use in massive reconstruction of the limbs especially when the grafts are subjected to weight bearing.

Biomechanical Strength of Demineralised Cortical Bone Allografts

With demineralised cortical bone allografts, where the calcium of the bone is removed, the biomechanical strength would be weaker compared to lyophilised and deep-frozen cortical allografts. Itoman and Nakamura (1991) showed that in rats, using compression testing, deep-frozen cortical allografts were stronger than lyophilised cortical allografts, which in turn were stronger than demineralised cortical allografts.

Summary

The types of bone grafts in the order of decreasing strength are:

- autografts,
- deep-frozen cortical allografts,
- lyophilised and gamma-irradiated cortical allografts and
- demineralised cortical allografts.

References

Bright R and Burchardt H (1983). The biomechanical properties of preserved bone graft. In Friedlaender GE, Mankin HJ and Sell KW (eds.), *Bone Allografts: Biology, Banking and Clinical Applications*, Little, Brown and Co., USA, pp. 223–232.

Delloye C (1999). The use of freeze-dried mineralized and demineralized bone. In Phillips GO, Von Versen R, Strong DM and Nather A (eds.), *Advances in Tissue Banking*, Vol. 3, World Scientific, Singapore, pp. 45–65.

Enneking WF and Shirley PD (1977). Resection-arthrodesis for malignant and potentially malignant lesions about the knee using an intramedullary rod and local grafts. *J Bone Joint Surg Am* **59**: 223–236.

Hamer AJ, Strachan JR, Black MM, *et al.* (1996). Biomechanical properties of cortical allograft bone using a new method of bone strength measurement. A comparison of fresh, fresh-frozen and irradiated bone. *J Bone Joint Surg Br* **78**: 363–368.

Itoman M and Nakamura S (1991). Experimental study on allogenic bone grafts. *Int Orthop* **15**: 161–165.

Komender A (1976). Influence of preservation on some mechanical properties of human Haversian bone. *Mater Med Pol* **8**: 13–17.

Mankin HJ, Doppelt S, and Tomford W (1983). Clinical experience with allograft implantation. The first ten years. *Clin Orthop Relat Res* **174**: 69–86.

Nather A (1999). Use of allografts in spinal surgery. *Ann Transplant* **4**: 7–10.

Nather A, Balasubramaniam P, and Bose K (1990a). Healing of non-vascularised diaphyseal bone transplants. An experimental study. *J Bone Joint Surg Br* **72**: 830–834.

Nather A, Goh JCH, and Lee JJ (1990b). Biomechanical strength of non-vascularised and vascularised diaphyseal bone transplants. An experimental study. *J Bone Joint Surg Br* **72**: 1031–1035.

Nather A and Goh JCH (2000). Biomechanical strength of large diaphyseal deep-frozen allografts. *Cell Tissue Bank* **1**: 201–206.

Nather A, Thambyah A, and Goh JCH (2004). Biomechanical strength of deep-frozen versus lyophilized large biomechanical cortical allografts. *Clin Biomech* **19**: 526–533.

Pelker RR, Friedlaender GE, and Markham TC (1983). Biomechanical properties of bone allografts. *Clin Orthop Relat Res* **174**: 54–57.

Pelker RR, Friedlaender GE, Markham TC, *et al.* (1984). Effects of freeze-drying on the biomechanical properties of rat bone. *J Orthop Res* **1**: 405–411.

Schneider J and Bright RW (1976). Anterior cervical fusion using freeze-dried bone allografts. *Transplant Proc* **8**(suppl. 1): 73–76.

Sedlin E (1965). A rheological model for cortical bone. *Acta Orthop Scand* **36** (suppl. 83): 1–77.

Spence KF, Jr, Bright RW, Fitzgerald SP, and Sell KW (1976). Solitary unicameral bone cyst: Treatment with freeze-dried crushed cortical-bone allograft. A review of 144 cases. *J Bone Joint Surg Am* **58**: 636–641.

Triantafyllou N, Sotiorpoulos E, and Triantafyllou J (1975). The mechanical properties of lyophilised and irradiated bone grafts. *Acta Orthop Belg* **41**: 35–44.

Wolfinbarger L, Zhang Y, Bao-Ling T, *et al.* (1994). A comprehensive study of physical parameters, biomechanical properties and statistical correlations of iliac crest bone wedges used in spinal fusion surgery. II. Mechanical properties and correlation with physical parameters. *Spine* **19**(3): 284–295.

Zhang Y, Homski D, Gates K, *et al.* (1994). A comprehensive study of physical parameters, biomechanical properties and statistical correlations of iliac crest bone wedges used in spinal fusion surgery: IV. Effect of gamma irradiation on mechanical and material properties. *Spine* **19**(3): 304–308.

Chapter 14

Basic Principles of Transplantation Immunology

Aziz Nather* and Shushan Zheng*

Introduction

The immune system has evolved to protect us from pathogens via two types of responses. The innate response is the first line of defence and is not specific for the invader. This response is primarily mediated by phago-cytic leukocytes: granulocytes and macrophages. The adaptive immune response is a specific response to the invader. It is mediated primarily by lymphocytes. An adaptive response is characterised by memory. The response to repeat exposures to the same invader is more vigorous than the first, and this memory is for life. Although the response to transplanted tissue contains elements of both types of immunity, this discussion will focus on the adaptive immune response.

The Concept of Antigen

Antigens are substances that are recognised by the adaptive immune response. Antigens are so called because they were initially thought of as antibody generators. However, this definition has since been widened to include all substances that can prompt an immune response. Antigens can

*NUH Tissue Bank, Department of Orthopaedic Surgery, Yong Loo Lin School of Medicine, National University of Singapore, Singapore.

be proteins or polysaccharides, and less frequently, lipids or nucleic acids combined with either proteins or polysaccharides.

The Adaptive Immune Response

The presentation of antigen is the first step to triggering the adaptive immune system. Using the analogy of soldiers guarding a castle, when the patrols spot an invader, they would defeat the invader and alert other soldiers to the attack. All castle soldiers would then be on guard against the enemy. Likewise, white blood cells defend the body against pathogens or foreign material. When antigen-presenting cells (APCs) spot a pathogen, they would ingest and digest the pathogen internally and display the resulting peptides (antigen) on the surface of the cell. This alerts the T-lymphocytes to the presence of the pathogens, and the cellular and humoral immune responses are activated to eliminate the pathogen (Fig. 1).

Antigen Presentation — The Induction of Immune Response

The processing and presentation of antigens from the environment outside of the cell are functions of certain groups of leukocytes: dendritic cells, macrophages, and B-lymphocytes (Fig. 2). These APCs function by ingesting the material and presenting it on human leukocyte antigen (HLA) class II molecules.

The process of antigen presentation starts as such (Fig. 3):

- The APC phagocytoses the pathogen or endocytoses material bound to antibodies on the cell surface.
- The material is enclosed in a series of endosomal vesicles of decreasing pH to which the cell adds digestive enzymes.
- The final vesicle in the series, which now contains peptides of the digested protein, coalesces with a vesicle from the Golgi body containing HLA class II molecules. In this acidic compartment, the peptide-binding groove of the HLA molecule is opened, allowing binding of a peptide.

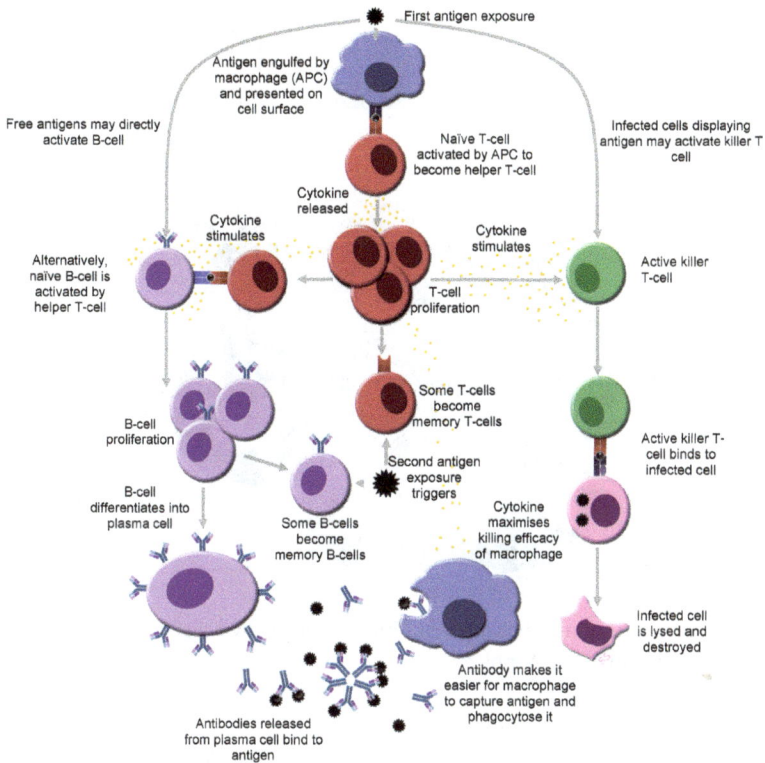

Fig. 1. Diagram showing pathway of immune response, with the first antigenic exposure leading to T-lymphocyte activation, which in turn triggers the humoral immune response (left half of the diagram) and the cellular immune response (right half of diagram). Some T-lymphocytes and B-lymphocytes become memory cells that would be swiftly activated upon second antigenic exposure.

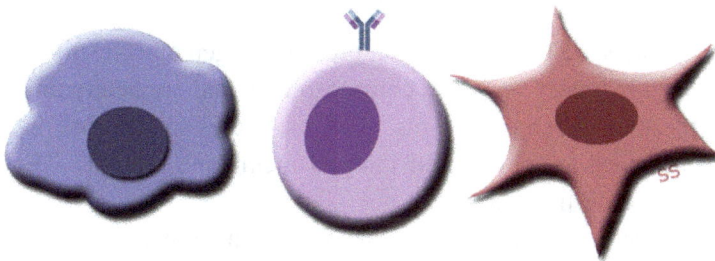

Fig. 2. Antigen-presenting cells. From left: macrophage, B-lymphocyte and dendritic cell.

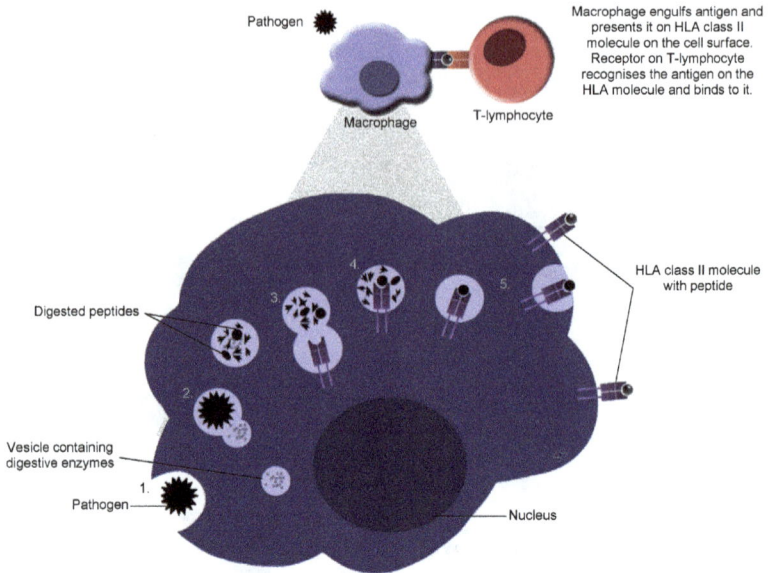

Fig. 3. The process of macrophage engulfing a pathogen by phagocytosis and presenting the antigen (peptide) on HLA molecules.

- Not all peptides fit; a peptide must have the appropriate amino acids at defined positions to fit into the binding grooves of the HLA molecule.
- The HLA molecule, now loaded with peptide, is transported to the surface of the cell and integrated into the membrane.

Human Leukocytes Antigens (HLA)

Structure and Characteristics of HLA

HLA molecules are the proteins that present antigen on cell surface. They are encoded by genes found on chromosome 6 and are heterodimers, i.e. made up of two different subunits. HLA molecules span across the membrane and hence are considered integral membrane proteins. They also possess a cytoplasmic tail.

HLA molecules share structural motifs with other members of the immunoglobulin (antibody) supergene family (Fig. 4). This enables them to bind to antigen. Each HLA molecule has a peptide-binding groove

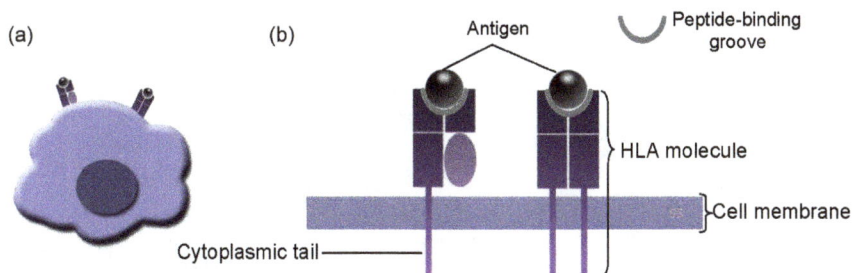

Fig. 4. (a) HLA molecules on the surface of an antigen-presenting cell, in this case, a macrophage. (b) basic structure of HLA molecules.

to allow the presentation of peptides (antigen). A peptide must have the appropriate amino acids at defined positions before it can fit into the binding grooves of the HLA molecule.

There are two main classes of HLA molecules concerned with the immune response: Class I and Class II.

HLA Class I Molecules

HLA class I molecules are expressed on the majority of nucleated cells in the body. They consist of three alpha domains and a β_2-microglobulin molecule. HLA class I molecules present peptides from inside the cell. The peptides in the binding groove of HLA class I molecules are generated in the cytosol of the cell from internal proteins and loaded into the HLA molecules in the endoplasmic reticulum (ER).

Through the presentation of internal peptides, HLA class I molecules can present both native antigens and foreign antigens. The latter is important in fighting infection. In virally infected cells, viral peptides are loaded into HLA class I molecules, identifying infected cells for destruction by killer T-cells (Fig. 5).

HLA Class II Molecules

HLA class II molecules are expressed on APCs, including B-lymphocytes, macrophages and dendritic cells. HLA class II molecules also consist of

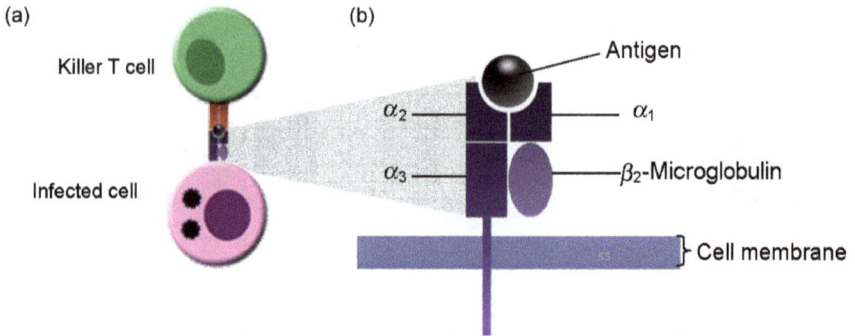

Fig. 5. (a) Virally infected cell presents antigen (viral peptide) on HLA class I molecules for identification by killer T-cell. (b) HLA class I molecule is made up of three alpha domains and a β_2-microglobulin molecule. The peptide-binding groove is located between α_1 and α_2.

two chains, both encoded by genes in the HLA complex. While the three-dimensional structure of class II molecules is very close to that of class I molecules, the peptide-binding groove differs in being open on both ends to accommodate longer peptides.

HLA class I molecules present peptides from outside the cell. The peptides in the binding groove of HLA class II molecules are produced by the ingestion and digestion of foreign material by APCs, and loaded into the HLA molecules in vesicles (Fig. 6). This presentation of external peptides accounts for the role of HLA class II molecules in antigen presentation in APCs for the induction of adaptive immunity.

High Variability in HLA Loci

The genes encoding HLA class I and class II molecules exhibit a high level of polymorphism: there are literally hundreds of alleles for some loci. Due to this polymorphism, HLA molecules are unique to each individual. Thus, the immune system uses HLAs to recognise and discriminate between self cells and non-self cells. Cells that display the individual's HLA type are recognised by the immune system as self cells. In contrast, cells that display foreign HLA types are recognised as non-self cells, i.e. invaders.

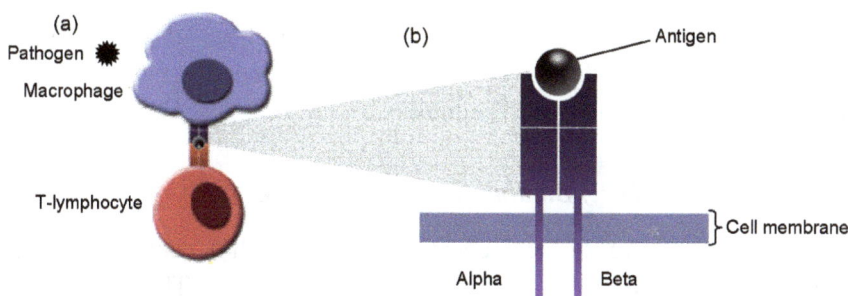

Fig. 6. (a) Antigen-presenting cell presents antigen on cell surface for recognition by T-lymphocyte using HLA class II molecule. (b) HLA class II molecule is made up of two chains — alpha chain and beta chain.

The high level of polymorphism has arisen and has been maintained in order to allow presentation of peptides from any foreign invader that might prove dangerous to our survival. As expected, this polymorphism differs among people whose ancestors lived in different areas of the world and were exposed to different environmental pathogens. This is important when considering the chances of selecting HLA-matched donors and recipients for transplantation.

Recognition of Alloantigen

In transplantation, the HLA antigens of the donor are the targets or antigens recognised by the T-lymphocytes. The T-lymphocyte receptor can recognise the donor HLA antigens on the donor cell and respond — a direct recognition of alloantigen. Alternatively, the recipient's APCs ingest, digest and present peptides derived from donor tissue. When T-lymphocytes respond to the peptide, it is called an indirect recognition of alloantigen.

Indirect recognition is the main mode of sensitisation after tissue transplantation. Phagocytes are recruited to the surgery site as part of innate immunity. They ingest donor tissue in the process of repairing the injury and present the peptides to the recipient's T-lymphocytes.

Direct recognition is possible with transplants of fresh bone, cartilage, heart valves or corneas, as these tissues may contain viable cells expressing HLA antigens and co-stimulatory molecules. Treatment of tissue to remove viable cells eliminates direct recognition of alloantigen. Processing tissues further to denature or remove donor HLA antigens also reduces chances for indirect recognition.

T-Cell Receptor (TCR) Binding to HLA–Peptide Complex

T-lymphocytes have surface receptors called T-cell receptors (TCRs), which interact with the HLA proteins and their peptides. The presence of a foreign peptide triggers a response in a T-lymphocyte with the corresponding receptor. Most T-lymphocytes can only recognise foreign antigen when it is processed as peptides and presented by an HLA molecule. The TCR then attaches to the HLA–peptide complex, fitting in it much like a key fits in a lock.

The TCR is made up of two chains and like HLA molecules, shares motifs with other members of the immunoglobulin supergene family. On the cell surface, the TCR is associated with the CD3 complex of proteins. Polymorphism associated with the TCR is extensive, generated from germ-line genes by some of the same mechanisms that create different specificities for antibodies.

If the binding of the TCR with the HLA–peptide structure is of sufficient affinity, a signal is transduced through the CD3 structure to give the first signal for T-lymphocyte activation (Fig. 7).

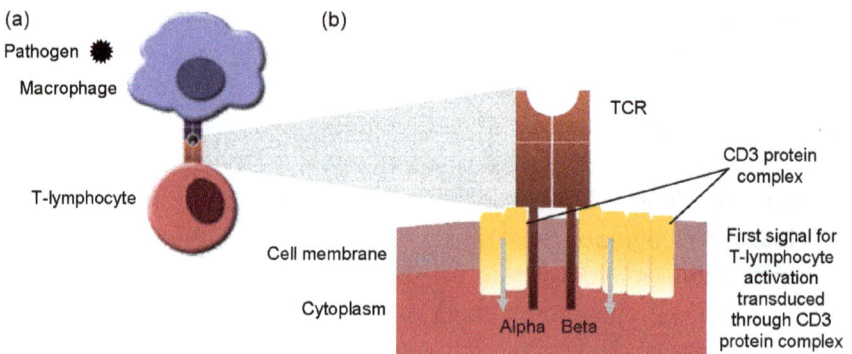

Fig. 7.   (a) Binding of TCR on T-lymphocyte with HLA–peptide complex on antigen-presenting cell. (b) TCR, made up of two chains — alpha chain and beta chain, is associated with the CD3 complex of proteins.

Co-Stimulation — The Second Signal for T-Lymphocyte Activation

The T-lymphocyte requires a second signal from the APC to continue to respond. This second signal, called co-stimulation, is provided by the interaction of other receptors on the T-lymphocyte with co-stimulatory molecules on the APC. The co-stimulatory molecules are identified by their cluster of differentiation or CD number. They include CD40, CD58, CD80 and CD86. The most potent co-stimulatory molecules on the APC are CD80 and CD86. They interact with CD28 and CD152 receptors on the T-lymphocyte. CD40 on APCs interacts with CD154 on the T-lymphocyte — this interaction is important for T-cell activation of B-lymphocytes. For the second signal to be delivered, CD80 or CD86 on the APC interacts with CD28 on the T-lymphocyte.

The second signal is essential for the proliferation and maturation of T-lymphocytes (Fig. 8). The interaction of co-stimulatory molecules with their receptors sets off an intracellular cascade of signals required for transcription of growth factors or cytokines, which is in turn required for further development of the T-lymphocyte and for development of other T-lymphocytes. If the second signal is blocked, the intracellular cascade is halted and the T-lymphocyte is left in a state of immune unresponsiveness (anergy).

The interaction of co-stimulatory molecules with their receptors on T-lymphocytes is a target for novel approaches to immunosuppression in transplant recipients. Monoclonal antibodies have been generated to bind to the co-stimulatory molecules on the APCs and prevent delivery of the

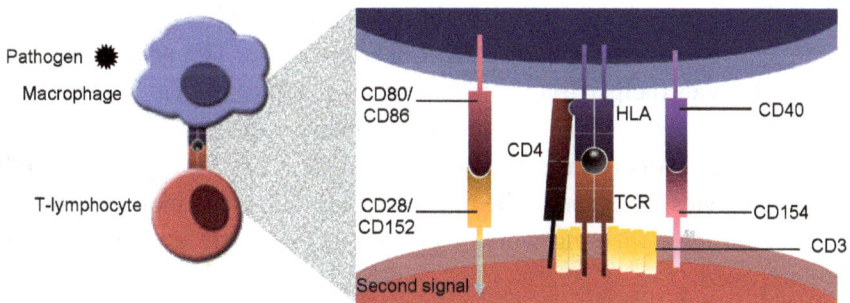

Fig. 8. Co-stimulation provides the second signal for the full activation of T-lymphocyte.

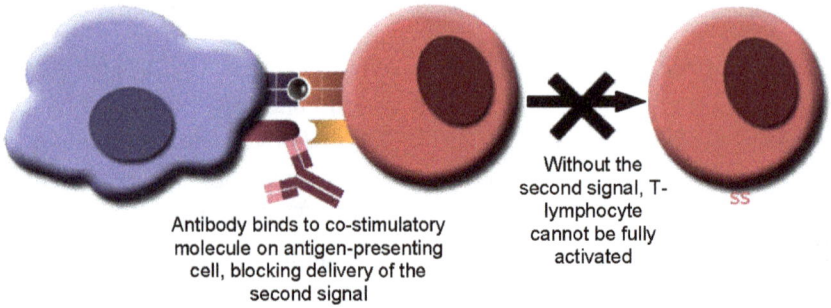

Fig. 9. Blocking of second signal by antibody.

second signal to T-lymphocytes that are interacting with the HLA class II–peptide complex on that APC (Fig. 9). The dendritic cell is emerging as the most potent APC due to its expression of high numbers of co-stimulatory molecules.

The Cellular Immune Response

The T-lymphocyte plays a central, orchestrating role in both the cellular and humoral responses. There are different populations of T-lymphocytes with different functions. These populations can be distinguished by the expression of different CD antigens.

CD4 T-lymphocytes are the first to recognise foreign antigen. Their TCRs interact with HLA class II and peptides. The CD4 molecule becomes part of the CD3–TCR complex, binds to a site on the HLA class II molecule and helps stabilise the complex, thereby prolonging the engagement between APC and CD4 T-lymphocyte. The activated CD4 T-lymphocytes (helper T-cells) then produce cytokines and growth factors to orchestrate the cellular and humoral immune responses. The cytokines and growth factors released enable other APCs, CD8 T-lymphocytes and B-lymphocytes to respond. The cytokine production also induces inflammatory responses, during which elements of the innate immune system are activated.

Mature, activated CD8 T-lymphocytes (killer T-cells) contain cytotoxic granules. They can kill infected cells or cells bearing donor HLA

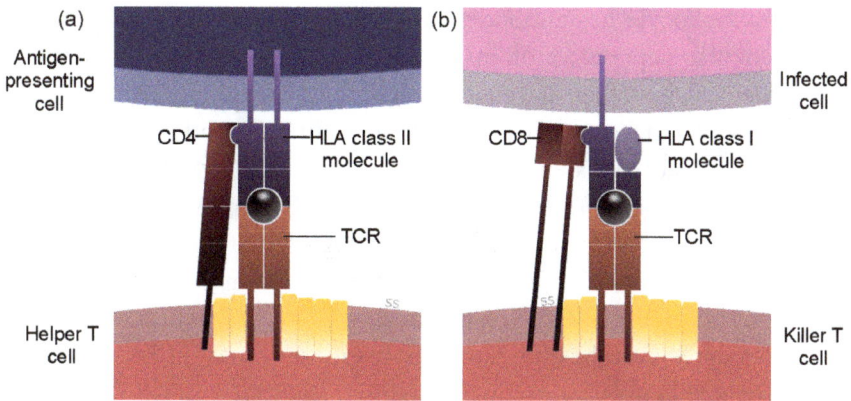

Fig. 10. (a) CD4 binds to HLA class II molecule on the antigen-presenting cell to stabilise the binding between HLA–peptide complex and TCR. (b) CD8 binds to HLA class I molecule on the infected cell to stabilise the binding between HLA–peptide complex and TCR.

antigens. The TCR of CD8 T-lymphocytes interacts with HLA class I molecule and peptides. The CD8 molecule becomes part of the CD3–TCR complex, binds to a site on the HLA class I molecule, and helps stabilise the complex (Fig. 10).

The cellular immune response is marked by CD4 and CD8 T-lymphocytes trafficking into the transplant site. The activation of the cellular response most likely occurred in lymph nodes close to the transplant site or in the spleen. Donor cells may have migrated from the graft. Recipient dendritic cells or macrophages (APCs) may have ingested material at the graft site and then migrated to lymph nodes or spleen. During activation, T-lymphocytes undergo multiple rounds of cell division to increase the number of T-lymphocytes with TCRs specific for donor antigens. These T-lymphocytes then mature to become cytokine producers or killer cells. Monocytes and other lymphocytes such as natural killer cells also traffic into the transplant site due to recruitment by cytokines made by the T-lymphocytes (Fig. 1).

The Humoral Immune Response

The humoral immune response is also activated in lymph nodes or spleen. Activation requires CD4 T-lymphocytes, APCs and B-lymphocytes.

The B-lymphocyte has antibodies as surface receptors for antigen, and can recognise peptides, proteins, lipids, nucleic acids, polysaccharides and chemical moieties. It requires signals or co-stimulation from the T-lymphocyte to become activated: to proliferate and mature into a plasma cell that makes antibody (Fig. 1). The antibody molecule has a part called the variable region that binds to the antigen inducing the response, and a part called the constant region that can combine with other molecules or cells. The first antibody made in a response has an IgM constant region. If the response continues, interaction with other T-lymphocytes switches the antibody response to use an IgG constant region (Fig. 11).

Antibody can interact with transplanted tissue in at least three ways. First, the variable region can bind to the HLA antigens on donor cells and the constant region can interact with plasma proteins called complement to destroy the donor cells. Second, antibody can bind to donor cells and mark them to be destroyed by lymphocytes and monocytes. Receptors on the lymphocytes or monocytes bind to the constant region of the antibody. Third, antibody can bind to donor cells and cause them to be "eaten" by phagocytes.

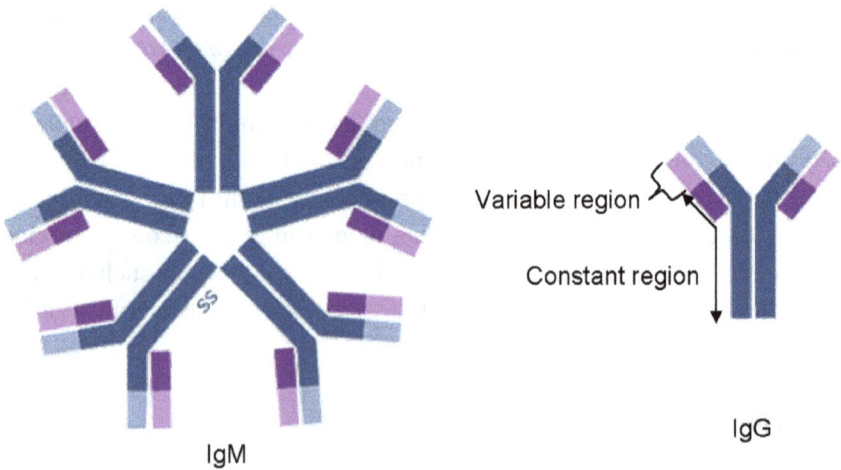

Fig. 11. Antibodies produced by plasma cell include IgM (left) and IgG (right).

Immunological Memory

The induction or activation phase of the adaptive immune response lasts from one to seven days. During this time, the T-lymphocyte and B-lymphocyte populations are expanding by cell division and maturing into cells that make cytokines, killer granules or antibody. Towards the end of this period, lymphocytes appear at the site of transplant as described above, and the first antibody, IgM, appears in the blood.

Both T-lymphocytes and B-lymphocytes can mature into memory cells. After activation and proliferation, some lymphocytes become memory cells rather than effectors of the immune response. These memory lymphocytes reside in lymph nodes, spleen or bone marrow. If the same pathogen or donor HLA antigens are reintroduced, these cells rapidly respond by proliferating and maturing to provide effector T-lymphocytes and IgG antibody (Fig. 12).

Recipients of tissue transplants may have been exposed to other people's HLA antigens as a result of blood transfusion, pregnancy or other transplants. If the HLA antigens expressed by the new donor are the same or similar to those in the previous exposure, there is a possibility of

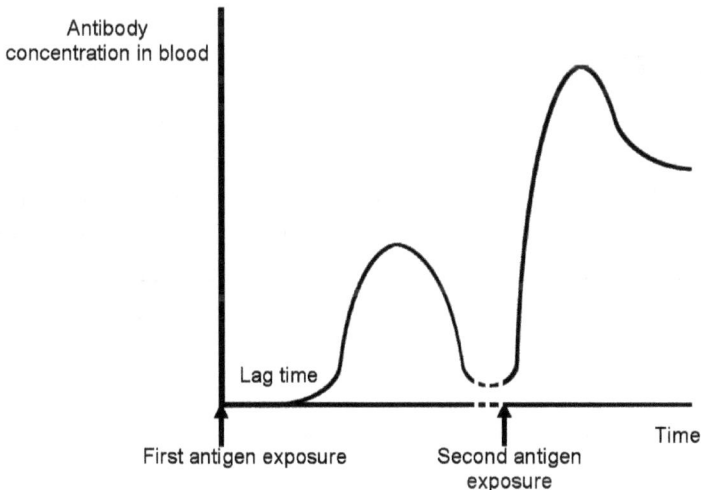

Fig. 12. Graph showing antibody response following first antigen exposure and second antigen exposure.

reactivating a memory immune response. Memory responses are more difficult to control than primary responses. Immunosuppressive drugs used in organ transplantation target the early activation and proliferation phases of the immune response. These drugs are less effective in controlling a memory response.

HLA Typing

As the immune system uses HLAs to differentiate between self cells and non-self cells, a close match between donor and recipient's HLA types would reduce the risk of graft rejection and graft-versus-host disease (GVHD). The latter occurs commonly following allogeneic bone marrow transplantation, when functional immune cells in the transplanted marrow recognise the recipient as "foreign" and mount an immunologic attack against it. In order to select a close match between the HLA types of donor and recipient, HLA typing is used.

HLA typing identifies the alleles of the HLA genes that were inherited by an individual from each parent. In transplantation, immune responses have been found to be different in class I and class II loci as follows (Fig. 13):

Class I Loci HLA-A, HLA-B, and HLA-C
Class II Loci HLA-DR, HLA-DQ, and HLA-DP

Typing of all six loci is important for bone marrow or stem cell transplantation, in order to select a new immune system that will function well in the recipient. For organ or tissue transplantation, typing of HLA-A, HLA-B and HLA-DR is considered sufficient. These molecules are present at a higher density on the cell surface, and are thought to be the target of the majority of immune responses to donor antigens. Most of the

Fig. 13.   Position of class I and class II loci on gene encoding HLA molecules.

differences between alleles are substitutions of the amino acids that form the peptide-binding groove. Typing of these differences is difficult using the serological techniques that have been the gold standard for HLA typing. The majority of HLA typing is now performed using techniques that detect differences in DNA sequence corresponding to the amino acid substitutions (Nelson, 2001a).

Testing for Transplantation

Potential recipients of organ allografts are tested for humoral or cellular immunity to donor HLA antigens to avoid memory responses. The HLA class I and class II phenotypes are determined for the patient and donor. Other genetic polymorphisms may also be tested to determine the capacity of the recipient to mount an inflammatory response (Wilson *et al.*, 1997). Testing for humoral immunity involves assays looking for IgG or IgM antibody that can bind to HLA molecules. These assays use lymphocytes from HLA-typed individuals or soluble HLA antigens immobilised on microbeads or ELISA trays. Techniques vary in sensitivity. The most sensitive techniques use fluorescence-activated flow cytometry for a quantitative measure of antibodies to HLA antigens. Testing with a panel of HLA antigens is performed during the time the patient is waiting for an allograft; the result is given as %PRA, meaning percent panel-reactive antibody, which is the number of antigens that reacted positively with the patient serum divided by the total number of antigen tested. It indicates the number of panel members to whom the patient has antibody and estimates the percentage of potential donors who would be ruled out for that patient due to prior exposure to HLA antigens of those donors. When a donor is identified, the antibody test is repeated using lymphocytes from the donor to ensure that the patient does not have immunological memory to that donor; this test is called cross-match. Figure 14 displays the results of a cross-match of donor T-lymphocytes and serum with no antibody compared to serum with anti-donor antibody. The human IgG molecules binding to donor HLA antigens are detected by means of an anti-IgG reagent that is coupled to the fluorochrome fluorescein isothiocyanate (FITC). The flow cytometer histogram displays the population of T-lymphocytes by the amount of

T-lymphocytes   +   Serum   Flow Cytometer Histograms

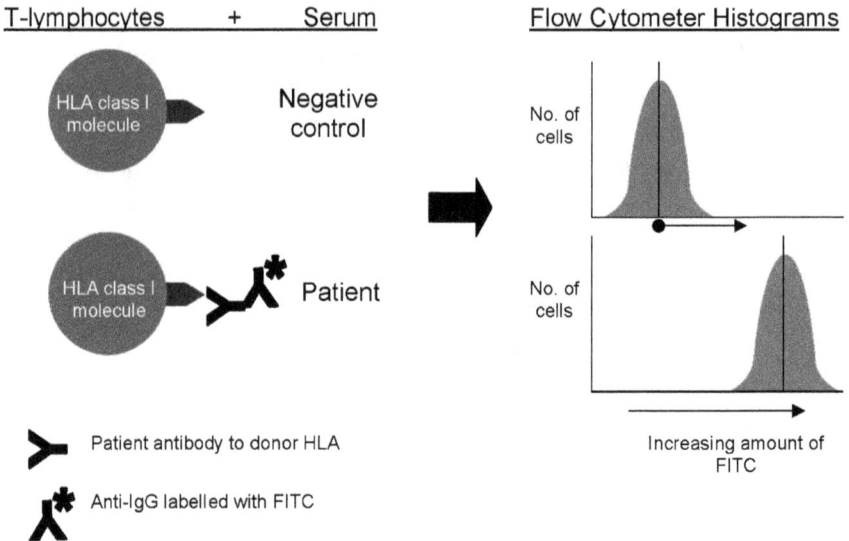

Fig. 14.   Flow cytometry cross-match.

FITC bound to their surface. The population of T-lymphocytes incubated with patient serum is compared to the population incubated with the negative control serum. This flow cytometry cross-match is a very sensitive measure of anti-donor antibody.

After transplant, the appearance of antibody to donor antigens is highly correlated with incipient rejection. Assays to monitor patient sera for these antibodies alert clinicians to institute biopsy and anti-rejection protocols (McKenna *et al.*, 2000). Recipients of large bone allografts who make antibody to their donor's class I HLA antigens also have poorer outcomes as compared to antibody-free recipients (Friedlaender *et al.*, 1999).

Measures of cellular immunity are used less frequently prior to transplant, but are used after transplant to identify patients who have lost T-lymphocytes capable of responding to donor antigens and are candidates for protocols to reduce immunosuppressive medication. Recent advances include using fluorescence-activated cell analysis to identify phenotypes of T-lymphocytes responding to donor cells (Wells *et al.*, 1997) or using tetramers of HLA and peptides to identify T-lymphocytes with TCR specific for donor HLA (Mutis *et al.*, 1999).

Glossary

Allele	Any of the alternative forms of a gene that may occur at a given locus.
Cytosol	The internal fluid of a cell, where a portion of cell metabolism occurs.
Endocytosis	A process where cells absorb material from the outside by engulfing it with their cell membrane.
Endoplasmic reticulum	An organelle in eukaryotic cells whose function includes synthesising protein and the folding and transporting of proteins to be used in the cell membrane.
Gene	A specific sequence of nucleotides in DNA or RNA that is located usually on a chromosome and that is the functional unit of inheritance controlling the transmission and expression of one or more traits by specifying the structure of a particular polypeptide and especially a protein or controlling the function of other genetic material.
Golgi body	An organelle in eukaryotic cells whose primary function is to process and package the macromolecules such as proteins and lipids that are synthesised by the cell.
Locus	The position of a gene on a chromosome.
Phagocytosis	A specific form of endocytosis where cells absorb solid particles from the outside by engulfing them with their cell membrane. It is used either for the acquisition of nutrients or for removing pathogens or cell debris.
Polymorphism	Existence of a gene in several allelic forms.

Suggested Readings

Abbas AK, Lichtman AH, and Pober JS (eds.) (2000). *Cellular and Molecular Immunology*, 4th edn. WB Saunders Co., Philadelphia.

Janeway CA, Travers P, Walport M, and Capra JD (eds.) (1999). *Immunobiology: The Immune System in Health and Disease*, 4th edn. Garland Publishing, New York.

Parham P (ed.) (2000). *The Immune System*, Garland Publishing, New York.

Roitt I, Brostoff J, and Male D (eds.) (1998). *Immunology*, 5th edn. Mosby, London.

References

Friedlaender GE, Strong DM, Tomford WW, and Mankin HJ (1999). Long-term follow-up of patients with osteochondral allografts. A correlation between immunologic responses and clinical outcome. *Orthop Clin North Am* **30**: 583–588.

McKenna RM, Takemoto SK, and Terasaki PI (2000). Anti-HLA antibodies after solid organ transplantation. *Transplantation* **69**: 319–326.

Mutis T, Gillespie G, Schrama E, Falkenburg JH, Moss P, and Goulmy E (1999). Tetrameric HLA class I-minor histocompatibility antigen peptide complexes demonstrate minor histocompatibility antigen-specific cytotoxic T lymphocytes in patients with graft-versus-host disease. *Nat Med* **5**: 839–842.

Nelson KA (2001a). HLA typing. In Rich R, Fleisher TA, Shearer WT, Schroeder HW Jr and Kotzin B (eds.), *Clinical Immunology: Principle and Practice*, Mosby, England, pp. 126.1–126.9.

Nelson KA (2001b). Basic principles of transplantation immunology. In Nather A (ed.), *The Scientific Basis of Tissue Transplantation*, World Scientific, New Jersey, pp. 553–565.

Wells AD, Gudmundsdottir H, and Turka LA (1997). Following the fate of individual T cells throughout activation and clonal expansion. Signals from T cell receptor and CD28 differentially regulate the induction and duration of a proliferative response. *J Clin Invest* **100**(12): 3173–3183.

Wilson AG, Symons JA, McDowell TL, McDevitt HO, and Duff GW (1997). Effects of a polymorphism in the human tumor necrosis factor alpha promoter on transcriptional activation. *Proc Natl Acad Sci USA* **94**: 3195–3199.

Chapter 15

Principles of Sterile Techniques

Siti Zubaidah Mordiffi*, Mary Tan† and Aziz Nather‡

Introduction

All tissue-bank operators must be trained on aseptic technique in order to perform sterile procurement of tissues from living and deceased donors. It is vital that each technologist be attached to the operating room for at least two or three months to learn and apply the principles of sterile techniques and to practise sterile procedures. The technologist must learn the proper technique of scrubbing and gowning in a sterile fashion (manner). He or she must also learn how to maintain sterility in the operating room as well as the methods of sterilising equipment, instruments and materials to be used in the operating rooms.

Basically, the principle of sterile technique covers the following areas:

- monitoring of sterility in the operating room,
- methods of sterilisation of equipment and materials and
- sterile technique in operating theatre.

Monitoring of Sterility in the Operating Room

Monitoring of sterility in the operating room is crucial in controlling the propagation of infection that can have undesirable outcomes to the

*Evidence Based Nursing Unit, Nursing Department, National University Hospital, Singapore.
†Nursing Education Unit, Nursing Department, National University Hospital, Singapore.
‡NUH Tissue Bank, Department of Orthopaedic Surgery, Yong Loo Lin School of Medicine, National University of Singapore, Singapore.

surgical patient who is immunologically compromised and highly susceptible to infections. Furthermore, the skin; a natural barrier protecting against the invasion of pathogenic microorganisms, is incised and the protecting barrier is broken (Woodhead *et al.*, 2005). Thus, the open surgical wound is a portal of entry for the invasion of pathogenic microorganisms. Sources of contamination in the operating room include human factors, air circulation, environmental controls and supplies used during surgical intervention.

Human Factors

The main source of contamination comes from humans because they harbour the most pathogenic microorganisms and are the means by which microorganisms are introduced into the wound (Phillips, 2004).

Health and Hygiene

Personnel working in the operating room must be free from transmissible bacterial infection such as upper respiratory tract infection (URTI) carbuncles, dermatitis and unhealed wounds and infection of mouth, eyes and ears (Phillips, 2004; Groah, 1990). Most infections are transmitted via direct contact from personnel to patient or via indirect contact from personnel to inanimate objects to patient. Handwashing procedure eliminates most of the pathogenic microorganisms that are present on the hands. It is one of the most effective methods to halt further propagation of infection (AORN, 2009). Thus, proper technique of handwashing must be performed consistently before and after each patient contact, before performing clean or aseptic procedure, after exposure to body fluids, and after touching patient's surroundings (WHO, 2009).

Attire

The operating room suite is a clean environment. The use of clean attire reduces contamination of the operating room from extraneous infection. The operating room suite is sectioned into three zones (Earl, 1996).

The unrestricted zone is the "grey area" where personnel or public may mingle with personnel in operating theatre outfit. The grey area refers to the reception and recovery area and the top-up store, where consumable and pharmaceuticals are replenished from external source.

The semi-restricted zone refers to the corridors, administrative office, storage rooms and the theatre sterile supply unit (TSSU) in the operating room suite. All personnel entering the operating room suite must change into freshly laundered operating theatre outfit (AORN, 2009; Phillips, 2004). Changing into fresh attire reduces particle count of microbial shedding from the body (Phillips, 2004). Hair covers are worn to prevent dandruff and hair from being shed onto the scrub clothing and wound and clean footwear is worn in the operating theatre suite (AORN, 2009).

The restricted zone is the operating room suite that includes the operating room where surgery is performed, preparation room, scrub room and induction room. In addition to the operating room attire, a mask must be worn prior to entering this zone.

Air Circulation

Opening of doors results in mixing of clean operating room air with that of the outside corridor air, which has a higher microbial count. Traffic in and out of the operating room must be kept to a minimum during surgery. In the construction of an operating theatre suite, it is preferable that doors should open from a clean room (double-door concept). Only authorised personnel and surgical team are allowed within the operating rooms and the number of people within the operating room must be limited. Movement within the operating room creates air turbulence, which can circulate lint, dust, human's skin shedding and microbes into the wounds (Phillips, 2004).

Environment

Methods employed in controlling the environment in the operating theatre suite are aimed at reducing the microorganism count in the operating rooms.

The design of the operating theatre suite includes considerations for:

- The traffic flow of personnel and equipment within the operating room complex. Separate movements of "clean" and "dirty" items prevent contaminated items from mixing with clean and sterile items.
- Designated locations of the TSSU, storage rooms, pantry, decontamination area and disposal rooms to reduce human and equipment movements, and workflow and process.
- The size of operating room must be adequate to allow the surgical team to move freely within the operating room and for surgical procedures to be performed without untoward contamination.
- Toilet must not be located within the vicinity of the operating rooms suite.
- Flooring and walls should allow easy cleaning, washing and mopping.

Considerations for mechanical and engineering control of the environment include air-conditioning and ventilation system, air exchanges, temperature control, and humidity.

- Dust and lint from fresh air that is sucked into the ventilation system is removed. Recirculation with fresh air takes place. Just before the air enters the operating room, it passes through a high-efficiency particulate air (HEPA) filter. Phillips (2004) stated that the use of HEPA filters can remove as much as 90% of particles larger than 0.3 μm. Air sampling is said to improve the bacteriological quality of air during surgery. It assists in limiting the colony-forming units and the bacteria-carrying particle counts (Edmiston *et al.*, 1999). However, it has not been shown to directly affect the infection rate. Holton *et al.* (1990) and Humphreys (1992) recommended its use only during commissioning and following major refurbishment of the operating theatre. The purpose of doing air sampling is to identify engineering faults.
- Air conditioning in each operating room is supplied from the air-conditioning ducts situated on the ceiling and is exhausted via vents, which is located on the wall, near to the floor. This system creates positive pressure within the operating rooms and air is forced out of the operating room via the vents. Air exchanges in the operating room

are maintained at 15 air exchanges per hour (Humphreys, 1992). Relative humidity should be maintained at 50–60% and temperature at 20–23°C (Phillips, 2004).

- Housekeeping is essential to maintain cleanliness of the operating-room environment. Before the beginning of the day's operation list, inspection of the general cleanliness and dryness of the floors of the operating rooms and suite are conducted by the operating-room personnel. It is essential to perform housekeeping audit of the operating room to ensure that standards of cleanliness are maintained (Phillips, 2004; NUH, 2006e).

  o Daily routine cleaning is performed with detergent and warm water. Tabletops, equipment, fixtures and stools are damp dusted to control airborne microorganisms that are settled on dust and lint. Corridors and operating room floors are mopped.

  o Standard operating procedures for housekeeping must also be followed (NUH, 2006e) to ensure that the operating room environment is maintained. Floors are mopped in between cases and washed at the end of the day's operating list, with recommended detergent and water.

  o Thorough cleaning of the operating theatre environment and equipment is also performed on a weekly basis, including damp dusting of walls and ceiling, with soap and water, and scrubbing and cleaning of trolleys and drip stands (NUH, 2006e).

  o Monthly cleaning schedule includes chemical stripping and polishing of the floor, and cleaning air-conditioning vents with changing of filters at regular intervals by housekeeping and technical personnel (NUH, 2006e).

Methods of Sterilisation of Equipment and Materials

The term "sterile" means free from any microorganism, including spores (Phillips, 2004). Sterilisation is the process by which an item is rendered sterile. In a hospital setting, the sterilisation process occurs in the theatre sterile supply unit (TSSU). The TSSU is an important functional part of an operating room setting. The processes of sterilisation include decontamination, preparation, packaging, loading, sterilisation and sterile storage of operating theatre sterile supplies. All criteria must be fulfilled at every stage in order to attain complete sterilisation.

Decontamination

Prior to any method of sterilisation, items must be cleaned to reduce the bioburden (NUH, 2006b). The sterilisation process is time based. Therefore, the more the microorganisms there are on an item, the longer it takes to destroy them completely or near completely. Furthermore, presence of blood or soil on an item will form a barrier that prevents the surface of an item from direct exposure to the sterilising agent. All instruments are unclamped and removable parts are disassembled before placing them into the open basket for loading into the washer steriliser (Phillips, 2004; NUH, 2006b). Figure 1 shows personnel arranging utensils into an open basket to allow water and detergent to access the surfaces of the items in the washer steriliser.

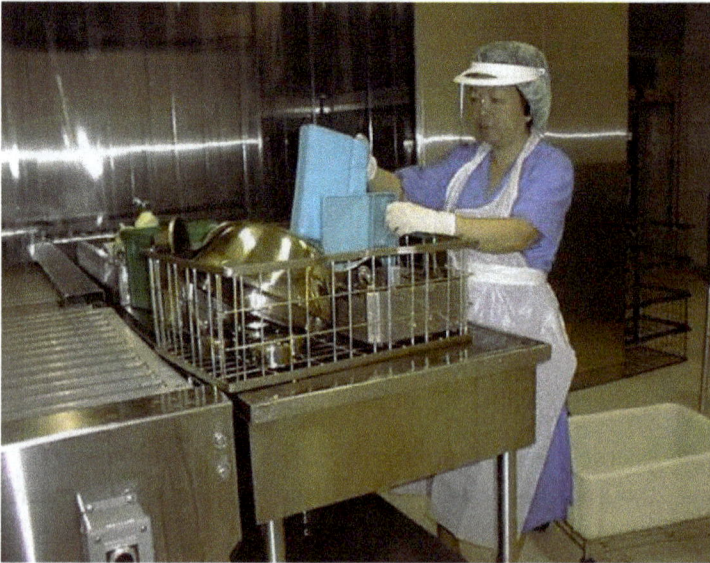

Fig. 1. Decontamination — arranging utensils and instruments into the washing basket to ensure that water and detergent can get to all surfaces.

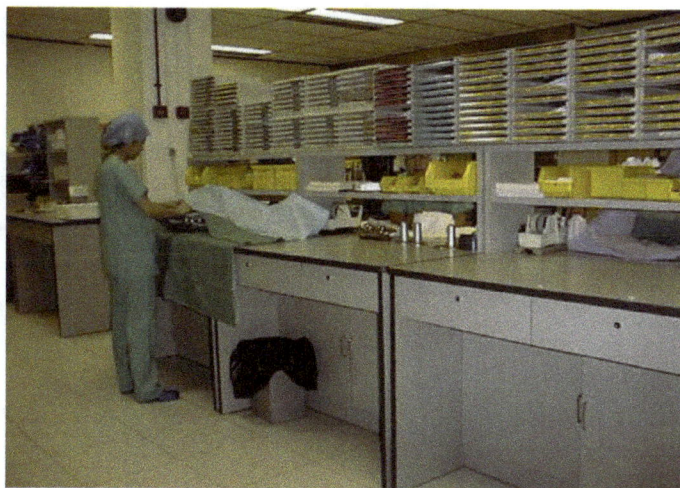

Fig. 2. Assembling and packaging area.

Assembly

When assembling instruments (Fig. 2), quality check of instruments is performed; instruments are inspected for integrity and function. All instruments in a set are counted for the correct quantity and arranged in sequence in a metal tray with perforated base (NUH, 2006f).

Packaging

Appropriate packaging materials and proper technique of wrapping enhance the effectiveness of the sterilisation process. All linen or paper-wrapped items used for surgery should be double wrapped. Linen and hydra paper wrappers require careful inspection for minute holes. Special tapes that are used for securing the wrappers must be sufficient in length to prevent the packages from being loosen during the process of sterilisation and handling. Container packaging system with lid may be used to pack instrument sets (Phillips, 2004).

Loading

All surfaces of an item must come in direct contact with the sterilising agent for effective sterilisation to occur (Phillips, 2004). Systematic

Fig. 3.   Arrangement of packed items ready for loading into the steriliser.

arrangement of items on the trolley shelves before loading into the steam
steriliser chamber determines adequate contact with the sterilising agent
(Fig. 3). Placement of an item in a chamber is dependent on the material,
the type of equipment employed and the sterilisation method. Most items
can be processed using the steam steriliser (NUH, 2006h). Other methods
of sterilisation, physical and chemical sterilisation methods, are also
available for items that cannot be processed using the steam sterilisation
method (Fig. 4a–f).

Sterilisation

Sterilisation can be classified into two groups — physical and chemical
methods. Physical sterilisation includes steam and hot air (Figs. 4a and b).
Chemical sterilisation methods include hydrogen peroxide, ethylene
oxide, paracetic acid and ortho-phthalaldehyde (OPA) (Figs. 5c–f). The
description for each type of sterilisation methods that are in use at this
hospital is displayed in Tables 1 and 2.

(a)

(b)

(c)

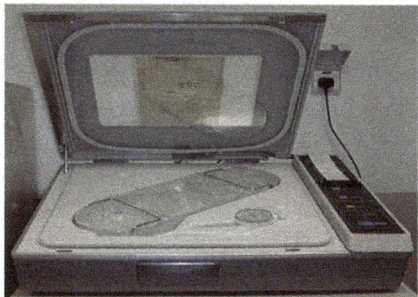

(d)

Fig. 4. Types of sterilisers: (a) steam steriliser; (b) hot-air oven; (c) hydrogen peroxide steriliser; (d) paracetic acid steriliser; (e) ortho-phthalaldehyde (OPA) steriliser and (f) ethylene oxide steriliser.

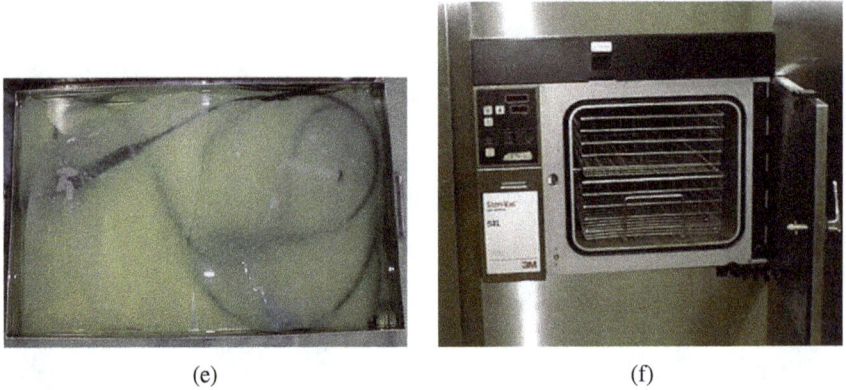

(e)   (f)

Fig. 4. (*Continued*)

TSSU is responsible to ensure that items sent out for use are sterile. Mechanical, physical and chemical monitoring of the sterilisation process ensure that all the criteria are met before the items are considered sterile.

Unloading

Upon completion of the sterilisation process, the door of the steriliser is opened slightly (cracked). Opening of the door widely will allow gush of cool air to enter into the chamber quickly, resulting in condensation. An item that is wet allows microorganisms to travel freely, thus rendering the item non-sterile (Phillips, 2004). When the temperature in the chamber has cooled down, the load of sterile items is removed by locking onto the trolley base and the trolley is pushed directly into the designated sterile storage area.

Storage

The load is allowed to cool further and the chemical indicators are checked to ensure that the items have undergone successful sterilisation process before storing them into the designated shelves. For items that are packed in see-through peel, pouches are inspected for presence of water droplets. If present, the items are returned to the assembly area for

Table 1. Physical sterilisation methods.

Method	Process of microbial destruction	Agent	Process of sterilisation	Temperature	Exposure time	Cycle time	Biological indicator	Items indicated
Steam (prevacuum) (NUH, 2006h)	Coagulation	Saturated steam at high temperature	1. Air removed 2. Steam injected	134°C @ 28–30psi	4 min	1 hour	*Bacillus stearothermophilus*	Instruments, utensils, linen, heat-resistant materials
Steam (gravity) (NUH, 2006h)	Coagulation	Saturated steam at high temperature	1. Saturated steam enters chamber 2. Air displaced downwards	121°C @ 16–19 psi	30 min	1 hour	*Bacillus stearothermophilus*	Instruments, laparoscopic instruments, heat-resistant plastic, e.g. suction tubing, diathermy cable
Hot air (NUH, 2006g)	Oxidation (slow burning process)	Heating process	Chamber heated to high temperatures	160°C	1 hour and 10 minutes		*Bacillus stearothermophilus*	Insoluble items, sharp-edged instruments, e.g. cataract set, scissors set

Table 2. Chemical sterilisation methods.

Method	Process of microbial destruction	Agent	Process of sterilisation	Temperature	Exposure time	Cycle time	Biological indicator	Items indicated
Hydrogen peroxide (STERRAD) (Johnson, 2005)	Disrupting cell metabolism	Hydrogen peroxide plasma	1. Air removed 2. H_2O_2 vial punctured 3. H_2O_2 vapourised and diffused 4. Radiofrequency applied 5. Plasma activated	55–60°C	16–20 min	55 min	Bacillus stearothermophilus	Heat-labile plastics, endoscopes and laparoscope instruments, camera system
Paracetic acid (Steris System 20)	Inactivates critical microbial cell system causing death	Paracetic acid 2%	1. Liquid buffer drained into chamber 2. Lid closed 3. Chamber filled with sterile water 4. 35% paracetic acid aspirated into chamber	50–56°C	12 min	<30 min	Bacillus stearothermophilus	Immiscible items, endoscopes and laparoscope instruments/ equipment
Ethylene oxide (no longer used in this hospital)	Alkalysis	Ethylene oxide gas at low temperature and high pressure	1. Air removed 2. ETO gas channelled into chamber 3. ETO gas diffuse into items	55–60°C	2 hours	18–24 hours	Bacillus subtilis	Heat-labile plastics, endoscopes and laparoscope instruments, camera system

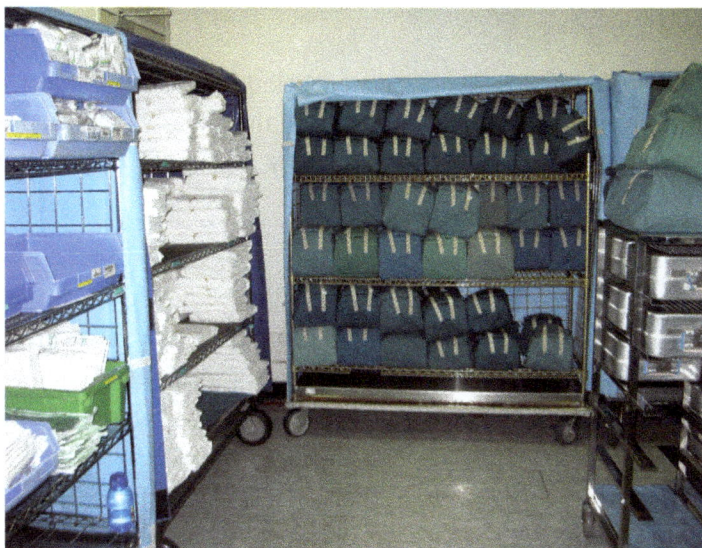

Fig. 5.   Sterile storage area.

re-processing (Phillips, 2004). The sterile items are distributed and stored in a cool, clean, dry and low humidity area (Fig. 5).

Monitoring of Sterilisation Process

Monitoring of the sterilisation process is an important aspect to ensure effective sterilisation has taken place. Described below are various methods of monitoring to establish that effective sterilisation has taken place.

Mechanical Monitoring

Modern sterilisers record the pressure and temperature at every stage of the sterilisation process (cycle). At the end of the sterilisation process, TSSU personnel will check the printout for the correct pressure, temperature and exposure time. The printouts are filed for the purposes of quality assurance. Mechanical monitoring is performed at every load.

Physical Monitoring

TSSU and operating theatre personnel check the items for presence of wet load, the change of colour on the strips of the adhesive tape and loosen packaging in steam-sterilised items. These items are rejected and re-processed.

Chemical Monitoring

Bowie Dick's test is performed daily to test the efficacy of steam steriliser (NUH, 2006d). Chemical indicator is placed on every item. TSSU personnel check for the change in the colour of chemical indicators to confirm that the item has undergone the process of sterilisation.

Biological Monitoring

Biological indicator is the ultimate test and the gold standard to test the efficacy of sterilisers and sterility of an item (Phillips, 2004). For steam sterilisers, the vial, which contains *Bacillus stearothermophilus* strip, is placed on the bottom shelf. For other types of sterilisers, it can be placed anywhere within the chamber. After the sterilisation, the vial is cracked, dated and placed in the incubator at a temperature of 37°C (ASP, 2004). A non-sterilised vial is cracked and marked "control". The results are read 48 hours later. The control will change colour, indicating growth of microbes whereas the sterilised vial is not affected, indicating conditions for sterilisation has been met (NUH, 2006d).

Sterile Technique in the Operating Room

Aseptic technique prevents contamination of microorganism by ensuring a sterile item remains sterile during storage and handling, and at the time of use for the patient. Sterile technique is employed to maintain sterility when working in a sterile environment (Phillips, 2004). Non-sterile item carries pathogenic microorganisms, which may cause wound infections; thus, the surgical team is responsible and must remain diligence to ensure items used during surgery must be sterile (Phillips, 2004; NUH, 2006a). The activities of the surgical team where sterile technique are applied include scrubbing, preparation of trolley, antiseptic cleaning of patient, draping procedure, surgical procedure, cleaning and dressing of wound after surgery and post-surgical procedures.

The concept of sterile technique is not easy to grasp. Therefore, concepts from Kolb's (Kolb, 1984) experiential learning theory and neurolinguistic programming theory (Lankton, 1980) will be adopted to aid in the visualisation and understanding of sterile technique. For effective visualising on the concept of sterility, colours will be used to differentiate non-sterile and sterile fields, and safety margin. Yellow denotes the non-sterile areas of a scrubbed person (person wearing a sterile operation gown), blue denotes the sterile areas and red denotes the safety margin. The safety margin is an area that is sterile, but is also considered non-sterile area. This safety margin is always present wherever there is a non-sterile and sterile field. This area provides an allowance of margin to protect and prevent crossing over the extension of the sterile field. Thus, in sterile technique, "blue" denotes sterile and can only come in contact with designated "blue" area, "yellow" which denotes non-sterile can only come in contact with designated "yellow" area and "red" denotes the safety margin, which can only come in contact with designated "red" areas.

As an example, the Association of Operating Room Nurse (AORN, 2009) recommends the practice for maintaining a sterile field; according to the recommendation that a sterile gown be considered sterile in front of the chest and from the chest to the level of the sterile field, and the sterile field of the sleeves is from two inches above the elbow to the cuff. In a scrubbed sterile person who is gowned up in sterile gown, these sterile areas are denoted with blue colour. The safety margin, which is denoted in red colour, would include areas at the neckline, shoulders, under the arms, back, the sleeve cuffs and below the waist level (trolley level). Other safety margins include areas 5 cm below the neckline of the gown and areas of the gown below the waistline. The hands and arms, which are denoted by blue colour, must remain within the sterile zone (blue zone) of the operating gown (Fig. 6). Unguarded sterile areas must be considered as non-sterile. The back of the gown, which is exposed to risk of contamination and cannot be monitored, becomes the "red" zone even though a wrap-around gown is used.

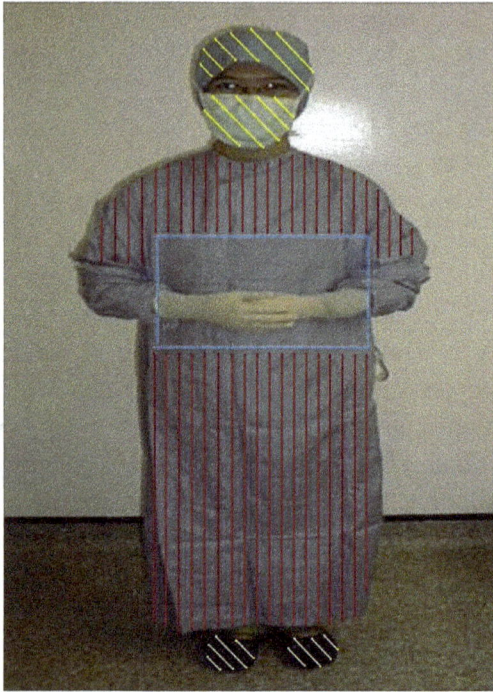

Fig. 6. Scrubbed person (person wearing sterile gown) with colours to denote sterile and non-sterile areas of the sterile gown.

Sterile Scrub (*Surgical Hand Scrub*)

A surgical hand scrub is performed prior to a surgical procedure. Surgical scrubbing is a process of removing and reducing transient flora from the nails, hands and forearm with the use of antiseptic agent and mechanical action (AORN, 2009). This technique may be a brush-stroke or time-based hand-rubbing technique. The skin harbours transient and resident microorganisms that are pathogenic when skin integrity is compromised. Thus, surgical hand scrubs effectively:

- remove debris and transient microorganisms from the nails, hands and forearms,
- reduce the resident microbial count to a minimum and
- inhibits the rapid rebound growth of microorganisms (AORN, 2009, p. 307).

Preparation Prior to Scrubbing

- Fingernails should be kept short and clean (WHO, 2009).
- Nail polish should not be used (WHO, 2009).
- Hands are inspected for cuts and abrasions.
- Jewellery should be removed from hands and arms (WHO, 2009).
- Hair must be well covered.
- Mask adjusted snugly and comfortably over nose and mouth.
- Fluid shield mask or goggles to be worn where blood splashing is expected.
- Disposable plastic aprons are worn where blood, fluid or water splashing is expected (NUH, 2006j).

Antiseptics

Common antiseptics used for scrubbing at this centre include 4% chlorhexidine gluconate and 7.5% povidone-iodine. Choice of antiseptic is dependent on the user and whether there is any user allergy. Antiseptics should never be used in combination. Chlorhexidine has been shown to be more effective in decreasing the bioburden and is effective against Gram-positive and some Gram-negative microorganisms. Upon application, chlorhexidine is effective for approximately eight hours. Chlorhexidine also has substantive effect. A study by Paulson (1994) found chlorhexidine gluconate to be the most effective antimicrobial agent with regard to its immediate, persistent and residual effect. Povidone-iodine is effective against Gram-negative and some Gram-positive microorganisms (Paulson, 1994). Once applied, povidone-iodine is effective for not more than six hours (Phillips, 2004).

Surgical Hand Scrub

There are two methods of surgical hand scrub; the brushing method and the hand-rubbing method. The WHO (2009) guidelines reported that almost all studies that were reviewed discouraged the use of brushes for surgical hand hygiene. Thus, the hand-rub method will be described in this chapter. The hand-rub method can be divided into three parts. The whole procedure should take about three minutes (NUH, 2006i).

Part I (Fig. 7)

- Adjust the tap to provide water with a suitable temperature and moderate flow.
- Wet hands and forearms up to 5 cm above the elbow.
- Dispense approximately 5 ml of hospital-approved antiseptic agent onto the palm.
- Perform hand rub to remove superficial soils. Using friction, rub the hands and arms in a backwards and forwards motion 15 times in the following manner:

   Step 1: Palm to palm.
   Step 2: Palm to palm with fingers interlaced.
   Step 3: Right palm over left dorsum with fingers interlaced and vice versa.
   Step 4: Back of fingers to opposing palms with fingers interlocked and vice versa.
   Step 5: Rotational rubbing of right thumb clasped in left hand and vice versa.
   Step 6: Rotational rubbing of centre of left palm with tip of right fingers and vice versa.
   Step 7: Virtually divide the arm into two sections; wrist to mid-forearm and mid-forearm to 5cm above the elbow. Clasp the left wrist with the right hand and rub in an upward and downward motion from the wrist to mid-forearm. Continue rubbing from mid-forearm to 5cm above the elbow. Do the same for the other arm. Ensure all areas of the arm are scrubbed.

- Rinse the left hand and forearm under running water from the fingertips to the elbow in one direction only. Do the same for the other arm. Keep the hands elevated above the elbow at all times (WHO, 2009).

Part II

- Dispense approximately 5 ml of hospital-approved antiseptic agent onto the palm.
- Reapply a thin film of antiseptic solution up to 5 cm above the elbow.

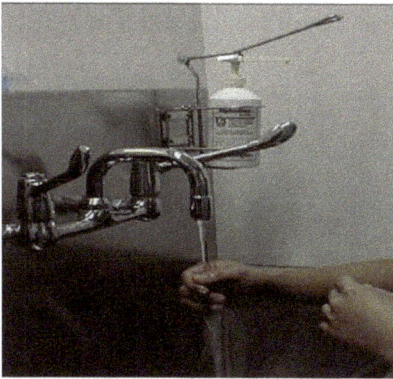

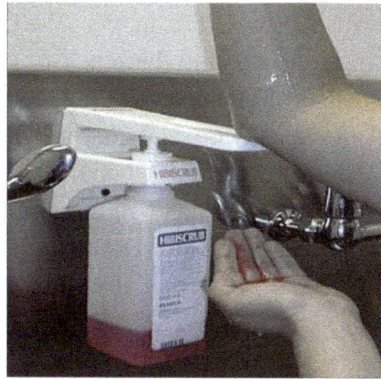

(a) (b)

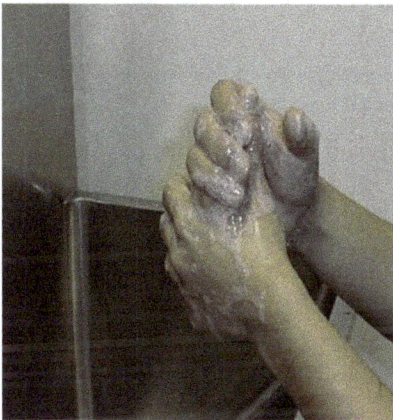

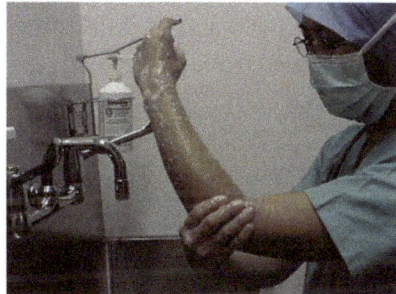

(c) (d)

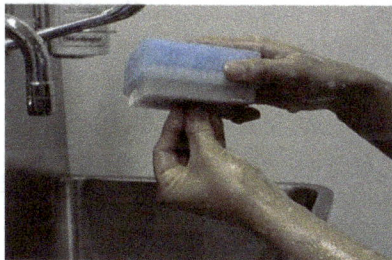

(e)

Fig. 7. Performing surgical hand rub: (a) wetting of hands; (b) obtaining the Hibiscrub; (c) scrubbing the hands; (d) scrubbing from mid-forearm to elbow and (e) scrubbing: start from finger nails and tips.

- Take a sterile brush, wet it and dispense sufficient amount of antiseptic solution onto the brush.
- Align fingertips and commence brushing ends of fingers and cuticles in a backward and forward motion for 30 seconds on each hand. Discard the brush when completed.
- Rinse the left hand and forearm under running water from the fingertips to the elbow. Do the same for the other arm. Keep the hands elevated above the elbow at all times.

Part III

- Reapply antiseptic solution to 5 cm below the elbow.
- Repeat steps 1 to 7 as per Part I.
- Rinse hands starting from the fingertips to the elbows with arms elevated at all times.
- Allow the water to drip from the flexed elbows before proceeding to the gown trolley.
- Proceed to the gown trolley.

Gowning

- Pick up the hand towel without contaminating any other sterile items on the trolley.
- Begin by drying the fingers, palms and dorsum of the hands.
- Continue to dab dry the left arm from the wrist to the elbow (Fig. 8a). Discard the hand towel. Perform the same sequences for the other arm.
- Stand about 30 cm away from the trolley (Phillips, 2004) and pick up the folded sterile gown at the top (collar).
- Lift the gown up from the trolley and away from the body. Holding the corners of the collar, allow the gown to unfold (Fig. 8b).
- Slip both arms simultaneously into the armholes and stretch the arms outward momentarily. Do not lift and spread arms higher than shoulder level. Maintain fingers within the sleeves of the gown.
- Circulating nurse pulls back the gown from within to fasten the gown.

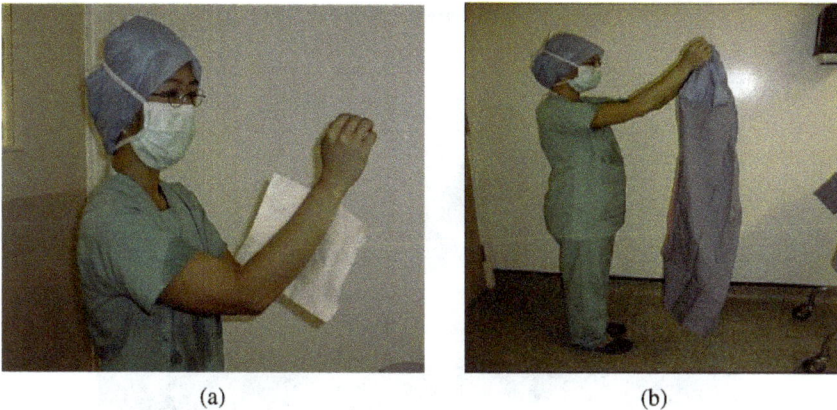

(a)   (b)

Fig. 8. (a) Drying of hands: start from palm of hands to elbow and (b) unfolding sterile operating gown.

Gloving

The close-donning gloving method is recommended by AORN (2009) prior to surgery.

- Open and orientate the gloves (fingers must remain about 1–2 cm within the cuff of the gown during this procedure). Pick up the glove by the cuff and place glove onto the palm of the opposite hand. Orientate the glove thumb facing the palm and the opening of the glove cuff in line (apposition) with the opening of the gown cuff.
- Pinch the centre of the cuff of the glove. Flip the glove over the fingers. Simultaneously tug the glove and push fingers through the gown cuff and into the glove (Fig. 9). Don the glove for the other hand in the same manner.
- Untie the knot of the strings attached to the front or side of the sterile gown.
- Wrap the end of the string that is attached to the wrap-around flap of the gown with a sterile wrapper. Pass the wrapper with the string within to the circulating nurse taking care not to contaminate the string.
- Whilst the circulating nurse is holding onto the string, rotate on the spot to allow the wrap-around flap of the gown to wrap around the

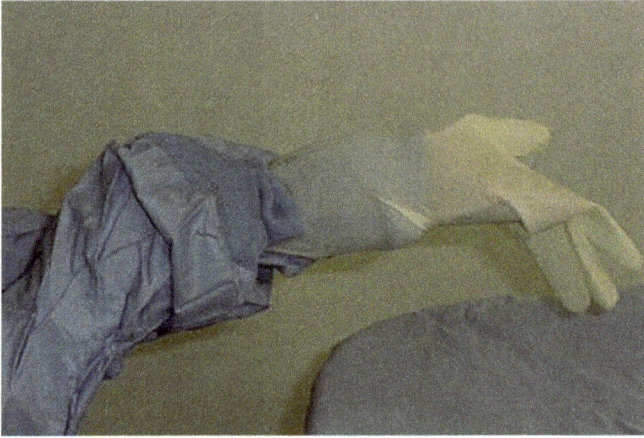

Fig. 9.   Close-donning gloving method.

back. Retrieve the string from the circulating nurse and leave the sterile wrapper with her, without touching the wrapper.

• Tie the strings together to secure the gown.

Preparation of Sterile Trolley

The sterile trolley should be prepared and set as close as possible to the time of surgery (Phillips, 2004). The operating-room nurse is responsible to ensure that the prepared items fulfills the criteria of sterility. Prior to opening any item, the non-sterile person checks for expiry date, change in colour of chemical indicator, and the packaging remains intact with no visible punctures, and items within the package remain dry. In some institutions, event-related sterility is practised, making expiry date not applicable.

The circulating nurse (non-sterile person) unwraps the outer wrapper of the draping pack, instrument and cleansing set. The scrub nurse (sterile-scrubbed person) unwraps the inner wrapper of the draping pack, making sure that her gloved hands do not go beyond and below the level of the trolley as she unfolds the wrapper. The scrub nurse proceeds to unfold the inner layers of other items, places the items on the sterile trolley and arranges receptacles, drapes and instruments onto the sterile trolley

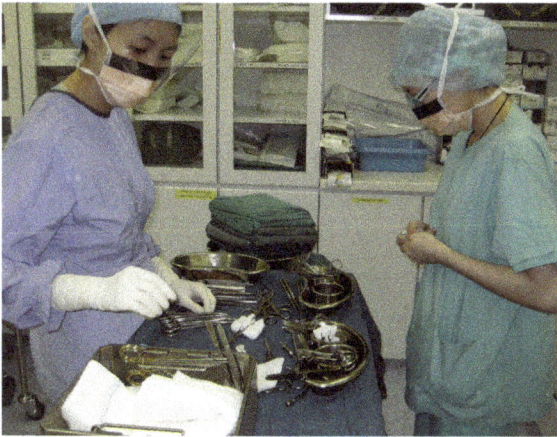

(a)

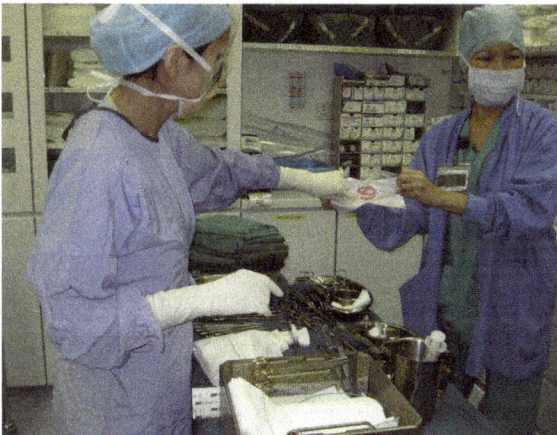

(b)

Fig. 10. (a) Preparation of sterile trolley. Circulating nurse maintains a distance of about 30 cm away from the sterile trolley (Phillips, 2004). (b) Scrub nurse retrieving sterile item from the circulating nurse.

(Fig. 10). The circulating nurse provides the necessary items for scrub nurse. The scrub nurse counts the instruments, blades and needles prior to the surgery witnessed by the circulating nurse (Fig. 10). The circulating nurse documents all instrument and swab counts; it must be verified correct by the scrub nurse to prevent discrepancy at the closing of the surgical wound.

Antiseptic Cleansing of Patient

The presence of hair may interfere with exposure, closure and dressing of surgical wound. If hair gets into the surgical wound, this may become a foreign body and exacerbate infection. Hair removal may be required within the perimeter of the incision and is best performed using either depilatory cream (if there is no allergy) or clippers. If shaving is used as a method for hair removal, it should be performed as close as possible to the time of surgery. Presence of nicks and abrasions becomes a source of infection.

Procedure

- Appropriate germicidal solution is used to decrease the microbial counts. A choice of chlorhexidine with 70% alcohol or povidone-iodine may be used. A sufficient volume of cleansing solution is poured into galipot. The gauze is folded and mounted onto the sponge-holding forceps. Additional gauze is placed into the galipot or into the kidney dish.
- The area is exposed only when the surgical team is ready to cleanse the patient. Modesty should be maintained and only a minimal area is exposed.
- Standing at the foot of the operation table, the technologist will assist to lift the lower limb by holding the heel.
- Place towels on the table beneath the limb to collect run-off solutions.
- Dip the gauze in the antiseptic solution, squeeze excess solution and commence cleansing from the surgical site outward towards the periphery in a circular motion. Sufficient pressure and friction should be applied to remove dirt and microorganisms from the skin and pores. Then the gauze is discarded. Repeat the procedure with a new piece of gauze in the same manner.
- The receptacles for the skin preparation are discarded.
- If run-off solution is expected, place towels along the incision site to prevent wetting of the underlying sheets. The non-sterile person, upon

completion of skin preparation, removes the side towels, as the cleansing solution can cause irritation to the skin. Pooling of alcohol-based solutions can cause chemical or diathermy burns or fire (Patient Safety Authority, 2005).

Draping

"Sterile drapes should be used to establish a sterile field" (AORN, 2009, p. 318). Sterility must be maintained during the draping procedure. The subsequent paragraph will describe the sterile draping technique in procuring the head of femur. The requisites include five protective drapes (Mackintosh or cellulose paper sandwiched between folded dressing towel), two abdominal towels, two trolley towels, one cotton bandage, one stockinette, one adhesive U-shape (3M) and one Ioban adhesive drape (iodine-impregnated drape).

The operating theatre technologist will continue to lift up the limb during this procedure.

- Arrange drapes in sequence of use.
- Place sterile protective drape beneath the limb, at the side at the hip region, covering the abdomen and the pubis area.
- Take the abdominal towel and stand on the other side of the OT table facing the surgeon. Carefully pass one end of the folded abdominal towel to the surgeon beneath the limb, grab the other end of the folded abdominal towel. Simultaneously pull the towel taut while supporting the mid-portion of the towel to prevent the towel from touching the non-sterile table. Unfold the towel by gathering the folds in the hand. Spread the towel and cover the lower portion of the table.
- Perform the same motion but this time cover the upper part of the patient. Two towel clips, one on each side, will be clipped to secure the two towels together.
- Take the trolley towel and again stand on the opposite side of the OT table. Unfold the towel. The technologist will lower the limb onto the towel. Wrap the limb with the towel. Secure with cotton bandage, bandaging from the foot to slightly beyond the proximal end of the towel. Tie the bandage to secure the towel in place.

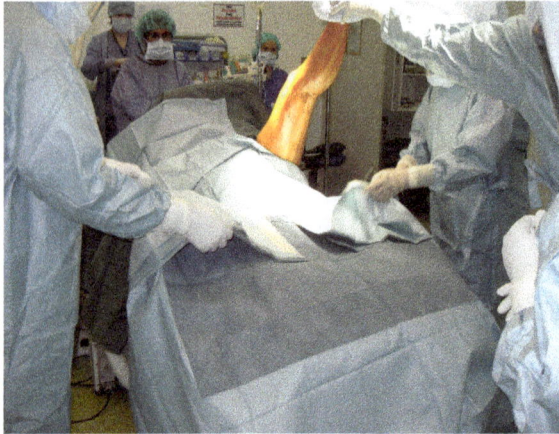

Fig. 11.   Sterile draping technique during surgery.

- A sterile member lifts the limb. Scrub nurses places an unfolded trolley towel beneath the limb to reinforce the drapes.
- A U-drape adhesive drape may be used. This drape should be placed beneath the limb. The two slits are to go around the proximal limb and cross at the anterior (NUH, 2005) (Fig. 11).

Sterility must be maintained throughout the surgical procedure. The surgical team must remain vigilant for any breaks in sterility. Items used during surgery must remain sterile (Phillips, 2004).

- Drapes must remain dry. Reinforce as soon as it is wet.
- Non-sterile items or persons must maintain a distance of about one foot away form the sterile field.
- Drapes below the table level are considered non-sterile.
- Sterile person always faces the sterile drapes. Never turn the back towards the sterile operating field and table. Referring to the principles of sterile technique, the practices of "blue touching or facing blue" (sterile), "yellow touching or facing yellow" (non-sterile) and "red touching or facing red" (safety margin) must be strictly applied and adhered to.
- Change gowns if it is contaminated.

- Surgical team must be constantly aware of sterile and non-sterile areas and be on the lookout for breaks in sterility at all times.

Cleaning and Dressing of Wound after Surgery

- The wound is cleansed with gauze soaked in saline after closure of the skin. Start cleaning from one end of the suture line to the other end. Use another gauze to clean around the wound.
- Drying of the wound is done in similar manner beginning at the wound, followed by drying of the surrounding area.

Post-Surgical Procedures

- Remove all instruments from the field.
- Discard soiled swabs into the biohazard plastic bag carrier.
- The drapes are removed from the patient and discarded into the appropriate linen carrier.
- Check that the underlying towel is clean and dry.
- Patient is covered to keep patient warm and maintain modesty (NUH, 2006c).

Acknowledgements

- Ms. Khor Pui Kwan, Shirin, Senior Nurse Educator, Operating Theatre Suite, Nursing Department, National University Hospital.
- Ms. Lee Yee Kew, Director of Nursing, Nursing Department, National University Hospital.
- Ms. Tan Soh Chin, Senior Manager, Nursing Administration, Nursing Department, National University Hospital.
- Mr. Mohd. Osman bin Hussain, Registered Nurse, Physician Assistant.
- Ms. Ong Hong Yar, Senior Nursing Officer, Theatre Sterile Supply Unit, Operating Theatre Suite, Nursing Department, National University Hospital.
- Ms. Leow Chai Choo, Senior Nursing Officer, Operating Theatre Suite, Nursing Department, National University Hospital.

- Ms. Helen Goh, Nursing Officer, Infection Control Nurse, National University Hospital.
- Ms. Ho Swee Fook, Wendy, Nursing Officer, Central Sterile Supply Department, National University Hospital.
- Ms. Ong Ah Hua, Senior Lecturer, Nursing Division, Nanyang Polytechnic.

References

AORN (2009). *Perioperative Standards and Recommended Practices.*

Earl A (1996). Operating room: Dans. Olmsted RN (ed.), *APIC Infection Control and Applied Epidemiology: Principles and Practice*, St. Louis, USA: Mosby.

Edmiston C, Sinski S, Seabrook GR, Simons D, and Goheen M (1999). Airborne particulates in the OR environment. *AORN* **69**: 1169–1183.

Groah LK (1990). *Operating Room Nursing: Perioperative Practice.* USA: Appleton & Lange.

Holton J, Ridgway GL, and Reynoldson AJ (1990). A microbiologist's view of commissioning operating theatres. *J Hosp Infect* **16**: 29–34.

Humphreys H (1992). Microbes in the air — when to count! (The role of air sampling in hospitals). *J Med Microbiol* **37**: 81–82.

Johnson J (2005). STERRAD cyclesure biological indicator, certificate of performance. In J Johnson (ed.), *Ethicon* (Vol. Ref. 14324), Irvine.

Kolb DA (1984). *Experiential Learning: Experience as the Source of Learning and Development.* New Jersey: Prentice-Hall.

Lankton S (1980). *Practical Magic: A Translation of Basic Neuro-Linguistic Programming into Clinical Psychotherapy.* California: Meta Publication.

NUH (2005). Draping of surgical patient (general) (Vol. SOP-NSG-OTS-014, pp. 3): National University Hospital, Singapore.

NUH (2006a). Aseptic technique (Vol. NUH-SOP-NSG-OTS-034, pp. 3): National University Hospital, Singapore.

NUH (2006b). Decontamination of surgical instruments (Vol. NUH-SOP-NSG-OTS-056, pp. 3): National University Hospital, Singapore.

NUH (2006c). Immediate post operative care in recovery room (Vol. NSG-SOP-OTS-027, pp. 3): National University Hospital, Singapore.

NUH (2006d). Monitoring the effectiveness of the steam sterilizer (Vol. NUH-SOPNSG-OTS-059, pp. 3): National University Hospital, Singapore.

NUH (2006e). Operation room sanitation (Vol. NUH-SOP-NSG-OTS-003, pp. 3): National University Hospital, Singapore.

NUH (2006f). Packing of instruments in TSSU (Vol. NUH-SOP-NSG-OTS-057, pp. 3): National University Hospital, Singapore.

NUH (2006g). Processing of surgical instruments with hot air sterilization (Vol. NUH-SOP-NSG-OTS-060, pp. 3): National University Hospital, Singapore.

NUH (2006h). Processing of surgical instruments/equipment using steam sterilisation (Vol. NUH-SOP-NSG-OTS-030, pp. 3): National University Hospital, Singapore.

NUH (2006i). Scrubbing, gowning and gloving (Vol. NUH-SOP-NSG-OTS-008, pp. 3): National University Hospital, Singapore.

NUH (2006j). Standard precaution (Vol. NUH-HAP-NSG-021, pp. 3): National University Hospital, Singapore.

Patient Safety Authority (2005). Risk of fire from alcohol-based solutions, p. 2.

Paulson DS (1994). Comparative evaluation of five surgical hand scrub preparations. *AORN* **60**(2): 246–256.

Phillips N (2004). *Berry & Kohn's Operating Room Technique* (10th edn.). St. Louis: Mosby.

WHO (2009). *Who Guidelines on Hand Hygiene in Healthcare: First Global Patient Safety Challenge Clean Care Is Safer Care* (pp. 270): WHO Press.

Woodhead K, Wicker P, and Cumming IR (2005). *A Textbook of Perioperative Care*. London: Churchill Livingstone.

Chapter 16

Preparation of Tissue Grafts during Transplantation

Aziz Nather* and Choon Wei Lee*

Introduction

As allograft transplantation is increasingly used by more orthopaedic surgeons, it is important to ensure that surgeons know how to use the bone and soft tissue allografts. Professional education is required regarding the proper indications for allograft transplantation, the biological and biomechanical behaviour of different types of processed allografts, and the preparation procedures required for different types of processed tissues before they can be transplanted. Surgeons must understand that it is only by paying meticulous attention to such details that good clinical outcomes could be obtained with tissue transplantation and the complications that would otherwise result from such transplantation could be minimised or avoided.

Reception of Tissue Allografts for Transplantation

All allografts received from the tissue bank must be carefully checked before use. The donor number, graft number, description of type of graft and expiry date must be closely scrutinised. The condition of the jars for "double-jar" deep-frozen specimens must be checked to ensure that the jars are not broken and sterility not already breached. With "triple-wrap"

*NUH Tissue Bank, Department of Orthopaedic Surgery, Yong Loo Lin School of Medicine, National University of Singapore, Singapore.

deep-frozen specimens, the integrity of the outer plastic layer must be carefully inspected. In the case of freeze-dried specimens, the polyethylene packing must be closely scrutinised for loss of vacuum (with vacuum-sealed specimens), or for minute cracks or tears in the outermost plastic layer. The latter indicates a breach of sterility. Furthermore, the cerric-cerrous dosimeter must be verified to be red in colour, indicating that gamma-irradiation sterilisation has indeed been performed. Instructions accompanying all specimens sent by the tissue bank must also be carefully read and complied with.

Donor and Recipient Teams during Tissue Transplantation

To maintain proper discipline in using the allografts, there must be a donor team, who receives and prepares the graft on a donor trolley specially draped and reserved for this purpose. The donor team should be headed by a surgeon, at least a registrar and assisted by a resident. This team is responsible for carrying out all preparations for the tissue graft in a sterile manner before the allograft is ready to be passed to the recipient team for transplantation. The recipient team is headed by the consultant in charge of the recipient.

Preparation for Tissue Allografts

Different preparation techniques are required for different types of tissue allografts, which have been processed differently — small deep-frozen bone allografts, large deep-frozen cortical allografts, cryopreserved osteoarticular allografts, freeze-dried bone allografts and deep-frozen soft tissue allografts.

Small Deep-Frozen Bone Allografts

The femoral head in the "sterile double jar" is taken out of the freezer and allowed to thaw for at least one hour before the start of the operation. The circulating nurse opens the lid of the outer jar without contaminating the inner jar and passes the sterile inner jar to the scrub nurse. The scrub nurse then opens the lid of the inner jar and removes the allograft. This is transferred into a kidney dish containing normal saline with 500 mg of ampicillin and 500 mg of cloxacillin placed on a separate sterile trolley specially draped for preparation of the allograft (Fig. 1) by the donor team. Using a bone nibbler, the surgeon or his/her assistant meticulously removes all cartilage from the femoral head (Figs. 2 and 3). It is extremely

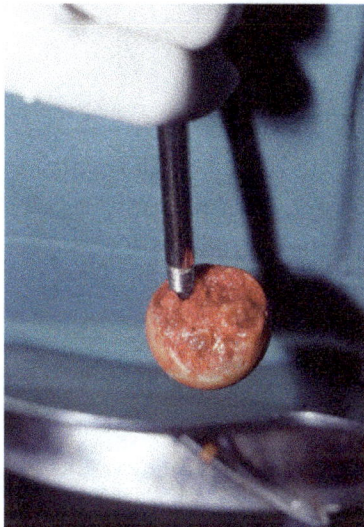

Fig. 1. Femoral head held with a corkscrew transferred to a kidney dish containing normal saline for thawing.

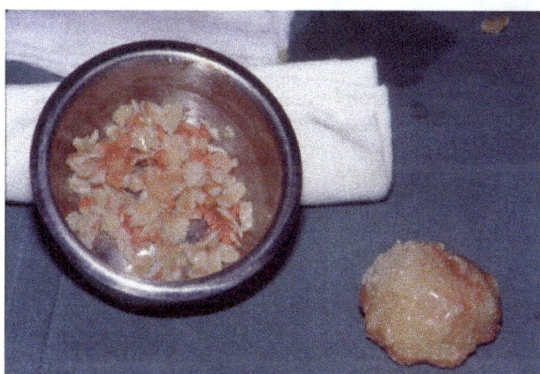

Fig. 2. All cartilage nibbled off from the femoral head. The cartilage pieces to be discarded are contained in the gallipot and the denuded femoral head is on the right side of the picture.

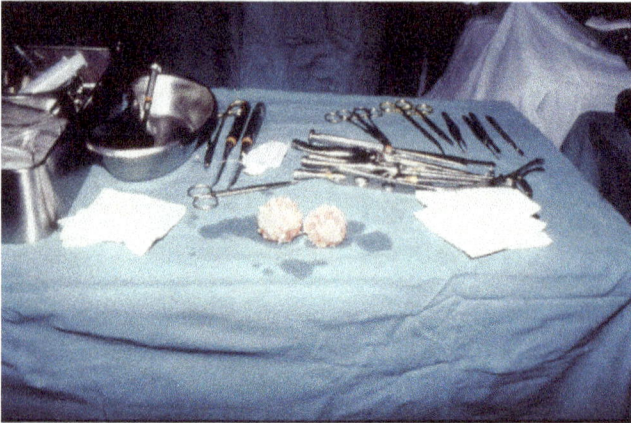

Fig. 3.   Two femoral heads denuded of cartilage on the donor trolley.

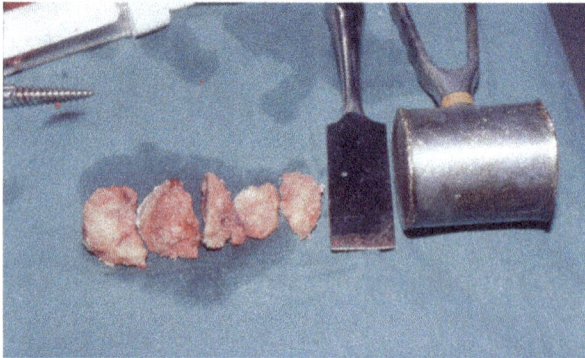

Fig. 4.   The femoral head osteomised into small pieces using a manual osteotome.

important to remove all cartilage since cartilage left behind will impair proper bone fusion and incorporation. All soft tissues must also be removed from the bone. Using a manual osteotome, the femoral head is then cut into small pieces (Fig. 4). As oscillating saw could also be used for this purpose. The bone pieces are then flushed with normal saline using jet lavage to remove all blood and fat globules (Fig. 5). A swab is taken and sent for culture and sensitivity testing (Fig. 6). The bone pieces are then soaked in another kidney dish containing 500 ml of normal saline and

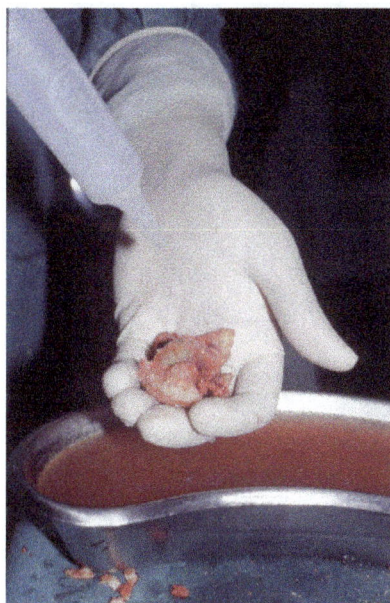

Fig. 5. The small pieces are flushed with sterile normal saline. The blood and fat globules removed in this lavage could be seen in the kidney dish holding the effluent wash.

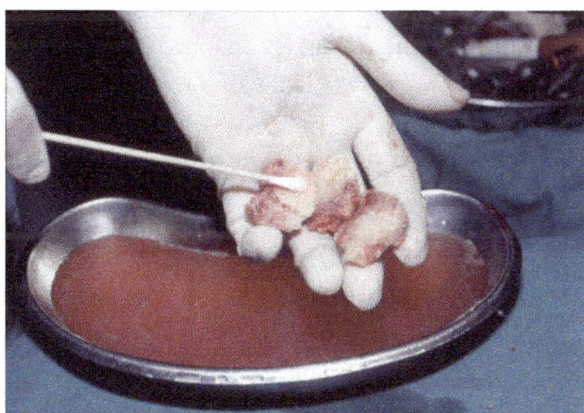

Fig. 6. A swab is taken from the washed grafts for culture and sensitivity testing.

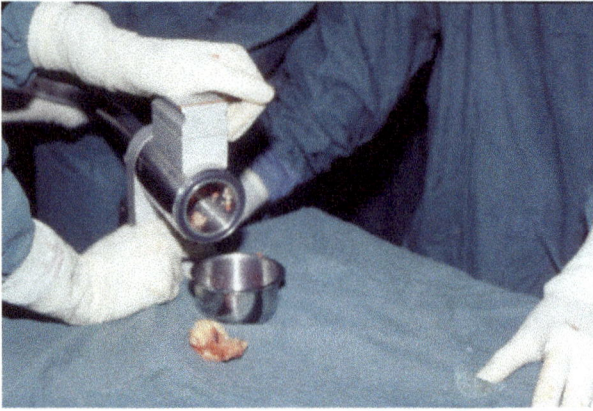

Fig. 7.   Grinding the bone pieces into smaller pieces using a sterile bone mill.

500 mg of ampicillin and 500 mg cloxacillin for about 30 minutes. (For patients allergic to penicillin, 500 mg of erythromycin is used instead.)

The bone grafts are now ready for use. If smaller bone pieces are required, the bone pieces are then ground into smaller pieces using a sterile bone mill (Fig. 7). On the recipient side of the operation, at the start of operation, the recipient is given a prophylactic dose of 1 g of cefazolin intravenously. A Redivac drain is usually inserted into the operation site before closure of the wound. For scoliosis cases (Fig. 8), no drain is inserted. Post-operatively, the patient is continued on intravenous cefazolin (1 g) six-hourly for 48 hours. This is then followed by oral antibiotics — Ceporex (500 mg) six-hourly for two weeks until the wound has healed completely (cyclosporin C is not used).

Long Deep-Frozen Bone Allografts

The long bone in the "sterile triple wrap" stored at −80°C must be taken out of the freezer and allowed to thaw for at least one hour before the start of the operation. The circulating nurse opens the outermost plastic layer without contaminating the sterile middle linen layer and passes the bone wrapped with middle linen layer and inner plastic layer to the scrub nurse. The latter places the graft on a separate trolley draped separately for preparation of the graft by the donor team.

Fig. 8. Bone grafts from two femoral heads inserted in the fusion bed on either side of the implant used for correction of scoliosis in a 12-year-old girl.

The scrub nurse removes the middle linen layer and the inner plastic layer and soaks the long bone in a basin containing two litres of normal saline. One gram ampicillin and 1 g of cloxacillin powder are added to this solution. After further thawing for half an hour, the surgeon who prepares the transplant places the bone on a sterile towel and dissects all soft tissues and periosteum from the bone. This is done meticulously using sharp cuts with a scalpel and further stripping using a periosteal elevator. On both ends of the bone, tough small parts of the periosteum are removed using a bone nibbler. Upon completion of this procedure, the bone is washed with sterile normal saline. It is extremely important to remove all periosteum and soft tissues since these are the tissues that render the allograft immunogenic.

The cleaned bone is then placed on a sterile towel and the required portion of the bone (intercalary) is marked with sterile methylene blue dye to mark the two sites for osteotomy (Fig. 9). The osteotomies are then made using an oscillating saw (Fig. 10). The large intercalary segment cut is then manually reamed using a manual reamer to remove all marrow contents from its medullary canal (Fig. 11). This must also be done meticulously, as the marrow is the most immunogenic part of the deep-frozen allograft. Upon completion of the reaming, jet lavage is done using normal saline to ensure that all marrow contents have been totally flushed out.

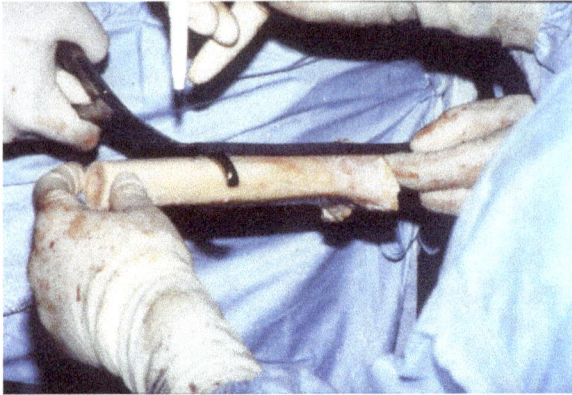

Fig. 9. Marking one osteotomy site with sterile methylene blue.

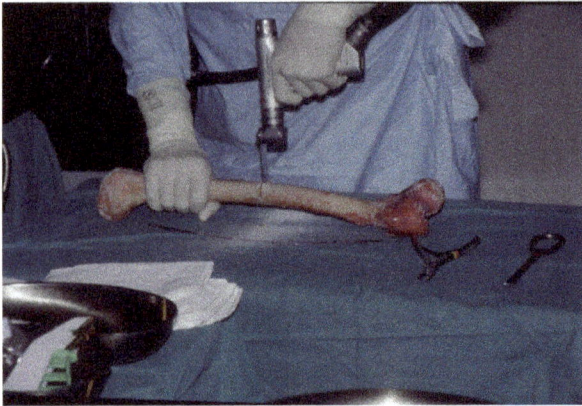

Fig. 10. Osteotomising the long bone using an oscillating saw.

The cut-bone allograft specimen is then finally soaked in a kidney dish containing one litre of normal saline (Fig. 12) with 500 mg of ampicillin and 500 mg of cloxacillin (Fig. 13) for at least 30 minutes.

The intercalary bone allograft is now ready for use by the recipient team (Fig. 14). The same antibiotic regime described for the use of small deep-frozen bone allografts is used. A drain must be inserted before closure of the wound. The same post-operative antibiotic regime is used (cyclosporin C is not used).

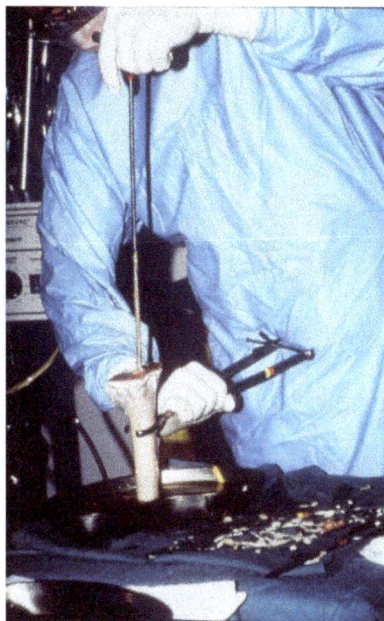

Fig. 11. Reaming the medullary canal of the long bone segment with a manual reamer.

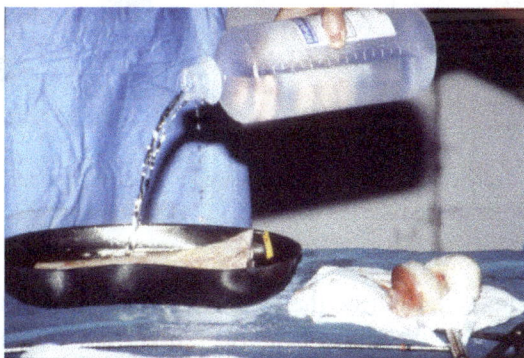

Fig. 12. Prepared allograft placed into new kidney dish and one litre of normal saline is poured.

Cryopreserved, Osteo-Articular Allografts

The long bone in the "triple wrap" cryopreserved at −160°C with 10% glycerol used as cryoprotectant for the articular surface of the long bone

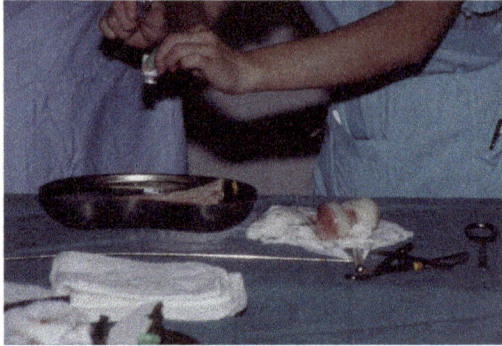

Fig. 13. Antibiotic powder being added to this solution.

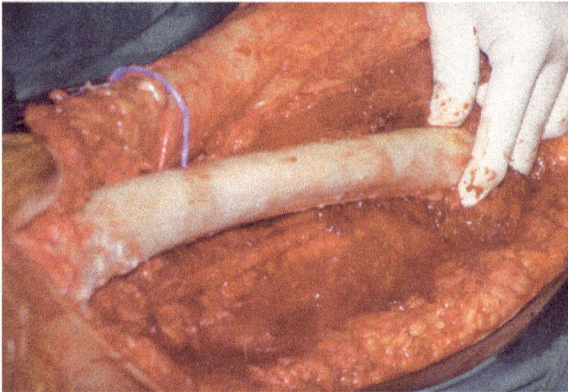

Fig. 14. Prepared intercalary allograft being placed into defect in a patient after resection of osteogenic sarcoma of the patient's femur.

must likewise be taken out of the freezer and be allowed to thaw for at least one hour before the start of the operation. In a similar fashion, the circulating nurse removes the outer plastic layer and hands over the graft to the scrub nurse who places it on a special trolley reserved for the preparation of the graft by the donor team.

The scrub nurse removes the remaining two layers and washes the articular end of the bone with running sterile normal saline until the gauze wrapping the end with 10% glycerol could be removed. The bone is then also soaked in a basin containing two litres of normal saline with antibiotics as described in the preparation of long deep-frozen bone allografts. After soaking for another 30 minutes, the bone allograft is placed on a sterile towel and a similar dissection of all soft tissues and periosteum from the bone is meticulously performed. The bone is again washed with sterile normal saline.

The cleaned bone is again placed on a new sterile towel and the site for osteotomy in the diaphysis of the bone is marked with sterile methylene blue. The osteotomy is then made using an oscillating saw.

Taking care not to damage the articular cartilage at the end of the bone, a manual reamer is used gently to ream out marrow contents of the diaphysis. The reaming must not enter the distal one inch of the end of the bone where the articular surface is. Jet lavage is again used to flush out all marrow contents.

The bone is then soaked in a kidney dish containing 500 ml of normal saline containing antibiotics for at least 30 minutes before it is ready for use by the recipient team for transplantation.

Freeze-Dried Bone Allografts

With freeze-dried cortical allograft (Fig. 15), the outer layer is removed by the circulating nurse and the graft contained in the sterile inner

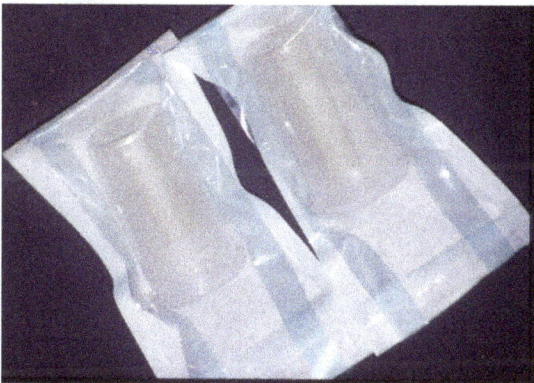

Fig. 15. Cortical femoral ring lyophilised allografts used for anterior spinal.

Fig. 16. Cancellous femoral head lyophilised chip allografts used for packing small bone cavities.

polyethylene layer is passed to the scrub nurse who places it on the donor preparation trolley. Then, the graft is soaked in a kidney dish containing 500 ml of sterile normal saline and 500 mg each of ampicillin and cloxacillin for about one hour before use. The exact length of the graft is cut using an oscillating saw by the donor team.

With small cancellous freeze-dried chip allografts (Fig. 16) it is best not to rehydrate the grafts for more than 10 minutes before they are used. Rehydration is done using the sterile normal saline containing antibiotics. Our experience showed that prolonged rehydration in such cases weakens the grafts substantially. It is common to find that the grafts have turned into powder if they were left to be rehydrated for more than one hour. Some surgeons prefer to use such grafts without rehydration. The serum from the recipient bed is sufficient to rehydrate such grafts.

Deep-Frozen Soft Tissue Allografts

Tibialis posterior and anterior tendons, patellar tendons, calcaneal tendons and fascia lata are processed deep frozen at −80°C without gamma irradiation and stored in "sterile double jars".

The same procedure is used in their preparation for small deep-frozen bones. The graft is allowed to thaw for at least one hour before

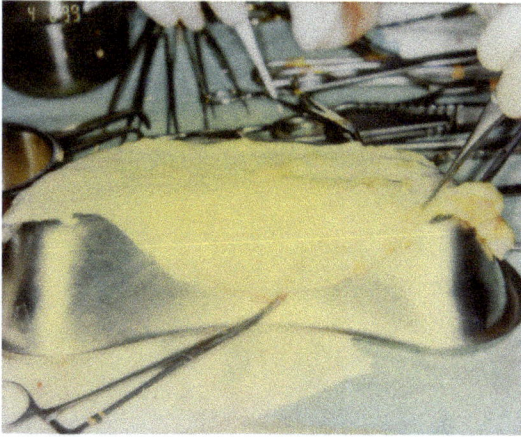

Fig. 17. Fascia lata held by tissue forceps at various edges and spread over an inverted kidney dish for dissection of all fats from its surface.

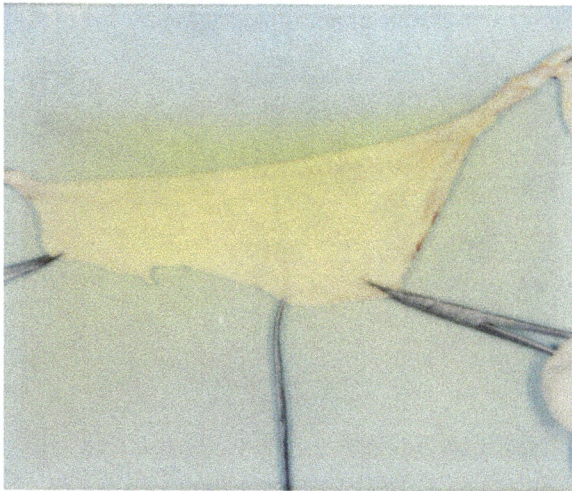

Fig. 18. Fascia lata gripped by an artery forceps at each end rolled to form a solid tube.

the start of the operation. The circulating nurse passes the inner sterile bottle to the scrub nurse who then soaks the graft in a kidney dish containing 500 ml of sterile normal saline and 500 mg each of ampicillin and cloxacillin on the donor preparation trolley. Using toothed

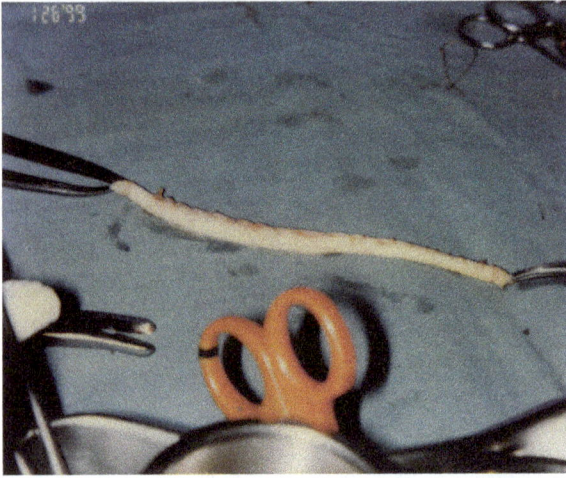

Fig. 19.   Fascia lata roll or "tendon" stitched with running 3-0 Dexon.

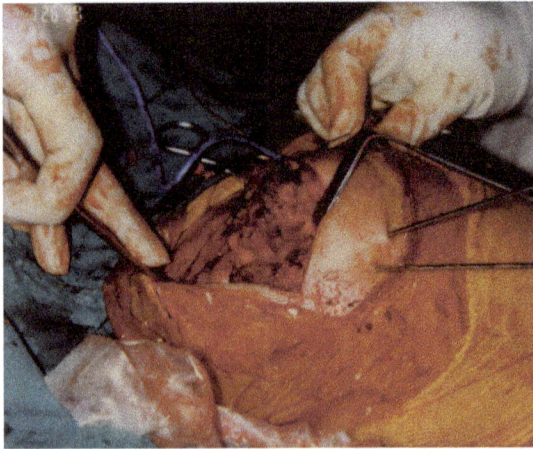

Fig. 20.   Fascia lata used as a "figure of eight" sling for reconstruction of achromic-clavicular dislocation of the shoulder.

forceps and dissecting scissors, all the fats and other soft tissues are meticulously dissected to leave only the tendon for use by the surgeon. The graft is then flushed with normal saline and finally soaked into a clean kidney dish containing sterile normal saline with 500 mg each of cloxacillin and ampicillin for at least 30 minutes. It is now ready for use by the recipient team.

In the case of fascia lata, several tissue forceps are clipped to the edges of the sheet to spread the fascia over an inverted kidney dish (Fig. 17) to allow dissection of all fats from the fascia. After completion of the dissection, the graft is rolled tightly using two artery forceps held at each end (Fig. 18). Upon completion of rolling this tissue, the solid tubular graft is maintained in this configuration by stitching the free end using a running stitch of 3-0 Dexon (Fig. 19).

The fascia lata "tendon" is then used for reconstruction (Fig. 20). This gives increased strength to an otherwise weak sheet of fascia.

Chapter 17

Transmissible Diseases

Xin Bei Chan* and Aziz Nather*

Features of Bacteria

Bacteria are unicellular microorganisms. Bacteria can be classified according to their shape. Cocci are spherical, bacilli are rod-shaped, and spirochaetes appear as corkscrew-like spirals (Fig. 1). Actinomycetes and *Nocardia* are filamentous and branching.

Bacteria possess general features such as the cell wall, cytoplasm, capsule, chromosomes, plasmids, fimbriae and pili (Fig. 2).

The rigid *cell wall* of a bacterium determines its shape. Without it, the cell would swell and burst due to high concentration of intracellular solutes. The cell wall is strengthened by mucopeptide (peptidoglycan), which can be damaged by certain antibiotics, e.g. penicillins. Gram's stain is used to classify bacteria according to the composition of their cell walls. Gram-positive organisms are blue-staining while Gram-negatives are red-staining. The difference is due to a greater amount of mucopeptide in Gram-positives and the presence of an outer cell membrane of lipopolysaccharides in the Gram-negatives' cell wall. Lipopolysaccharides liberated during the disintegration of cells result in Gram-negative sepsis, a cause of systemic immune response syndrome (SIRS) (Nystrom, 1998).

The membrane-bound *cytoplasm* forms the protoplasm. The cytoplasm comprises a watery sap packed with many tiny granules called ribosomes.

*NUH Tissue Bank, Department of Orthopaedic Surgery, Yong Loo Lin School of Medicine, National University of Singapore, Singapore.

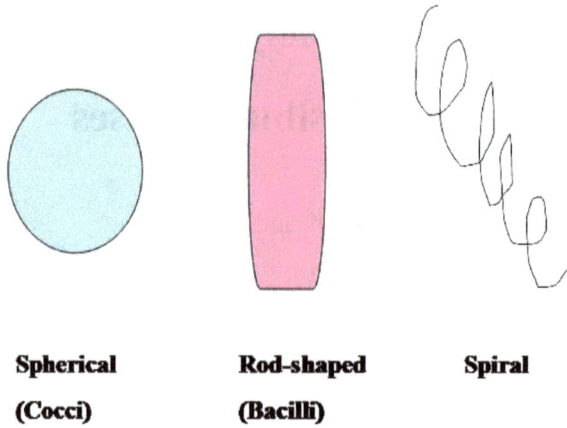

Spherical   Rod-shaped   Spiral

(Cocci)   (Bacilli)

Fig. 1.   Shapes of bacteria.

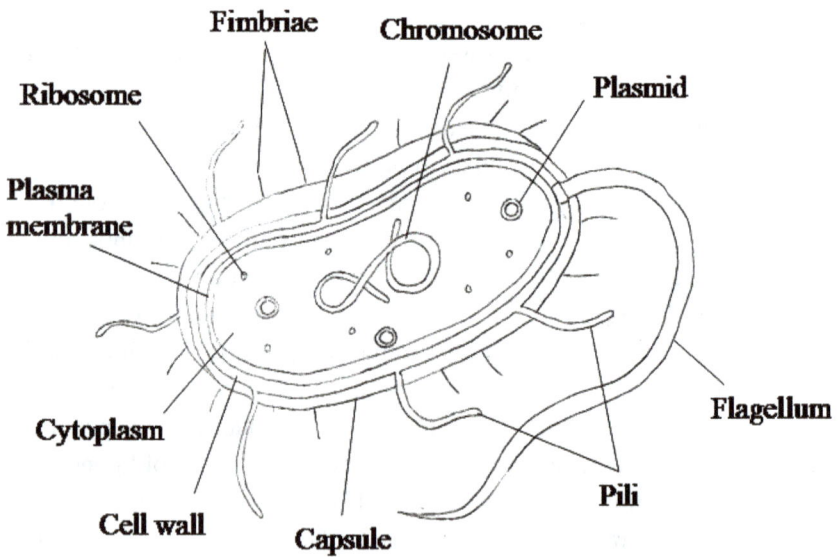

Fig. 2.   Structure of bacterium.

Ribosomes are the sites of protein synthesis and differ from those of human cells in size and composition. These differences explain the selective action of several antibiotics that inhibit bacterial protein synthesis more than that of humans.

The *bacterial chromosome/nuclear body* is the area in the cytoplasm where DNA is located and has little resemblance to the human cell nucleus. There is no nuclear membrane in prokaryotes.

Plasmids are circular DNA molecules that exist independently from the bacterial chromosomes. They may carry genes that confer properties of medical importance, such as antibiotic resistance. Multi-resistant bacteria found in hospitals often possess them.

Spores are produced by certain species of bacteria. They have thick, highly resistant cell walls. Spores can withstand adverse conditions, including extraordinarily high and low temperatures. Boiling cannot destroy spores, so autoclaving is essential to achieve a complete sterilisation process (Levinson and Jawetz, 1994).

Antimicrobial agents used to treat infections target various sites such as the cell wall, cell membrane, ribosomes, nucleic acids and folate synthesis (O'Grady *et al.*, 1997). There are structural and functional differences between these cellular targets and the human equivalents. This enables us to design drugs which damage bacterial systems while causing less or no harm to human ones.

Bacteria reproduce by binary fission; some are as quickly as every 20 minutes while others are slower (Levinson and Jawetz, 1994). Fast growers may contaminate specimens resulting in high counts and the misinterpretation of results. Slow growers, such as *Mycobacterium tuberculosis*, require 12–20 hours to divide. They may take several weeks to produce a visible growth on culture media (Volk *et al.*, 1996).

Features of Viruses

Viruses are not "living cells". They do not possess nuclei or organelles and cannot reproduce independently. Viruses replicate within host cells. They are too small to be seen with a light microscope but some can be detected and characterised by size and morphology using an electron microscope. The infectious particle of virus is known as the virion, which consists of nucleic acid and a protein coat (capsid). Some viruses possess viral envelopes that help them infect cells (Fig. 3).

Viruses can be rod-shaped, spherical, or icosahedral whereby the capsid has 20 triangular facets. Bacteriophages are viruses that infect bacteria.

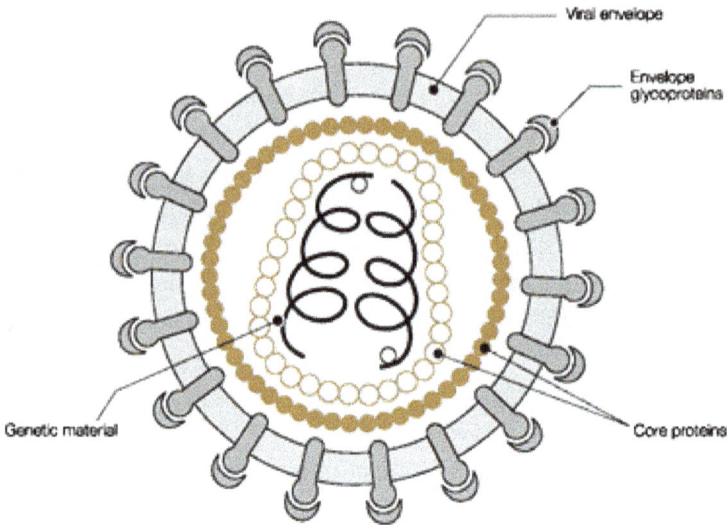

Fig. 3. Structure of virus.

Source: http://www.odec.ca/projects/2004/lija4j0/public_html/disease.htm.

They have elongated icosahedral heads and a tail with fibres for attachment to the bacterium (Fig. 4).

As viruses replicate only in living cells, and many of them are inactivated at room temperature, it is important to transport specimens in a suitable medium on ice and put them into cell cultures as soon as possible.

Transmission of Pathogens

Pathogens can be disseminated through direct contact (inadequate hand-washing, sexual activity etc.) or indirect contact (fomites such as inadequately sterilised instruments, door handles, and insect vectors such as flies and cockroaches).

Organisms that cannot directly invade the skin inoculate humans through penetrating wounds (e.g. tetanus), needle stick injuries (e.g. hepatitis B), biting insects and other arthropods — malaria, dengue and many others. Subcutaneous, intravenous, intramuscular, and intrasternal injections, which occur outside the alimentary tract, constitute parenteral transmission.

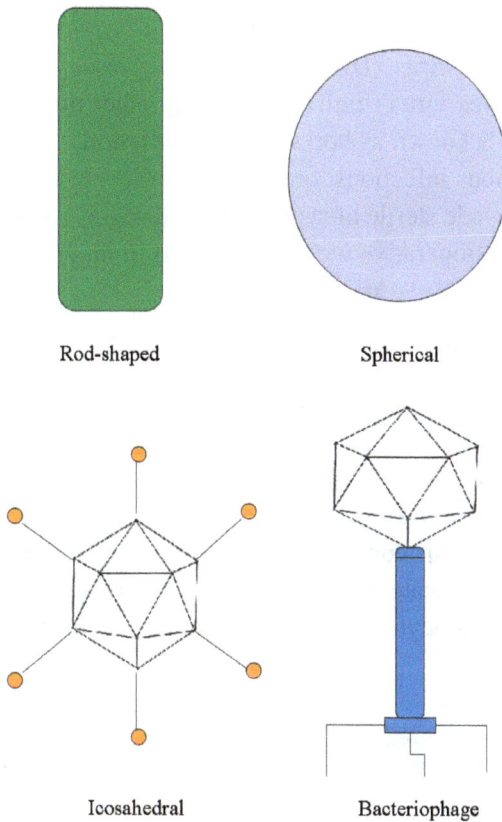

Fig. 4.   Shapes of viruses.

Microorganisms disseminated via ingestion are *Salmonella typhi*, *Shigella* spp., *hepatitis* A and E viruses and *Giardia intestinalis*.

Microorganisms can be transmitted by inhalation including the viruses that cause tuberculosis, chickenpox, mumps and measles. Staphylococci can be dispersed via skin in operating theatres. Each person can emit about 10 000 bacteria per minute at rest, increasing to 50 000 per minute during activity as the friction of clothing against the skin releases more skin scales (Howarth, 1985).

Vertical transmission occurs when the infection spreads from a mother to the foetus *in utero*. An acronym, PoRTHaTCH was devised to highlight the microorganisms using vertical transmission, excluding the vowels. They are parvovirus B19, rubella, *Toxoplasma gondii*, human immunodeficiency virus (HIV), *Treponema pallidum*, cytomegalovirus and herpes simplex virus (Inglis, 1996). Any other mode of transmission among humans is known as horizontal transmission.

In endogenous infections, organisms previously colonising the body surfaces may invade sterile areas and cause disease, e.g. intestinal organisms causing post-operative wound infections in large bowel surgery.

Human Immunodeficiency Virus (HIV) Infection

HIV is one of the leading causes of death in young individuals. In 2007, the number of people living with HIV globally was 33 million and two million died of acquired immune-deficiency syndrome (AIDS) (World Health Organization, 2008).

HIV is a retrovirus containing an enzyme called reverse transcriptase that allows it to replicate within new host cells (Fig. 5). HIV infects and kills CD4 T-lymphocytes. Other white blood cells that contain CD4

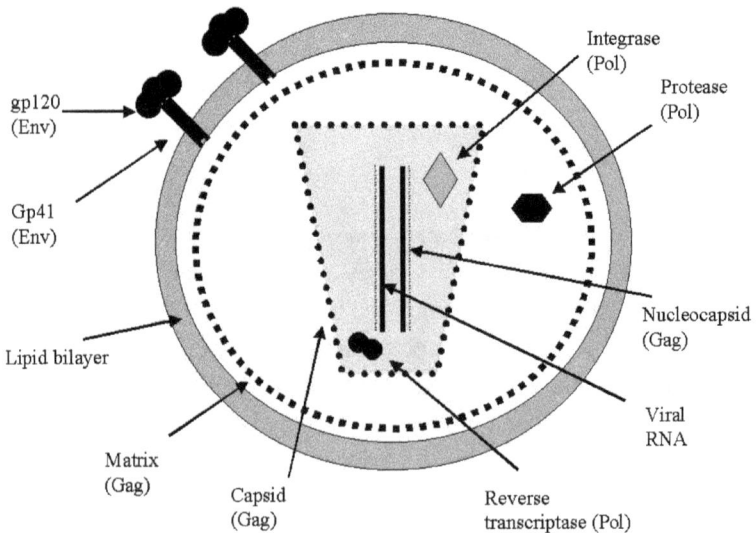

Fig. 5. Structure of HIV.

Source: http://www.soton.ac.uk/~ceb/teaching/2005/Image64.gif.

proteins on their surface can also be infected, e.g. macrophages and monocytes. The resulting loss of cell-mediated immunity makes the infected individual more likely to develop opportunistic infections.

There are two strains of HIV: HIV-1 and HIV-2. HIV-2 was first isolated in 1986 (Clavel *et al.*, 1986). Both strains have similar genomic organisations. Modes of transmission for both are the same but HIV-2 is less infectious. For instance, sexual and vertical transmissions of HIV-2 are around 5- to 9-fold and 10- to 20-fold reduced relative to HIV-1, respectively (Reeves and Doms, 2002).

HIV can be transmitted through sexual contact, transfusion with infected blood products, sharing infected needles and syringes, and perinatal transmission. Although a small amount of virus is present in the saliva, tears and urine, infection from these sources is unlikely. Occupational exposure among healthcare workers may occur from contaminated needle stick injuries. The risk of transmission is about 0.3% following such an injury (Tokars *et al.*, 1993).

HIV-1 and HIV-2 can be detected using the enzyme-linked immunosorbent assay (ELISA) to test for antibodies to HIV-1 and HIV-2 (anti-HIV$_1$ and anti-HIV$_2$ test). If ELISA is positive, the Western blot is used as a confirmatory test. The window period between the point of infection and time taken for anti-HIV tests to detect an infection is 22 days. The window period can be reduced to 16 days using p24 antigen assay and to 12 days using nucleic acid testing (see Chapter 10).

Acute or primary HIV infection manifests itself as a flu-like illness with swollen lymph nodes, fever, malaise and rash lasting two to three weeks. Signs of HIV-related illness usually develop within 5 to 10 years of infection (World Health Organization, 2008). The infection progresses to the asymptomatic stage, which can persist for many years. Viraemia may be low or absent, but the lymph nodes will continue to produce HIV during the latent state. In later stages, persistent generalised lymphadenopathy develops. The lymph nodes enlarge in the neck, underarms and groin for more than three months.

AIDS is the final stage of the disease caused by HIV. Systemic symptoms include fever, unexplained weight loss and diarrhoea. There is a gradual reduction in CD4 cells. When the HIV viral load is high in the advanced stages and the CD4 count falls below 200/mm^3, the infected individuals succumbs to infections an uninfected person can resist. Cancerous conditions may be precipitated by opportunistic infections, e.g. Kaposi's sarcoma, HHV8 (Human Herpesvirus 8), lympho-proliferative disorders and HHV6 (Luppi *et al.*, 1998). Various opportunistic infections seen in AIDS patients include viral infections, such as herpes simplex, herpes zoster and cytomegalovirus, fungal infections, and bacterial infections caused by *Mycobacterium tuberculosis* and *Salmonella* spp.

Antiviral therapy prolongs survival and reduces the number of opportunistic infections by inhibiting viral replication. However, the virus is not eliminated from the infected cells. Combination antiretroviral therapy suppresses the plasma HIV-1 RNA titer to less than 50 copies per millilitre (Hammer, 2005). This extends the period of viral suppression and prevents drug resistance.

Prevention can be achieved in healthcare workers by taking measures known as "universal precautions", to avoid exposure to blood and body fluids (Centers for Disease Control, 1991). It is mandatory to regard blood and body fluids of any person as hazardous. No vaccine effective against HIV infection has been developed.

Hepatitis B Virus Infection

Hepatitis B virus (HBV) is the prototype member of the Hepadnaviridae family (Fig. 6). The members of Hepadnaviridae have strong propensity to infect liver cells but small amounts of hepadnaviral DNA can be found in kidney, pancreas and mononuclear cells (Ganem and Prince, 2004).

Hepatitis B infection accounts for one million deaths worldwide from cirrhosis, liver failure and liver cancer each year (Dienstag, 2008).

The virus spreads through contact with blood and body fluids of an infected person. It can survive outside the body for at least seven days and has an incubation period of 90 days on average (World Health Organization, 2009). Although HIV is similarly transmitted, the infective dose required to cause an infection is lower for HBV: 0.001 ml of "e" antigen-positive HBV-infected blood compared to 0.1 ml HIV-infected blood (Shanson, 1999). HBV is 50 to 100 times more infectious than HIV (World Health Organization, 2009).

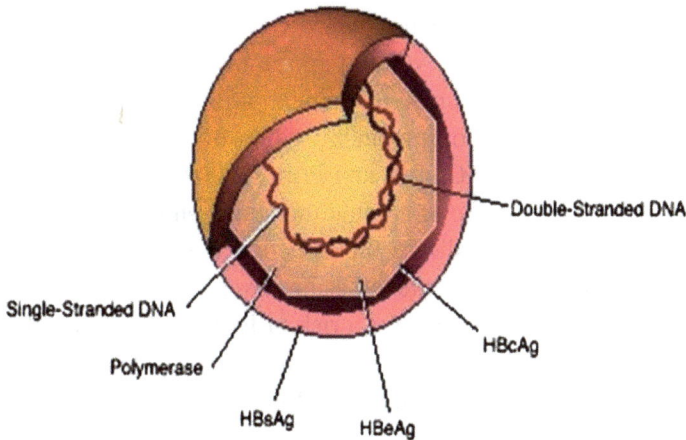

Fig. 6. Structure of HBV.

Source: http://medical-dictionary.thefreedictionary.com/hepatitis+B.

In Asia, where hepatitis B infection is often acquired perinatally; immunologic tolerance to the virus is established for life but incomplete. About 25% of adults who become chronically infected during childhood die from liver cancer or cirrhosis (World Health Organization, 2009). Hepatitis B infection is common in intravenous drug abusers, sex workers and homosexual men. Nosocomial infections among healthcare workers due to sharp injuries could occur. One of the major hazards to healthcare workers is hepatitis B infection (Shapiro, 1995).

An infection is detected using ELISA to screen for hepatitis B surface antigen (HBsAg) (Fig. 6). Anyone who has a positive test for HBsAg should be presumed to have some level of ongoing viraemia. A supplemental test uses ELISA to detect total hepatitis B core antibody (anti-HBc), which respond to hepatitis B core antigen (HBcAg) (Fig. 6) (see Chapter 10).

Hepatitis B carriers are asymptomatic and have minimal liver injury despite extensive viral replication in the liver. They have subclinical persistent infection and normal or nearly normal findings on liver biopsy.

Primary infection may be symptomatic or asymptomatic. The latter is more common, especially in young children. Symptoms include jaundice, dark urine, fatigue, nausea, vomiting and abdominal pain. Most primary infections in adults are self-limited, defined by the disappearance of viruses from the blood and liver and appearance of anti-HBs antibodies. They acquire lasting immunity to reinfection thereafter. Generally, less than 5% of primary infections in healthy adults do not resolve but advance into persistent infections (Ganem and Prince, 2004).

HBV persistence suggests an ongoing attack on infected hepatocytes, usually insufficient to eradicate the infection but does reduce the viral load. Chronic infections can develop into liver cirrhosis or liver cancer. Chronically infected subjects have a risk of liver cancer that is 100 times higher than that of non-carriers (Ganem and Prince, 2004).

Chronically infected patients should be screened twice a year. They can be treated with interferon and oral antiviral agents. Interferon inhibits viral replication. It is used in the treatment of chronic liver disease. HBV is suppressed but not eradicated by interferon; therefore, relapses occur when treatment is terminated.

A vaccine to prevent hepatitis B infection has been developed. Hyperimmune and anti-hepatitis B immunoglobulin are available for passive immunisation of non-immune persons, e.g. hospital staff who sustain inoculation injuries. Hepatitis B vaccine is 95% effective in preventing HBV infection and its chronic consequences (World Health Organization, 2009).

Hepatitis C Virus Infection

Hepatitis C virus (HCV) is an icosahedral RNA virus that belongs to the family of Flavivirudae (Fig. 7). The HBV and HCV behave similarly, with regard to epidemiology and clinical presentation. HCV infects an estimate of 170 million worldwide and represents a viral pandemic (Lauer and Walker, 2001).

The HCV is transmitted mainly through the parenteral route. Hepatitis C infection accounted for 80–90% of post-transfusion hepatitis in the United States, Europe and Asia before the introduction of anti-HCV screening (Cuthbert, 1994). Other parenteral risk factors include injection drug use, tattoos, needle stick accidents and transplantation of infected organs. Injection drug use is the most common risk factor worldwide. The virus can also be transmitted sexually and perinatally, though vertical

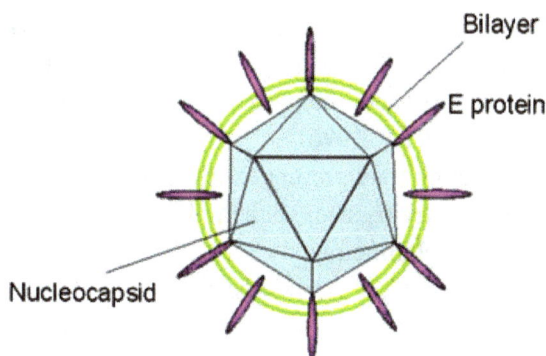

Fig. 7. Structure of HCV.

Source: http://www.usg.edu/ehs/training/pathogens/10.phtml.

transmission is uncommon. Healthcare workers represent 2% of acute hepatitis C patients. Risk of transmission is considerable as needle stick exposure may lead to transmission of HCV in up to 10% of instances (Cuthbert, 1994).

Hepatitis C infection can be diagnosed via serologic assays for HCV antibodies and viral antigens. In recent years, nucleic acid testing has been used to reduce the window period from 70 to 12 days (see Chapter 10). A liver biopsy is recommended to assess the activity of HCV-related liver disease in chronically infected patients.

Acute HCV infection has an incubation of seven weeks and is symptomatic in only one-third of patients (Hoofnagle, 1997). Clinically acute hepatitis C resembles other forms of acute viral hepatitis with onset of malaise, nausea, and right upper quadrant pain followed by dark urine and jaundice. About 85% of patients with acute HCV infection develop chronic infection (Hoofnagle, 1997).

In patients with chronic infection, approximately two-thirds have elevated aminotransferase levels (Hoofnagle, 1997). Such patients are mostly asymptomatic. Fifteen to 20% of chronically infected patients develop liver cirrhosis. Once cirrhosis is established, symptoms are more common and the risk of liver cancer is approximately 1–4% per year (Lauer and Walker, 2001). Liver cirrhosis can lead to end-stage liver disease marked by fatigue, muscle weakness and wasting, fluid retention, easy bruisability, upper intestinal haemorrhage, jaundice, dark urine and itching. The only solution to end-stage liver disease is liver transplantation. HCV is the most common cause of cirrhosis requiring liver transplantation (Detre *et al.*, 1996).

Hepatitis C infection may present extrahepatic manifestations such as arthritis, keratoconjunctivitis sicca, lichen planus, glomerulonephritis and essential mixed cryoglobulinemia.

Hepatitis C infection can be treated with monotherapy using interferon or combination therapy using pegylated interferon and ribavirin. So far, there is no vaccine against hepatitis C.

Syphilis

Syphilis is a sexually transmitted disease caused by the spirochaete *Treponema pallidum* (Fig. 8). An estimated 12 million new cases of syphilis occur each year (Hook and Peeling, 2004).

Syphilis is acquired by direct contact, usually sexual, with primary or secondary lesions. Lesions can occur on the external genitals, vagina, anus and rectum, and on lips or mouth. Vertical transmission is also possible.

To screen for syphilis, a blood test is done. Rapid plasma reagin (RPR) is used to detect non-specific treponemal antibodies. If RPR tests positive,

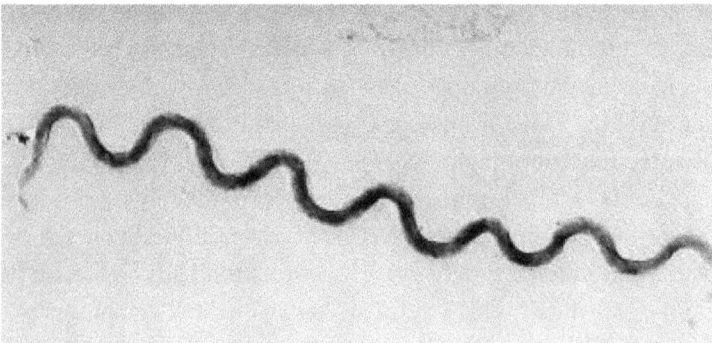

Fig. 8.   *Treponema pallidum.*

Source: http://www.shapelogic.org/images/particles/treponema_pallidum_bacterium_2333.jpg.

a *Treponema pallidum* haemagglutination assay (TPHA) is used as a confirmatory test to detect specific treponemal antibodies (see Chapter 10).

Infection occurs when *T. pallidum* penetrates dermal microabrasions or intact mucous membranes. At the primary stage, a primary chancre develops at the site on inoculation at an average of three weeks after exposure; the incubation period ranges between 10 and 90 days (Centers for Disease Control, 2007). The chancre usually hardens and progresses to ulceration but is typically not purulent. The primary chancre heals spontaneously within four to six weeks, but may still be discernible in about 14% of patients at the onset of secondary syphilis (Rompalo *et al.*, 2001). Moderate regional lymphadenopathy may be present.

Secondary syphilis occurs within three months of initial infection. Rashes develop on one or more areas of the skin. Mucous membrane lesions form. Systemic symptoms include fever, sore throat, headaches, malaise, muscle aches and weight loss. Generalised lymphadenopathy occurs in up to 86% of cases (Chapel, 1980). Secondary syphilis usually resolves spontaneously. Neurological complications such as meningitis and ocular disease may occur.

The latent stage of syphilis begins when primary and secondary symptoms disappear. Late latent syphilis is defined as asymptomatic infection of longer than one year or unknown duration. Serologic testing during the late latent stage is positive, but transmission is unlikely. Latent syphilis ends when curative antibiotic therapy is administered or when tertiary syphilis develops.

Tertiary syphilis is rarely seen today, as antibiotic therapy is common. Clinical manifestations often do not appear until, 20–40 years from onset of infection. Signs and symptoms include difficulty coordinating muscle movements, paralysis, numbness, gradual blindness and dementia. Damage to internal organs occurs and death may result.

A single intramuscular injection of penicillin antibiotic will cure a person who has had syphilis for less than a year (Centers for Disease Control, 2007). Additional doses are needed to treat patients infected with syphilis for a longer time. Other antibiotics can be used for patients allergic to penicillin.

Syphilis can only be prevented through lifestyle changes such as avoiding casual intercourse or practising safe intercourse with the use of condoms.

Creutzfeldt–Jakob Disease (CJD)

CJD is a rare, degenerative brain disorder. It belongs to a family of human and animal diseases known as the transmissible spongiform encephalopathies (TSEs). CJD is mostly sporadic, but can be hereditary or transmitted through exposure. CJD affects one in every one million people a year worldwide (National Institute of Neurological Disorders and Stroke, 2009). The onset of symptoms occurs at about age 60 and 90% of patients die within a year (National Institute of Neurological Disorders and Stroke, 2009).

CJD is characterised by rapidly progressive dementia. The incubation period may be as long as 40 years. Initially, patients experience problems with muscular coordination, personality changes and impaired memory, judgement, thinking and vision. Patients may also encounter insomnia, depression or unusual sensations. As the mental impairment becomes severe, sufferers often develop involuntary muscle jerks and may go blind. Eventually, they become immobile and comatose. Pneumonia and other infections may occur and cause death. In the new variant CJD (nv-CJD), the disease begins with psychiatric symptoms, affecting younger patients and having a longer duration from onset of symptoms to death.

CJD is believed to be caused by a type of protein called a prion. Prions do not contain any genetic information. They are much more resistant to inactivation by ultraviolet light, heat and formaldehyde. However, they are inactivated by higher concentrations of hypochlorite and sodium hydroxide and by autoclaving for a prolonged period of time at high temperature (Steelman, 1994).

Prion proteins occur in both a normal form, which is a harmless protein found in the body's cells, and in an infectious form, which causes disease. The harmless and infectious forms of the prion protein have the same sequence of amino acids (building blocks of proteins) but the infectious form of the protein has a different shape (Fig. 9). Abnormal proteins arise from mutations in the prion gene that code for these proteins. The abnormal proteins aggregate together. Researchers think these protein aggregates lead to neuron loss and brain damage.

Diagnosis can be established by a brain biopsy. Isolation precautions similar to those used for patients with HBV and HIV have been advised

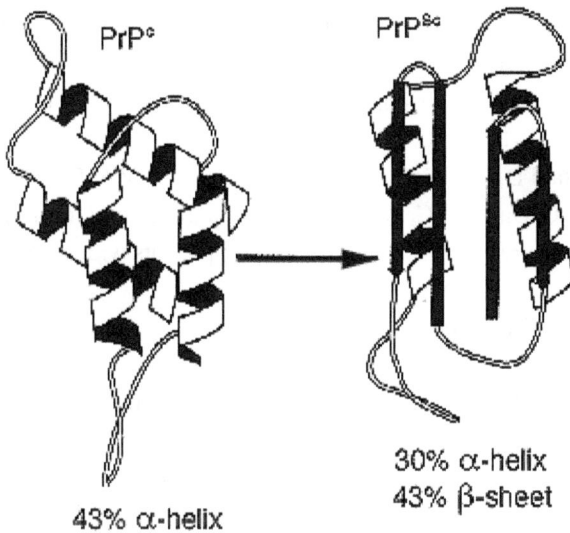

Fig. 9. Normal prion (left) and abnormal prion (right).

Source: Cann, 1997.

(Greenlee, 1982). Experiments have clearly demonstrated that prions are transmissible by successfully transmitting the agent to primates and other species. There are also reports of human-to-human transmission of CJD via human dura mater grafts, corneal transplants, intracerebral electrodes and as a result of injecting growth hormone derived from pituitary glands (Fradkin *et al.*, 1991). Studies have found that infectious prions from nv-CJD may accumulate in the lymph nodes, spleen and tonsils. These findings suggest that blood transfusions from people with nv-CJD might transmit the disease.

There is no treatment that can cure or control CJD. Treatment to alleviate symptoms can be provided. Currently, there is no vaccine to prevent CJD.

References

Cann AJ (1997). *Principles of Molecular Virology*. Academic Press, San Diego.

Centers for Disease Control (1991). Recommendation for preventing transmission of human immunodeficiency virus and hepatitis B virus to patients during exposure-prone invasive procedures. *Bull Am Coll Surg* **76**: 29–37.

Centers for Disease Control (2007). CDC fact sheet — syphilis. Retrieved from: http://cdc.gov/std/syphilis/STDFact-Syphilis.htm.

Chapel TA (1980). The signs and symptoms of secondary syphilis. *Sex Transm Dis* **7**(4): 161–164.

Clavel F, Guétard D, Brun-Vézinet F, Chamaret S, Rey MA, Santos-Ferreira MO, Laurent AG, Dauguet C, Katlama C, Rouzioux C, *et al.* (1986). Isolation of a new human retrovirus from West African patients with AIDS. *Science* **233**: 343–346.

Cuthbert JA (1994). Hepatitis C: progress and problems. *Clin Microbiol Rev* **7**(4): 505–532.

Detre KM, Belle SH, and Lombardero M (1996). Liver transplantation for chronic viral hepatitis. *Viral Hepat Rev* **2**: 219–228.

Dienstag JL (2008). Hepatitis B virus infection. *N Engl J Med* **359**: 1486–1500.

Fradkin JE, Schonberger LB, Mills JL, Gunn WJ, Piper JM, Wysowski DK, Thomson R, Durako S, and Brown P (1991). Creutzfeldt–Jacob disease in pituitary growth hormone recipients in the United States. *JAMA* **265**: 880–884.

Ganem D and Prince AM (2004). Hepatitis B virus infection — natural history and clinical consequences. *N Engl J Med* **350**: 1118–1129.

Greenlee JE (1982). Containment precautions in hospitals for cases of Creutzfeldt–Jacob disease. *Infect Control* **3**: 222–223.

Hammer SM (2005). Clinical practice. Management of newly diagnosed HIV infection. *N Engl J Med* **353**: 1702–1710.

Hoofnagle JH (1997). Hepatitis C: the clinical spectrum of disease. *Hepatology* **26**(3 Suppl. 1): S15–S20.

Hook EW 3rd and Peeling RW (2004). Syphilis control — a continuing challenge. *N Engl J Med* **351**: 122–124.

Howorth FH (1985). Prevention of airborne infection during surgery. *Lancet* **16**: 386–388.

Inglis TJJ (1996). Congenital and neonatal infections. In *Microbiology and Infection*, Churchill Livingstone, New York, Edinburgh, London, Madrid, Melbourne, San Francisco and Tokyo, pp. 137–143.

Lauer GM and Walker BD (2001). Hepatitis C virus infection. *N Engl J Med* **345**(1): 41–52.

Levinson W and Jawetz E (1994). (a) Structure of bacterial cells (b) Growth. In *Medical Microbiology and Immunology*, Appleton & Lange, USA, pp. 3–13, 13–14.

Luppi M, Barozzi P, Morris CM, Merelli E, and Torelli G (1998). Interaction of human herpes virus 6 genome in human chromosomes. *Lancet* **352**: 1707–1708.

National Institute of Neurological Disorders and Stroke (2009). Creutzfeldt–Jakob disease fact sheet. Retrieved from: http://www.ninds.nih.gov/disorders/cjd/detail_cjd.htm?css=print.

Nyström PO (1998). The systemic inflammatory response syndrome: definitions and aetiology. *J Antimicrob Chemother* **41**(Suppl. A): 1–7.

O'Grady F, Lambert HP, Finch RG, and Greenwood D (1997). Modes of action. In *Antibiotic and Chemotherapy*, Churchill Livingstone, London, pp. 10–22.

Reeves JD and Doms RW (2002). Human immunodeficiency virus type 2. *J Gen Virol* **83**(pt 6): 1253–1265.

Rompalo AM, Joesoef MR, O'Donnell JA, Augenbraun M, Brady W, Radolf JD, Johnson R, Rolfs RT, and Syphilis and HIV Study Group (2001). Clinical manifestations of early syphilis by HIV status and gender: results of the syphilis and HIV study. *Sex Transm Dis* **28**(3): 158–165.

Shanson DC (1999). AIDS and other diseases caused by retroviruses. In *Microbiology in Clinical Practice*, Butterworth-Heinemann, Great Britain, pp. 357–382.

Shapiro CN (1995). Occupational risk of infection with hepatitis B and hepatitis C virus. *Surg Clin North Am* **75**: 1047–1056.

Steelman VM (1994). Creutzfeld–Jacob disease: recommendations for infection control. *Am J Infect Control* **22**: 312–318.

Tokars JI, Marcus R, Culver DH, Schable CA, McKibben PS, Bandea CI, and Bell DM (1993). Surveillance of HIV infection and zidovudine use among health care workers after occupational exposure to HIV-infected blood. *Ann Intern Med* **118**: 913–919.

Volk WA, Gebhardt BM, Hammaskjold ML, and Kadner RJ (1996). (a) Microbial cells and their function: a review of cell biology. (b) *Mycobacterium*, (c) & (d) Normal flora, infections, and bacterial invasiveness. In *Essentials of Medical Microbiology*, Lippincott-Revan, Philadelphia, PA, pp. 3–13, 429–439, 315–328.

World Health Organization (2008). HIV/AIDS. Retrieved from: http://www.who.int/features/qa/71/en/print.html.

World Health Organization (2009). Hepatitis B fact sheet. Retrieved from: http://www.who.int/mediacentre/factsheets/fs204/en/print.html.

Part V

Procurement

Chapter 18

Bone Procurement: Living Donors

Aziz Nather* and Cui Lian Chong*

Introduction

Bones are procured from either living or deceased donors. However, before procurement, consent is sought from the donor or the next of kin. This legal practice is commonplace in Singapore and most Asia Pacific countries. The Medical (Therapy, Education and Research) Act 1972 serves as the legal framework in Singapore.

Consent must also be obtained for conducting serological and immunological tests. Blood samples are screened for anti-HIV1, anti-HIV2, hepatitis B, hepatitis C and syphilis. Once the relevant approval is acquired, bone procurement from living donors is carried out in the operating theatre (OT) under sterile conditions.

Potential Donors

The following people are potential living donors:

- Elderly patients with fracture neck of femur undergoing hemiarthroplasty (Fig. 1).
- Elderly patients with osteoarthritis of the hip undergoing total hip replacement (Fig. 2).

---

*NUH Tissue Bank, Department of Orthopaedic Surgery, Yong Loo Lin School of Medicine, National University of Singapore, Singapore.

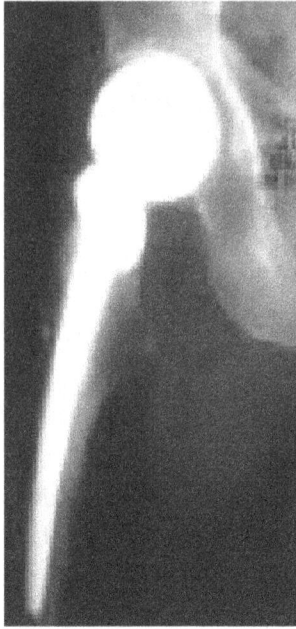

Fig. 1. Hemiarthroplasty.

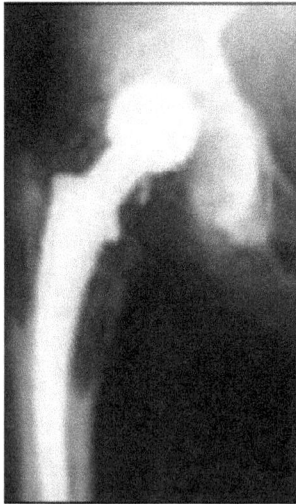

Fig. 2. Total hip replacement.

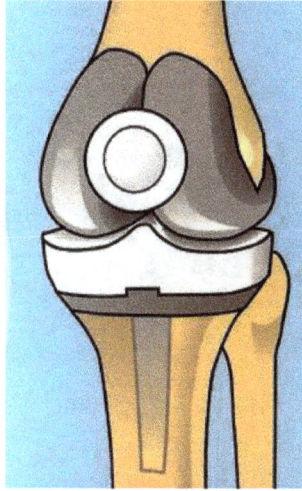

Fig. 3. Total knee replacement.

- Elderly patients with osteoarthritis of the knee undergoing total knee replacement (Fig. 3).
- Patients with vascular ischaemia of the lower limb without infection undergoing above-knee amputation (less common procedure).

One of the commonest bones procured is the femoral head from hemiarthroplasty for patients with fractured neck of femur. In Singapore, total hip replacement is not so commonly performed. In contrast, total knee replacement (TKR) is commonly carried out for osteoarthritis of the knees. However, the bone slices obtained from TKR are not as good as the femoral head.

Preparation of Sterile Trolley

A separate sterile trolley must be prepared in OT where the living donor is being operated for the collection of the procured bone. The trolley must have the following instruments and consumables:

- kidney dish (2), (Fig. 4),
- jug (1),

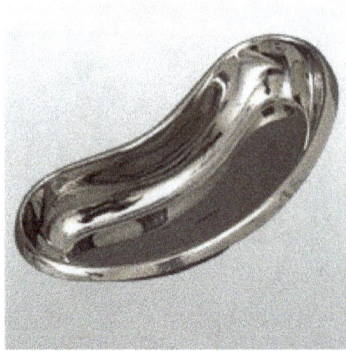

Fig. 4. Kidney dish.

Fig. 5. Squirt.

- squirt (1) (Fig. 5),
- bone nibbler (small),
- blade holder and sterile surgical blade,
- toothed dissecting forceps,
- curved Metzenbaum scissors,
- normal saline,
- ampicillin,

- cloxacillin,
- sterile inner jar,
- sterile outer jar and
- sterile culture bottle.

Femoral Head Procurement Procedure

The technologist must change into OT attire comprising of surgical cap, mask and theatre sandals before entering the OT along with sterile double jars, culture bottle, Living Donor Form, Consent Form and Laboratory Investigation Form.

The inner and outer sterile jars (Fig. 6) as well as a culture bottle are passed in a sterile manner to the scrub nurse to be placed on the procurement trolley. After procuring the femoral head, the surgeon performing the operation would measure its diameter for prosthesis sizing before passing it to the scrub nurse (Fig. 7). The scrub nurse would then flush the bone with normal saline until all blood and fat globules have been washed away (Fig. 8). A small piece of bone is nibbled from the neck of the femoral head to be sent for culture and sensitivity tests. The tissue is then soaked in a kidney dish containing 500 ml of normal saline with 500 mg each of ampicillin and cloxacillin.

Fig. 6.   Sterile autoclavable polyethylene double jar.

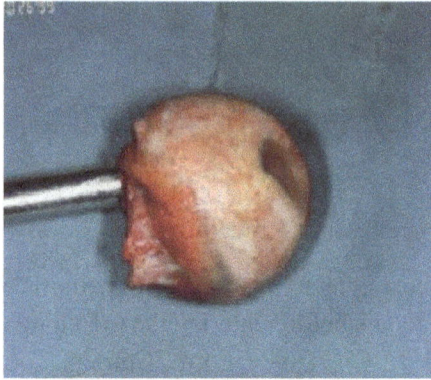

Fig. 7.   Femoral head procured with corkscrew instrument from patient.

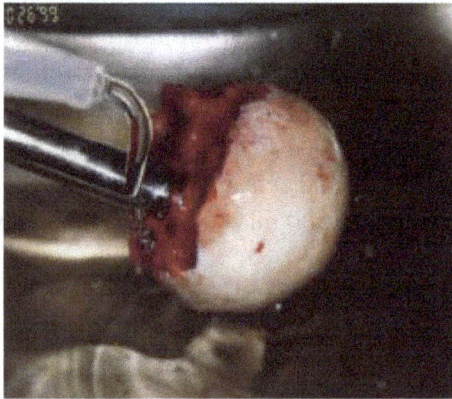

Fig. 8.   Femoral head washed with normal saline.

The femoral head is then placed inside the sterile inner jar and the lid is closed. Then the inner jar is in turn placed inside the sterile outer jar, closed and passed on together with the culture bottle to the tissue bank technologist waiting to receive the procured specimen and culture specimen.

Using adhesive labels, the technologist labels the sterile double jar (Fig. 9) and the culture bottle with the particulars of the patient. The technologist leaves the OT with the double jar and the culture bottle. Upon

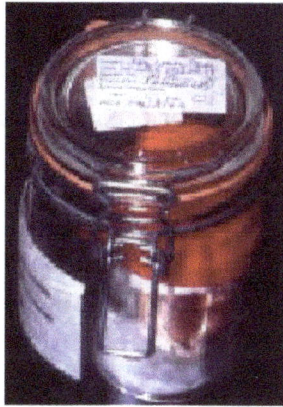

Fig. 9. Sterile double jar containing the femoral head.

Fig. 10. "Quarantine freezer" of deep freezer (−80°C) containing double jars.

returning to the tissue bank, the double jar is immediately stored in the "quarantine freezer" (Fig. 10). The culture bottle is sent as soon as possible to the Department of Laboratory Medicine for culture and sensitivity tests for both aerobic and anaerobic organisms.

Documentation

The technologist must document all procurement procedures meticulously and accurately using the Living Donor Form, which includes the following details:

- date of operation,
- donor number,
- donor particulars (name, identification number, sex, age etc.),
- type of tissue procured,
- name of the surgeon,
- name of the hospital,
- ward number,
- medical history,
- test results,
- clinical examination and
- status of the tissue procured:
  - o awaiting results,
  - o ready for use and
  - o to be discarded.

The results of all laboratory serological tests, including culture and sensitivity results, must be accurately recorded in the Living Donor Form. The status of the tissues procured must be decided by the director of the tissue bank. The Consent Form of the living donor must also be properly filled for proper documentation. Once test results are negative, the procured specimen are transferred to the "ready-for-use freezer" and recorded in the "ready-for-use freezer" log book.

Reference

Nather A (2001). Sterile procurement of bones and ligaments. In Nather A (ed.), *The Scientific Basis of Tissue Transplantation*, World Scientific, pp. 265–270.

Chapter 19

Bone Procurement: Deceased Donors

Aziz Nather* and Cui Lian Chong*

Introduction

In Singapore, bone procurement from deceased donors is carried out in the operating theatre (OT) under sterile conditions. However, some countries conduct the procurement under non-sterile, clean environment in the mortuary.

The bones are procured within 12 hours of death if body is not kept refrigerated and procured within 24 hours of death if body is kept refrigerated at 4°C.

Multi-Organ Tissue Donation System

In Singapore, bone procurement from living and deceased donors is part of a multi-organ and tissue procurement system coordinated by the Ministry of Health (MOH). Under this system, the organs and tissues procured include kidneys, liver, heart, cornea and bones. A national transplant co-coordinator is responsible for approaching the donor or the relatives for procurement and test screening consent. The aforementioned laboratory screening tests include blood grouping, HLA typing, anti-HIV1 and anti-HIV2, HbsAg, anti-HCV, RPR (TPHA) and anti-CMV.

*NUH Tissue Bank, Department of Orthopaedic Surgery, Yong Loo Lin School of Medicine, National University of Singapore, Singapore.

All potential donors with sudden but known cause of death are approached. About 10–20% of consenting donors who agree to donate both kidneys will also agree to donate their bones. In these cases, the consent form indicates the limb(s) from which the bones and ligaments can be procured.

Upon receiving consent, the national transplant coordinator activates the various teams concerned to perform the procurement. With bone procurement, the national transplant coordinator will notify the National University Hospital (NUH) transplant coordinator who will then assemble the bone team for procurement. The bone procurement team comprises of at least two orthopaedic surgeons, two residents and three technologists. Whenever possible, the director (orthopaedic surgeon) leads the team. If the director is not available, the deputy director (another orthopaedic surgeon) will lead. The orthopaedic surgeons and residents are provided by the Department of Orthopaedic Surgery in the NUH.

The manager of the NUH Tissue Bank will activate a team of medical, allied medical, non-medical, and counselling personnel to partake in the procurement procedure.

Medical Personnel

- transplant coordinators and
- surgeons.

Allied Medical Personnel

- nurses,
- medical technologist and
- surgical technologists.

Non-Medical Personnel

- administrators,
- legal counsels,
- religious counsels and
- social workers.

Counselling Personnel

Those involved in counselling should be acquainted with local customs and tradition. Alternative approaches to counselling include utilising the media and brochures. It is imperative that all forms of counselling take into account religious and socio-cultural beliefs and habits. The method and type of counselling depend on the country or community involved.

Preparations for Procurement

Deceased donors usually donate bones and soft tissues from both lower limbs. Some donors also donate one upper limb in addition to two lower limbs. It is important to plan what bones and soft tissues are to be procured from each donor. Furthermore, the number of soft tissues to be procured from each limb and total number of specimens to be obtained must be planned in advance.

To facilitate the planning and procurement, various sterile instrument sets and consumables must always be available for activation of bone procurement.

- Transportation (NUH Ambulance booked for transportation of staff, instruments and equipment if donor is in another hospital).
- Autoclaving of surgical instruments.
- Autoclaving of major cleansing sets.
- Preparing electric-driven portable osteotome (Fig. 1).
- Autoclaving of double jars.
- Autoclaving of triple wraps.
- Sterile surgical gowns/drapes.
- Other consumables.
- Plastic bones for reconstruction.
- Sticker labels for documentation.

Triple Wraps

It is important to note that polyethylene bags cannot be autoclaved, instead they are sterilised by ethylene oxide or by gamma irradiation.

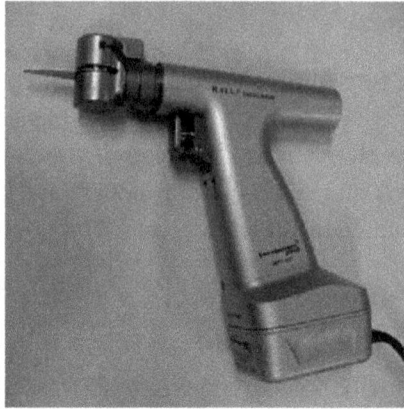

Fig. 1.   Powered portable osteotome.

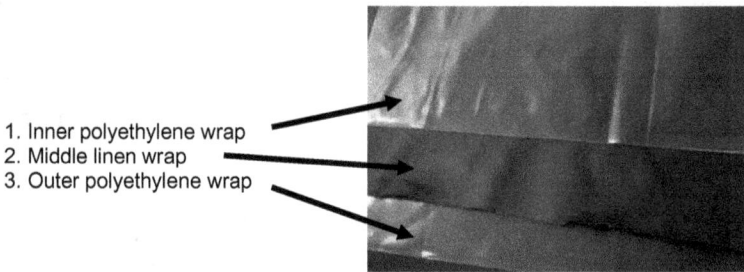

1. Inner polyethylene wrap
2. Middle linen wrap
3. Outer polyethylene wrap

Fig. 2.   Sterile triple wrap.

They are used for "sterile triple wrap technique" to package larger specimens — namely the long bones (Fig. 2).

Bone Procurement System

The tissue bank is fully equipped with operating instruments, trolleys, sterile disposable gowns, sterile disposable drapes and all consumables for performing sterile procurement of bones and ligaments from deceased donors.

As required in the procedure manual, the NUH Tissue Bank has all the necessary equipment and consumable items ready to be transported immediately to any hospital for procurement from any deceased donor at any time.

One major orthopaedic surgical instrument set including a powered portable osteotome, a major cleansing set and two trolleys are autoclaved and packed in polylite containers ready for mobilisation for bone team activation.

All consumable items prepacked in polylite containers and ready to use are sterile disposable drapes, sterile disposable gowns, sterile gloves, sterile penny towels, Raytec gauze, sterile disposable blades, povidone-iodine solution, normal saline solution and vials of ampicillin and cloxacillin.

All sterile double jars and linen, autoclaved cotton tapes and sterilised polyethylene bags as well as sterile culture and sensitivity bottles are also prepacked into polylite containers ready to be used. The normal protocol is to have *ready* at any one time all containers, bags and culture bottles to receive specimens from two lower limbs. This is because the majority of donors give consent for procurement from both lower limbs.

Time is the limiting factor when operating within the confines of a multi-organ and tissue procurement system. The kidney team procures the vital organs first, followed by the liver team and cornea team. Next, the skin team procures the skin. The bone procurement team is the last to operate on the donor and is given a maximum of two hours to complete all its procurement procedures. Even if the donor consents to give all bones, it is usually wise to restrict procurement from two lower limbs only. In addition, when a donor pledges two lower limbs, to save time and to perform the procurement efficiently, two surgical teams will scrub up to operate on the patient simultaneously; each team consisting of surgeon, a resident and a technologist.

The tissue bank also has ready all items required for proper reconstruction of the limbs to be operated on. This includes plastic bones (femur, tibia), sutures and Primapore dressings.

Adhesive labels are prepared beforehand to ensure more efficient labelling of the large number of specimens, which will be retrieved during the procurement. This preparation is started upon activation of the tissue bank team as soon as all laboratory test results of the donor have been received and the procurement confirmed. Donor number, date of procurement and type of tissue procured are printed or typed onto each adhesive label before the bone team moves to the hospital where the procurement is to be done.

Tissue Procurement Plan

As donors are scarce, it is important to plan in advance the number of specimen that will be retrieved to benefit the maximum number of recipients possible. With proper planning, one donor can benefit at least 50 to 60 recipients. For instance, it is wasteful to procure the whole femur as one bone thus the femur is procured in three parts — the femoral head, a proximal half and a distal half. The tibia is likewise procured in two portions — a proximal half and a distal half. The patellar tendon is procured in two longitudinal halves for two recipients. Likewise, the calcaneal tendon is also procured in two longitudinal halves. Other soft tissues procured include the fascia lata, menisci, tibialis anterior and tibialis posterior tendons. A large piece of the iliac crest is also procured from each iliac wing.

Bone Procurement Procedure

All procurement staff must change their attire into theatre attire, including theatre caps, masks and surgical boots or surgical sandals. There are two trolleys used: one containing the major orthopaedic surgical instrument set and the other containing the major cleansing set as prepared by the tissue bank technologists (who have been trained in sterile technique and nursing procedures).

The surgeons and residents that scrub up for the operation wear sterile gowns and double pairs of sterile gloves.

Each technologist lifts the lower limb by the toes for cleansing while each surgeon cleanses the lower limbs with povidone-iodine solution from the hind foot all the way to the groin and including the iliac crest region. The limbs are then carefully draped using disposable sterile drapes. The foot is then wrapped with sterile linen and then placed on the draped operating table. The technologists then scrub up to assist in the procurement by the two surgical teams.

Operative Procedure for Lower Limbs

A midline vertical incision is made over the front of the lower limb (A) from the middle of the groin above (Fig. 3) to the middle of the ankle joint below (Fig. 4).

The incision is deepened using sharp blades to reach the deep fascia, which is then incised. Before the operative procedure commences, it is of utmost importance that every member of each surgical team is reminded to use the *"no-touch technique"* to avoid accidental injuries. This means that during the operation, no one is allowed to touch or hold any tissue with his or her fingers; hence, minimising the risk of being accidentally cut by the surgeon. Instead, surgical instruments are used to hold the tissues.

The key to easy procurement of bones in the lower limb is to start operating on the knee joint. The patellar tendon–tibial tuberosity complex is procured first (Fig. 4). The portable electric-powered osteotome is used to split the patella and the tibial tuberosity into two equal halves.

This is best achieved *"in situ"* instead of dissecting the patellar tendon completely and then splitting the free patellar tendon into two halves

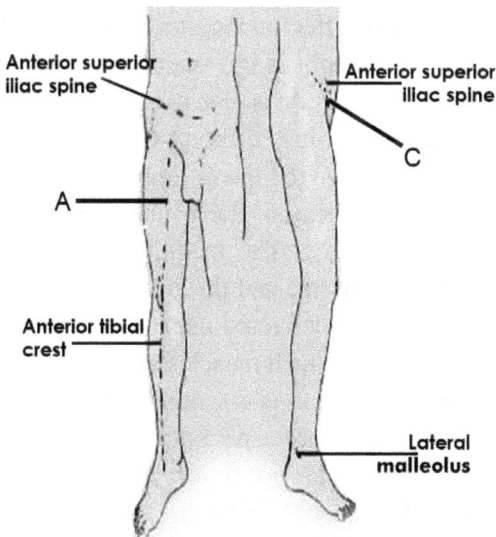

Fig. 3. (A) Incision made over front of thigh and leg; (C) Incision made over curved iliac crest.

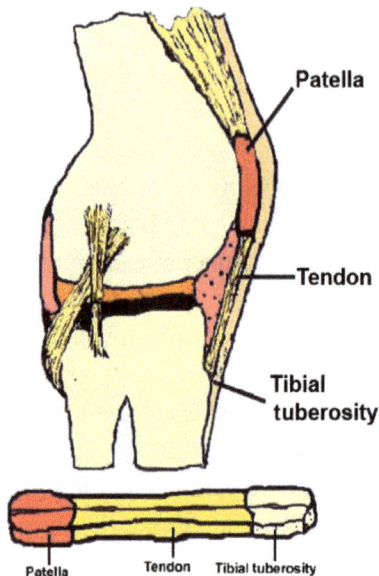

Fig. 4. Patella–tendon–tibial tuberosity.

(medial and lateral halves). After the longitudinal splits in the bones using the powered osteotome, the split in the tendon is completed using a large surgical blade. The patellar tendon is then dissected completely and is free to be removed for collection. It is difficult to split the patella specimen once it has been removed from the lower limb.

The menisci are then procured. Each meniscus is procured with the tibial articular surface attached. The medial and lateral collateral ligaments are divided at the joint line and this dissection enables the articular lower end of the femur to be delivered out easily. Using a periosteal elevator and sharp knife dissection, all muscles are stripped off the surface of the femur. Sharp knife dissection is essential to free the soft tissues from their attachment to the thick linea aspera posteriorly.

At the upper end, the ligamentum teres is divided and the femoral head dislocated from the acetabular socket.

Using the portable electric-driven osteotome, the femoral head is cut at the base of the neck and the femoral shaft is osteotomised in the middle to give an upper and a lower half (Figs. 6 and 7). The three specimens of the femur are then removed for collection (Fig. 5–7).

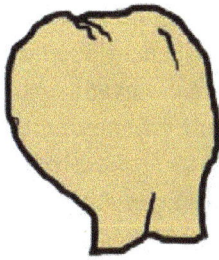

Fig. 5. Femoral head.

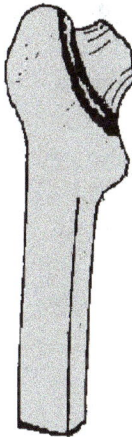

Fig. 6. Proximal femur.

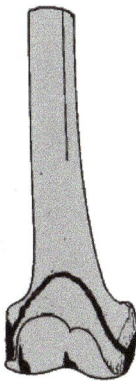

Fig. 7. Distal femur.

Below the knee, the dissection in the leg is likewise deepened to the deep fascia. The muscles are then stripped off the tibia and fibula bones, once again using the periosteal elevator and sharp knife dissection. For best dissection results, begin by removing the fibula intact as one bone. This is done by disarticulation at the proximal tibio-fibular joint above and cutting the inter-osseous ligament between the tibia and the fibula below as well as the calcaneo-fibular ligament inferiorly. The fibula is then removed, osteotomised into two halves and retrieved as two bone specimens (Fig. 8).

Once the fibula has been excised, it is then easy to remove the tibia. The tibia is osteotomised into two halves for collection (Figs. 9 and 10).

A small incision is then made posteriorly to expose the calcaneum (B) (Fig. 11). The tendinous part of the tendo-achilles is excised in continuity with a large piece of the calcaneal tendon (Fig. 12). This is first split into two longitudinal halves as for the patellar tendon for collection as two specimens. Again, the splitting is best done *in situ* before the calcaneum tendo achilles complex is removed from the hind leg.

Other soft tissues are procured at this stage. In the thigh, a large rectangular piece of fascia lata, about 24 cm × 6 cm, is procured. In the leg,

(a) (b)

Fig. 8. (a) Proximal fibula, (b) distal fibula.

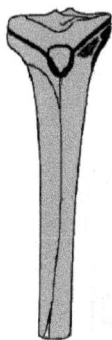

Fig. 9. Proximal tibia.

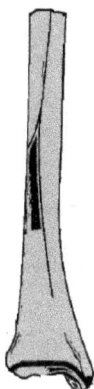

Fig. 10. Distal tibia.

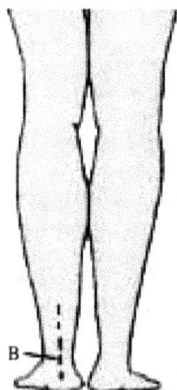

Fig. 11. Incision made over back of the heel (B).

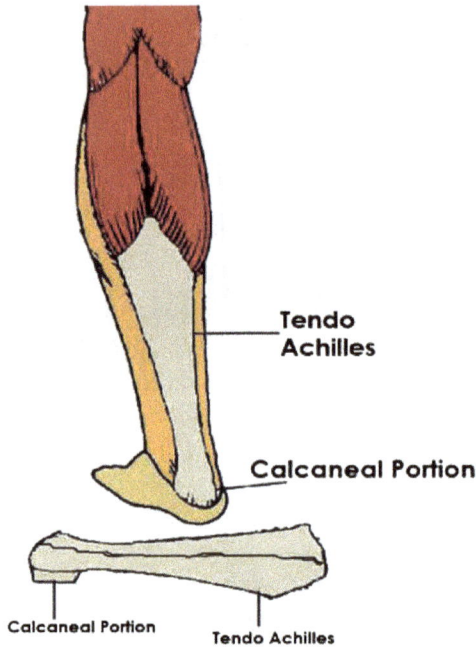

Fig. 12. Calcaneal tendon.

the tibialis anterior and tibialis posterior tendons are procured (as long a tendon as possible) (Fig. 13).

A separate curved incision is made over the crest of the iliac wing (C) (Fig. 14). The iliac bone is exposed and the muscles on its external and internal surfaces are stripped. Using the powered osteotome, a rectangular piece about 6 cm × 4 cm is procured (Fig. 14).

Operative Procedure for Upper Limbs

After cleansing and draping, the whole upper limb from above the axilla to the wrist below is exposed. A vertical midline incision is made over the front of the arm and extending over the middle of the forearm to the wrist below (D in Fig. 15). Proximally, the incision is curved upwards to enter the deltopectoral groove to the front of the shoulder joint.

To procure the humerus bone easily, start by dissecting at the elbow joint. The capsule of the elbow joint is cut. This helps to disarticulate the

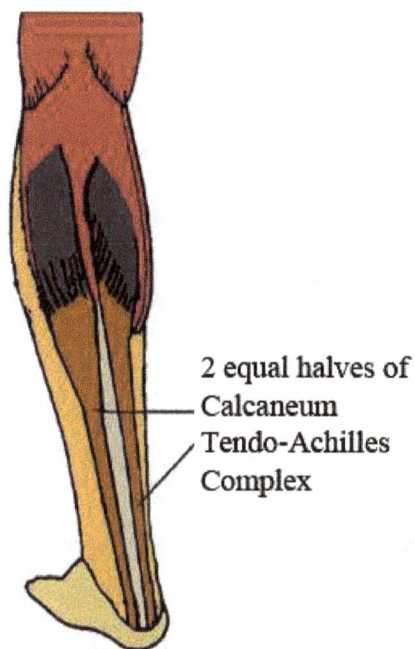

2 equal halves of Calcaneum Tendo-Achilles Complex

Fig. 13. Posterior and anterior tibialis tendons.

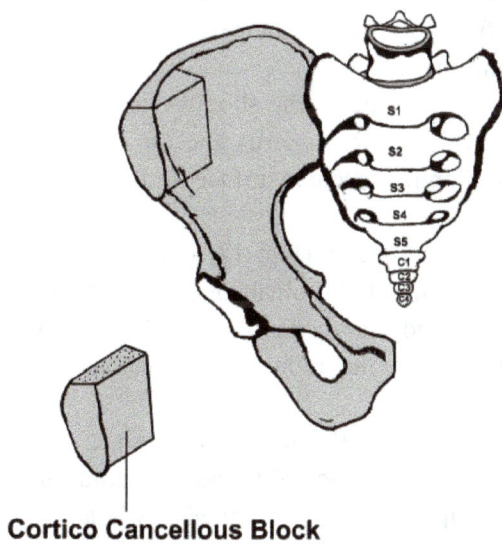

Cortico Cancellous Block

Fig. 14. Iliac crest.

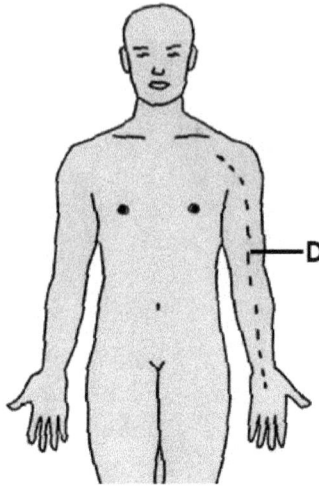

Fig. 15. Incision made over the front of the arm and forearm (D).

radius from the humeral articular surface on the lateral side, the ulna on the medial side and the olecranon from the humeral articular surface posteriorly.

The humeral bone (Fig. 16) is then delivered readily. All muscles and soft tissues are dissected from the humeral surface using periosteal elevator and sharp knife dissection.

Proximally, the capsule of the shoulder joint is cut to leave a cuff attached to the neck of the humerus and the humeral head dislocated from the socket (glenoid articular surface of the scapula) of the shoulder joint.

In the forearm, the radius (Fig. 17) is dissected free from the ulna (Fig. 18) at the proximal radio-ulnar joint. The inter-osseous membrane between the radius and ulna bones is incised. Likewise, all soft tissues are dissected free from the radius and ulnar surfaces in the similar fashion.

At the level of the wrist, the inferior radio-ulnar joint is dissected. The capsule of the wrist joint is incised to free the radius and ulna bones. All three bones in the upper limb are procured as whole bone specimens.

Usually consent is not obtained to extend the incision into the palm and therefore, usually flexor tendons could not be procured.

Fig. 16. Humerus.

Fig. 17. Radius.

Fig. 18. Ulna.

Specimen Collection

It is the onus of the director or manager of the tissue bank to ensure that specimens are collected in the most sterile and organised fashion. It is also important to ensure that they are labelled correctly.

This is done using a separate sterile trolley containing the major cleaning set. Specimens are received in kidney dishes on the trolley. Each specimen is flushed with copious amounts of sterile normal saline. After thorough cleaning, a small piece of tissue is obtained from each specimen and placed in a small bottle for culture and sensitivity tests. The specimen is then soaked in a washbasin containing a solution of two litres of sterile normal saline with four vials 2 gm each of ampicillin and cloxacillin.

After soaking in the antibiotic solution for about 10 to 20 minutes, the smaller specimens — femoral heads, patellar tendons, calcaneal tendons, tibialis anterior and tibialis posterior tendons, menisci and fascia lata — are collected in sterile jars using the "sterile double-jar technique" previously described for collection of small bone specimen from living donors.

A sterile metal ruler is first used to measure the length of the tendinous part of the patellar tendon and the calcaneal tendon and also the lengths of the tibialis anterior and tibialis posterior tendons. The length and breadth of all long bone specimens are also measured and documented as is also the side of the limb, where the long bone is procured from. This is meticulously and accurately documented by the third technologist who does not scrub for the operation. The role of this technologist is to receive the collected specimens in their proper containers and to label them properly and correctly using the adhesive labels already prepared beforehand.

The large specimens, such as femur, tibia and iliac crest, are collected using the "sterile triple-wrap technique" — inner polyethylene bag, a middle linen wrap tied with two sterile cotton tapes and an outer polyethylene bag (Figs. 19–21).

All specimens including the culture bottles accompanying each specimen are labelled immediately to avoid mistakes and stored in the polylite container specially reserved only for collection of specimens. Care is taken to make sure that each specimen is accompanied by a corresponding culture bottle containing tissue from that specimen.

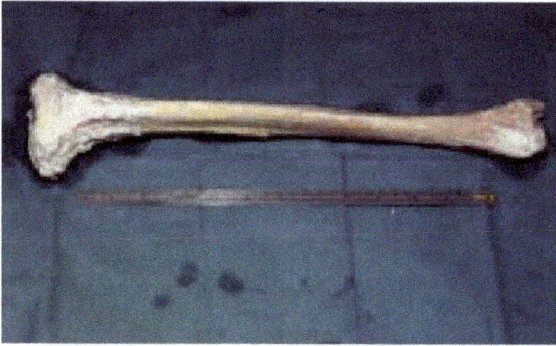

Fig. 19. Tibia specimen being measured with a ruler.

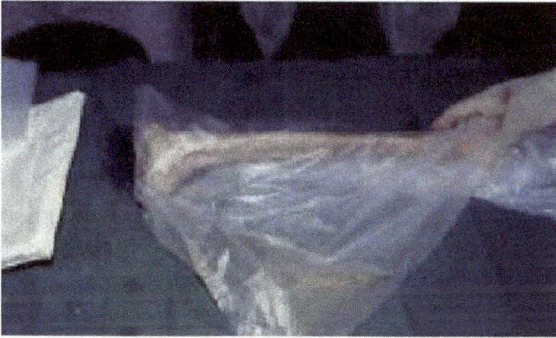

Fig. 20. Tibia placed in sterile polyethylene bag (first layer of triple wrap).

Fig. 21. Sterile triple-wrap technique used to procure long bones from a deceased donor.

Reconstruction

Religious and cultural considerations dictate that respect for the deceased body should be upheld for all donors. Hence, all efforts should be directed to restore the body to its natural appearance. Among the tissues donated, cornea and long bones require reconstruction. After skin procurement, a dressing should be applied.

Bone

A plastic bone or a wooden stick approximating the size of donor bone procured is used to replace the harvested bone (Fig. 22).

The split muscles are approximated using appropriate surgical sutures. This is followed by fascia, subcutaneous tissue and skin. One should make sure that the natural anatomic contours of the body are restored. Burial arrangements after tissue donation may affect the extent of reconstruction. If cremation is planned, just the gross contours of the body are reconstructed. The body is then cremated and the ashes given to the next of kin. When open casket or coffin burial is planned, meticulous attention should be made to restore the external appearance of the body. Exposed areas should appear like their natural appearance. Dignity and respect for the deceased body must be observed at all times.

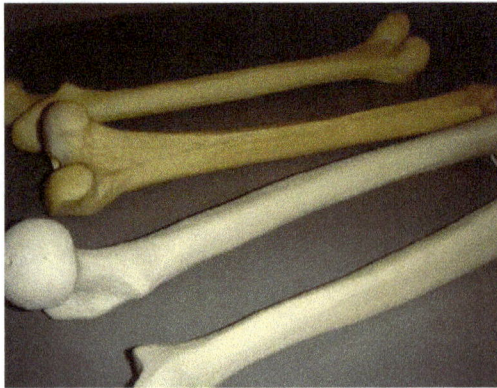

Fig. 22. Plastic bones.

Lower limb

Upon completion of retrieving the last specimen from each lower limb —
usually the iliac crest — reconstruction of both lower limbs begun. A plastic femur and a plastic tibia arthrodesed at the knee by two rolls of 4-inch
plaster of paris (Fig. 23). The length of this construct may have to be
trimmed using the powered osteotome to ensure a snug fit. The tissues in
the incision are then closed in two layers, the muscles and fascia lata with
1-0 Prolene or 1-0 Vicryl and the skin stitched with 1-0 prolene using a
well-knitted running stitch. The wounds are then dressed with Primapore
dressings.

Care must be taken that all bleeding has stopped and that a clean
dressing remains.

Upper Limb

For reconstruction, plastic humerus, radius and ulna bones are used. The
soft tissues are again closed in two layers, the muscles and fascia lata with
1-0 coated Vicryl or 1-0 Prolene and the skin with similar suture using a
well-knitted running stitch. Primapore dressing is then applied over the
whole incision wound.

It is important to procure tissues in a manner showing the utmost
respect for the deceased donor. Every cadaveric body must be treated with
respect. In the case of a deceased donor, we must take further into account
the good deed he/she has done by donating tissues. It is therefore very

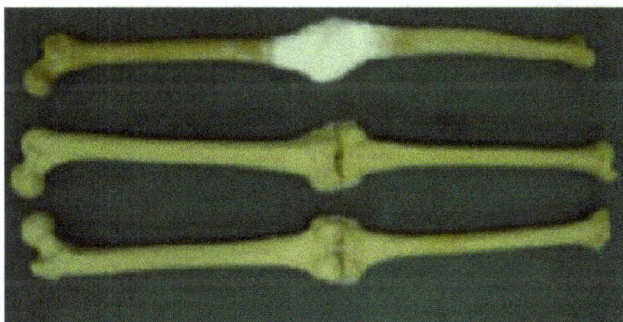

Fig. 23. Plastic femur and tibia arthrodesed at the knew with plaster of paris.

important to reconstruct the limb as best as can be done to give a cosmetically acceptable reconstructed upper limb to allow for open casket funerals.

At the end of the procurement, the leader of the team should thank the relatives who are usually waiting outside the operating room chanting prayers with the Buddhist monk whilst waiting to start funeral procedures. (So far all donors in Singapore are Buddhists.)

Documentation

All specimens are documented in separate deceased donor forms, that is, one deceased donor form for each specimen retrieved. All results (including the culture and sensitivity tests) are meticulously and accurately recorded in each form. A copy of the consent form is also filed for documentation.

In the case of deceased donors, detailed documentation must be conducted using the deceased donor form, which includes the following information:

- donor number,
- donor particulars (name, identification number, sex and age),
- date of death,
- cause of death,
- name of hospital/ward number,
- medical history,
- serological test results,
- details of tissue procurement (date, time, surgeons, residents and technologists),
- Description and size measurements of allograft and
- Status of tissues:
  - o awaiting quarantine,
  - o ready for use and
  - o to be discarded.

The status of the tissues must be decided by the director of the tissue bank.

Reference

Nather A (2001). Sterile procurement of bones and ligaments. Nather A (ed.), *The Scientific Basis of Tissue Transplantation*, World Scientific, pp. 270–290.

Chapter 20

Amnion Procurement

Aziz Nather* and Cui Lian Chong*

Introduction

Foetus is covered by two membranes. The outer membrane is the chorion whilst the inner membrane is the amnion (Fig. 1). The amnion wall is a tough, transparent, nerve-free and non-vascular membrane. It contains the amniotic fluid and has a stromal matrix, a thick collagen layer and an overlying basement membrane with a single layer of epithelium.

The membrane has unique characteristics, which include anti-adhesive effects, bacteriostatic properties and epithelialisation effects. It also demonstrates wound protection, pain reduction and exhibits a lack of immunogenicity.

Functionally, the amnion and its fluid protect the foetus from injury while allowing it to move freely during the later stages of pregnancy.

Procurement Procedure for Placenta

The donor's serum is first screened for anti-HIV1, anti-HIV2, syphilis, hepatitis B virus and hepatitis C virus. Only seronegative donors are accepted.

---

*NUH Tissue Bank, Department of Orthopaedic Surgery, Yong Loo Lin School of Medicine, National University of Singapore, Singapore.

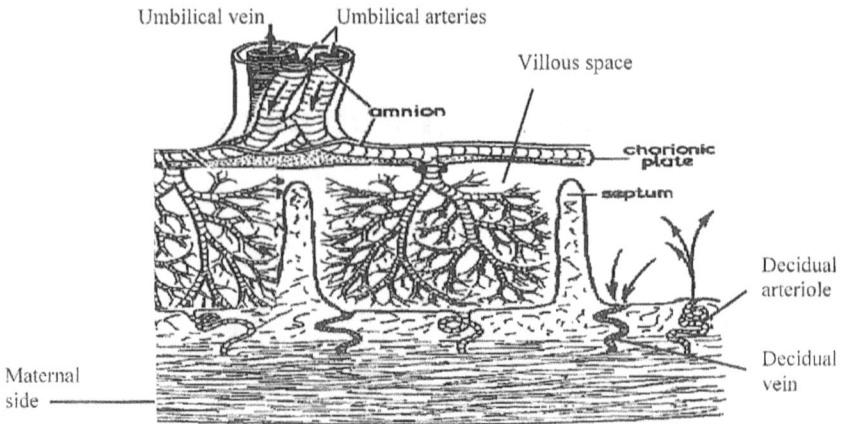

Fig. 1. Amnion (Mohamad, 2001).

Vaginal Delivery

Fresh placentae are procured from the delivery room under clean conditions after vaginal delivery. Meconium-stained placentae are disposed. The placenta is kept in a sterile basin or filled with normal saline solution. The placenta is flushed with normal saline to remove as much blood clots and debris as possible. The clean placenta is then transferred into a clean polyethylene bag containing one litre of normal saline with ampicillin and cloxacillin. This is transferred to the tissue bank as soon as possible for storage in 4°C freezer. Processing should be performed on the next day.

Caesarean Delivery

Some tissue banks prefer to procure the placenta under sterile conditions from pregnancies undergoing caesarean sections. In this situation, the technologist must be scrubbed, gowned and gloved to observe sterile technique. After sectioning the umbilical cord and the baby delivered, the placenta is separated carefully from the uterus and transferred to the technologist. The placenta is first flushed with normal saline to remove as much blood clots and debris as possible. The sterile placenta is then transferred to a sterile polyethylene container containing one litre of normal saline with 500 mg of ampicillin and 500 mg of cloxacillin.

To ensure that there is no leakage, this plastic bag is placed into a second sterile polyethylene bag. The specimen is then transferred to the tissue bank in the 4°C refrigerator as soon as possible.

Packaging

The amnion should be kept completely immersed in a sterile normal saline solution in a plastic container or two layers of plastic bags. It is then stored in a refrigerator until the final processing.

Reference

Mohamad H (2001). Anatomy and embryology of human placenta, amnion and chorion. Nather A (ed.), *The Scientific Basis of Tissue Transplantion*, World Scientific, p. 145.

Chapter 21

Skin Procurement and Processing

Aziz Nather*, Cui Lian Chong* and Xin Bei Chan*

Basic Anatomy of the Skin

Skin is the largest organ in the human body (Fig. 1). It serves as a protective shield, regulates body temperature and detects pain and heat whilst ensuring a balance of intra-environmental conditions. There will be a great risk for the body if this protective layer is injured, destroyed or lost. When skin injury is severe and extensive, such as in burn injuries, fatality often ensues.

Skin consists of the epidermis, dermis and subcutaneous tissue. The epidermis is the outermost layer and is about 1.5 mm on the palms and soles. The dermis comprises of three types of tissues, namely collagen, elastic tissue and reticular fibres. The bottommost layer is the subcutaneous tissue which is primarily made up of fat and connective tissue.

Each layer serves a different function. The epidermis is impermeable to the entry of microorganisms while the pigment cells at the dermal–epidermal junction (melanocytes) afford protection from the harmful effects of the sun and the reticular deep dermal layer is responsible for thickened keloidal/hypertrophic scars in some individuals.

Skin Procurement System

Once a suitable donor is found, the national transplant coordinator activates the various procurement teams across Singapore. The skin team

*NUH Tissue Bank, Department of Orthopaedic Surgery, Yong Loo Lin School of Medicine, National University of Singapore, Singapore.

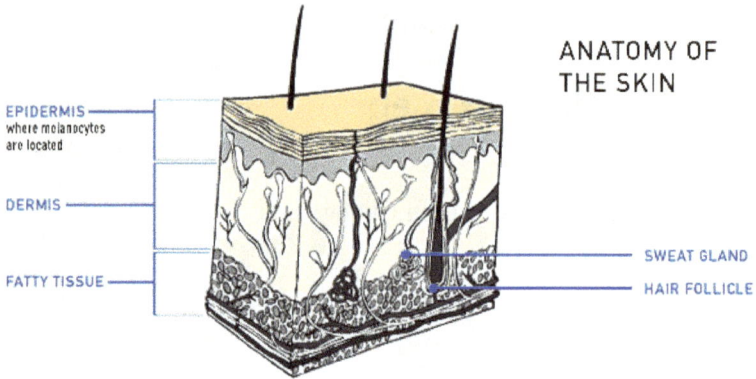

Fig. 1. Skin.

Source: http://www.antigenics.com/diseases/melanoma.html.

is contacted after the kidney, heart, liver and cornea teams. Skin donation is primarily carried out by the Singapore General Hospital (SGH) Skin Bank located in the SGH Burn Centre of the Department of Plastic Surgery. Once a suitable donor has been found, the SGH Burns Centre will be activated.

The skin procurement team consists of two duty doctors (a registrar and a medical officer), a skin bank officer and a staff nurse. For multi-organ donation cases, procurement takes place in the sterile operating theatre. In the event of tissue (skin and cornea) donation only, a clean room within the SGH Burn Centre is available for skin harvesting. For such cases, skin preparation, back table set-up and draping are conducted in an aseptic manner, adhering to OT requirements (Chua *et al.*, 2004).

Skin Procurement Procedure

Upon passing the relevant immunological and serological tests, the transplant coordinator will indicate to the procurement team that the donor has met the necessary criteria for harvesting. The procurement team has standard equipment, which is ready for use at any one time. Disposable materials will be utilised where possible to minimise contamination and to aid in expediency.

Skin procurement from a deceased donor is restricted mainly to the back, buttocks and the lower limbs. This is to ensure that concerns over "raw" appearance on donor's exposed areas will not be an issue for the donor family in an open casket funeral.

Paraffin oil (Fig. 2) is first applied to lubricate the area and the skin retrieval is then conducted using an electric dermatome (Fig. 3). It is important to apply constant pressure and to move down the surface of the skin with a consistent speed. Normally, skin of 0.38 mm in depth is procured.

The procured skin (Fig. 4) with an average area of 100 cm^2 is placed in a 1000-ml sterile container filled with Dulbecco's modified eagle medium (DMEM) as the transport medium. An antibiotic/fungicide mixture composing of penicillin (50 IU/ml), streptomycin (50 μg/ml) and amphotericin B (0.125 μg/ml) is also added into the container.

During procurement, a random skin sample of about 0.5 cm^2 is isolated in a small sterile bottle for three standard microbiology tests — aerobic, anaerobic and fungal cultures. A blood sample from the donor is tested for HIV, hepatitis B, hepatitis C and syphilis.

Procured skin tissues are transported to the skin bank in a container filled with ice. An average of 2000 cm^2 is obtained from each deceased donor.

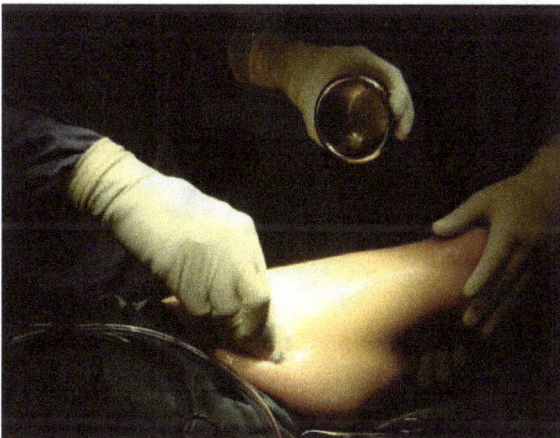

Fig. 2. Application of paraffin oil.

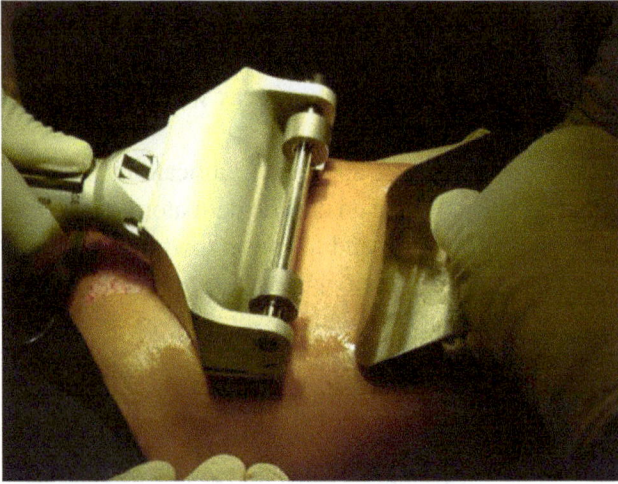

Fig. 3. Procuring skin with an electric dermatome.

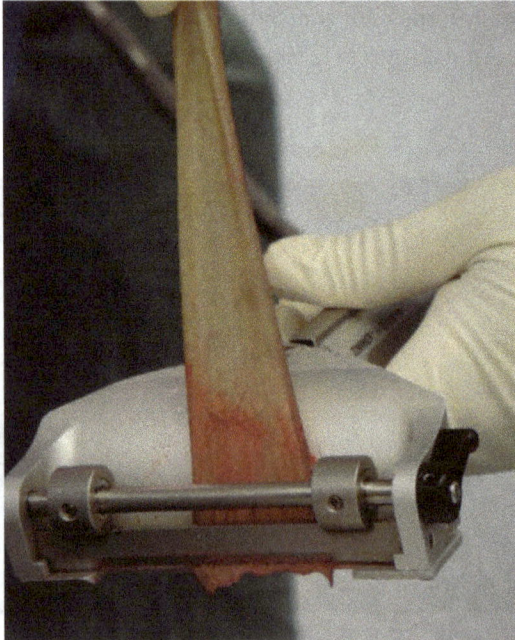

Fig. 4. Procured skin.

Skin Reconstruction

The raw defect is then covered with tulle-gras dressing followed by gauze dressing. Skin is not usually obtained from areas that will not be covered by clothes.

Skin Processing

The procured skin is rinsed with a mixture of 0.025% sodium hypochlorite and phosphate buffer saline to remove excess paraffin oil and dead skin cells. A skin sample is isolated for microbiology tests.

The skin tissue is divided into sizes of 200 cm^2 and placed into 150-ml sterile containers filled with DMEM transport medium, antibiotics and fungicide. Labels indicating the donor's identity, date and size of tissue are placed on each container. The containers are stored at 4°C for 10 days, with the medium changed every three days. On the 10th day, the fresh skin tissues are meshed, double packed and cryopreserved. They are trimmed at the edges and meshed with a Brennan mesher in a 2:1 ratio (Fig. 5).

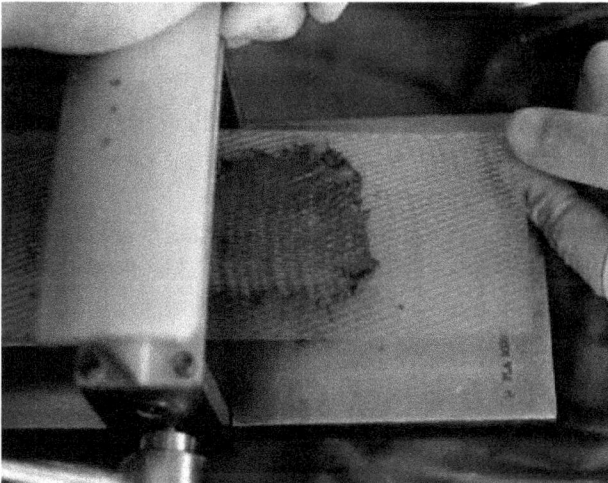

Fig. 5. Skin perforated by meshing machine.

Source: http://www.airahospital.org/?page_id=576.

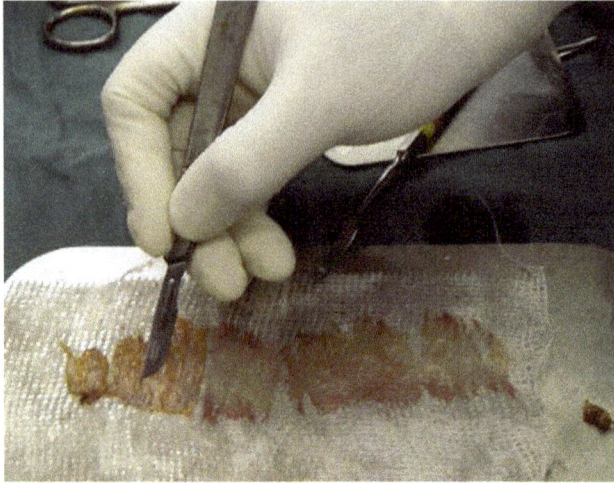

Fig. 6. Spreading skin onto cotton bandage.

Tissues may be used to apply to burn wounds before ten days are up, if microbiology tests are negative three days after rinsing and serology tests are also negative. Consent from the surgeon-in-charge is required.

The skin sheets are spread onto a 4-inch cotton bandage with the dermis facing upwards (Fig. 6). The cotton bandage is folded such that it can fit into a 6 inch × 8 inch plastic pouch. Each cotton bandage can accommodate up to four skin sheets. Skin samples are taken for microbiology tests again.

The cotton bandages containing the skin sheets are soaked in DMEM with 10% dimethyl sulfoxide for 20 minutes. This allows cryoprotectants to penetrate the skin tissues. Excess medium is then drained from the tissues. The tissues are double-packed and placed into the plastic pouch. The plastic pouch is placed in a larger foil pouch (8 inch × 10 inch) using a medical-grade heat sealer.

The pouches are placed in freezer programmed for uniform freezing of skin tissues at a controlled rate of −1°C to −5°C per minute. At −100°C, the pouches are transferred to a −150°C CFC-free ultralow freezer, where they can be stored for up to five years.

Before skin allografts are used, they are thawed and rinsed with warm saline at 37°C to remove any cryoprotectant present.

Reference

Chua A, Song C, Chai A, Chan L, and Tan KC (2004). The impact of skin banking and the use of its cadaveric skin allografts for severe burn victims in Singapore, *Burns* **30**: 696–700.

Part VI

Processing

Chapter 22

Processing of Bone and Musculoskeletal Soft Tissue Allografts

Aziz Nather* and Li Min Tay*

Introduction

All bones procured from living and deceased donors must be processed by methods that have been validated to ensure tissue quality, safety and efficacy for safe transplantation into recipients. Processing of bones is a very important function of every tissue bank. Bones can be processed as:

- deep-frozen and gamma-irradiated bone allografts,
- cryopreserved and non-irradiated osteoarticular bone allografts,
- freeze-dried (lyophilised) and gamma-irradiated bone allografts and
- demineralised and gamma-irradiated bone allografts.

Type of Allografts Processed

The type of processed bone allografts produced must be tailored to the needs and demands of orthopaedic and maxillo-facial surgeons who use these grafts. The range varies from massive deep-frozen and gamma-irradiated long-bone allografts and cryopreserved and non-irradiated long-bone allografts for large defect structural reconstruction to small or

*NUH Tissue Bank, Department of Orthopaedic Surgery, Yong Loo Lin School of Medicine, National University of Singapore, Singapore.

morsellised (powder, granules, chips, cubes, etc.) bone grafts for packing small bone defects in orthopaedic and maxillo-facial surgery.

The range of products processed by the North West Tissue Centre in Seattle, USA (Figs. 1 and 2) is shown as an example of the range of products that is processed by other tissue banks.

Processing by Deep Freezing

Bone allografts processed by deep freezing come from two sources:

- living donors and
- deceased donors.

Bone allografts from living donors include mainly:

- femoral heads from hemi-arthroplasty following displaced (Garden's Grade III and IV) fracture neck of femur in the elderly,
- femoral heads from total hip replacement in patients with osteoarthritis of the hip and
- resected bone slices from total knee replacement in patients with osteoarthritis of the knee.

Bone allografts from deceased donors include:

- cortical bones (long bones):
  - femur, tibia, and fibula
  - humerus, radius, and ulna
- Osteo-articular bones:
  - proximal half femur and distal half femur
  - proximal half tibia and distal half tibia
- Corticocancellous bones:
  - iliac crest

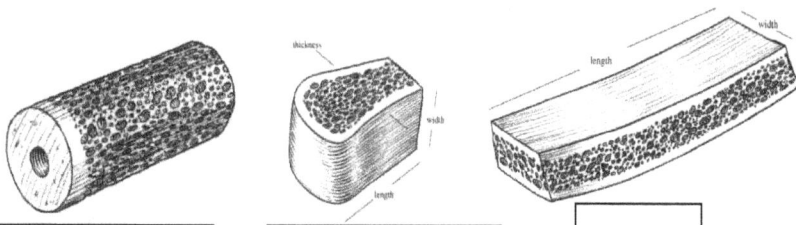

| Cloward dowel | Tricortical ilium block | Ilium strip |

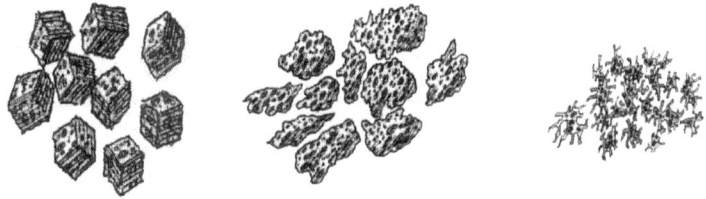

Cancellous chips (cubic, rongeured and fine)

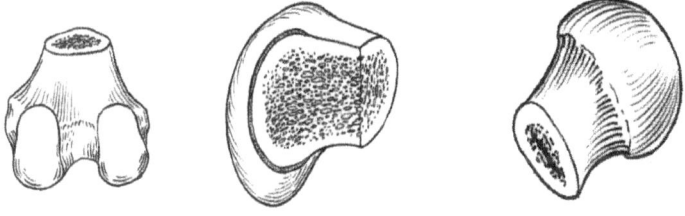

| Femoral condyle, whole and hemi | Femoral head |

| Cancellous block | Cortical powder |

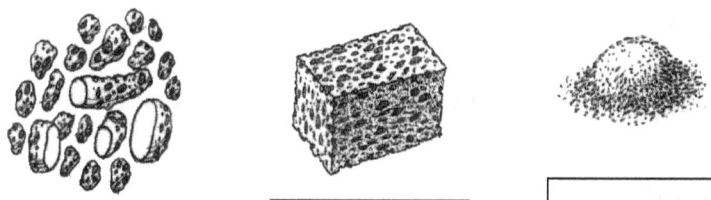

Cortical-cancellous chips

Fig. 1.   Catalogue of bone grafts processed in North West Tissue Centre.

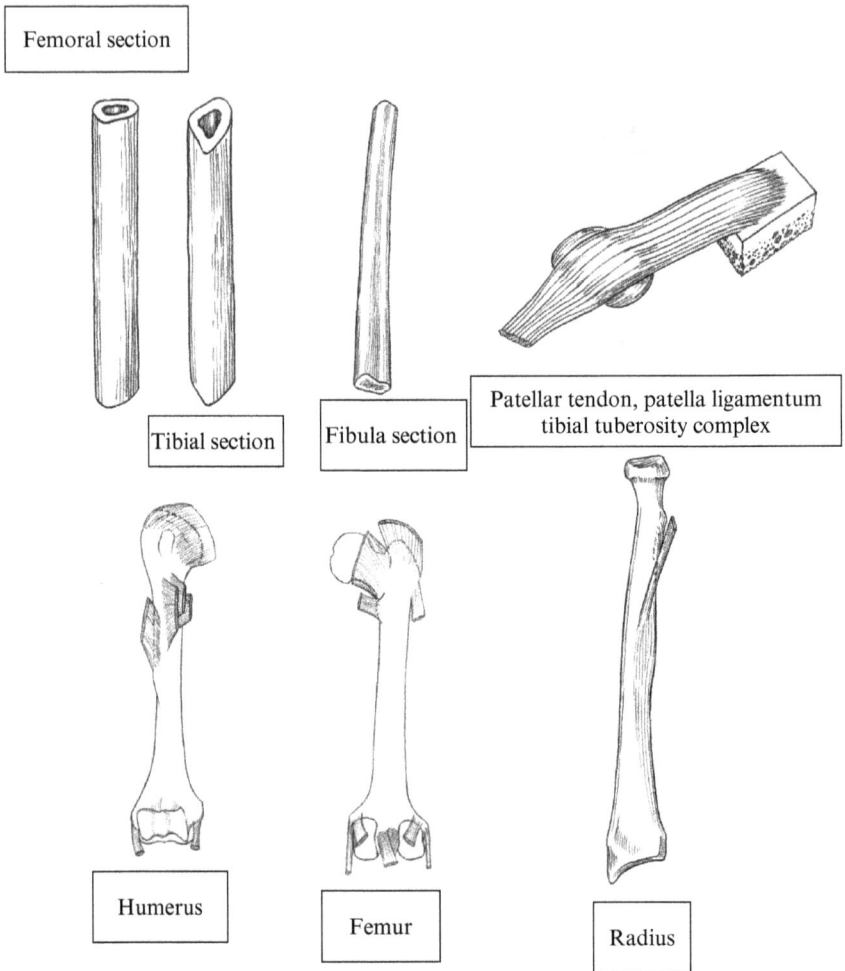

Fig. 2. Catalogue of bone grafts processed in North West Tissue Centre.

Triple Wraps

The long bones are sterile procured and packed in "triple wraps" (Fig. 3) for storage in the deep freezer at −80°C (electrical freezer). The triple wraps are:

• inner polyethylene layer,

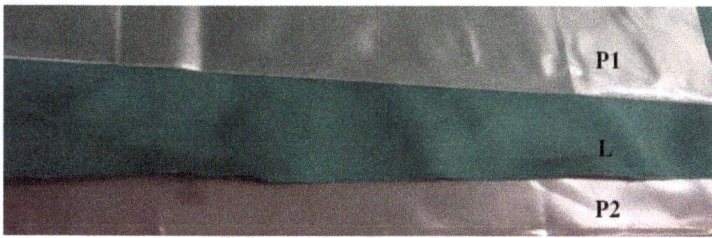

Fig. 3. Triple wrap technique: inner polyethylene layer (P1), middle linen (M) and outer polyethylene layer (P2).

- middle linen and
- outer polyethylene layer.

For packing and storage of osteoarticular grafts, "triple wrap technique" is used. However, the articular ends of the bones are additionally processed with 10% glycerol by wrapping gauze containing 10% glycerol over the articular ends before it is wrapped with the first polyethylene layer. The triple wrap is stored by cryopreservation at −160°C in a liquid nitrogen cryofreezer.

Double Jars

The smaller bones such as femoral heads and resected total knee replacement slices (living donors) and pieces of iliac crest (deceased donors) are also sterile procured and packed in sterile double jars (Fig. 4).

- Specimen received in sterile polyethylene jar (autoclavable), which is now placed into an outer polyethylene jar container (autoclavable).

Deep-Freezing Processing

Both triple wraps and double jars are placed in the quarantine freezer in the wet-processing laboratory of the tissue bank whilst awaiting the results of the laboratory tests. Once all tests are negative, they are transferred to the "ready-for-use freezer".

These tissues must be processed by deep freezing for at least two weeks to reduce significantly the antigenicity of the allograft. Cells in both soft tissues and bones die since they do not have blood supply. As the cells undergo necrosis — the process takes at least two weeks before the cells disappear completely on macroscopic examination (Catto, 1965) — the antigenicity reduces significantly. To ensure that this process is completed and double secured, the deep-freezing process must be carried out for at least one month before the tissue grafts are allowed to be released for transplantation.

Radiography of Long Bones

X-rays of the long bones (Fig. 5) are performed using a calibrated ruler to:

- give a measurement of the dimensions of the bone specimen produced,
- confirm the nature of the long bone specimen:

  o femur or tibia,

Fig. 4. Sterile double jar technique.

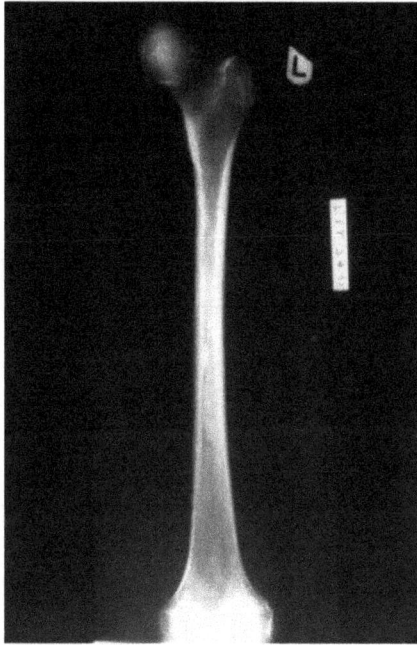

Fig. 5.   X-ray of left femur.

○   proximal half or distal half.

•   confirm the side of the long bone specimen:

○   left or right.

End-Sterilisation of Bones

During the deep-freezing period, the bones are packed into polylite container packed in dry ice (for 5 kg of bone, put 5 kg of dry ice) and sent to a cobalt-60 plant for gamma irradiation at a dosage of 25 kGy.

Figure 6 shows the flowchart of activities involved in the deep-freezing process.

Freeze-Drying Processing for Bones

There are various methods used for freeze-drying by individual tissue banks. A well tried and -tested method developed by Professor Frank Dexter

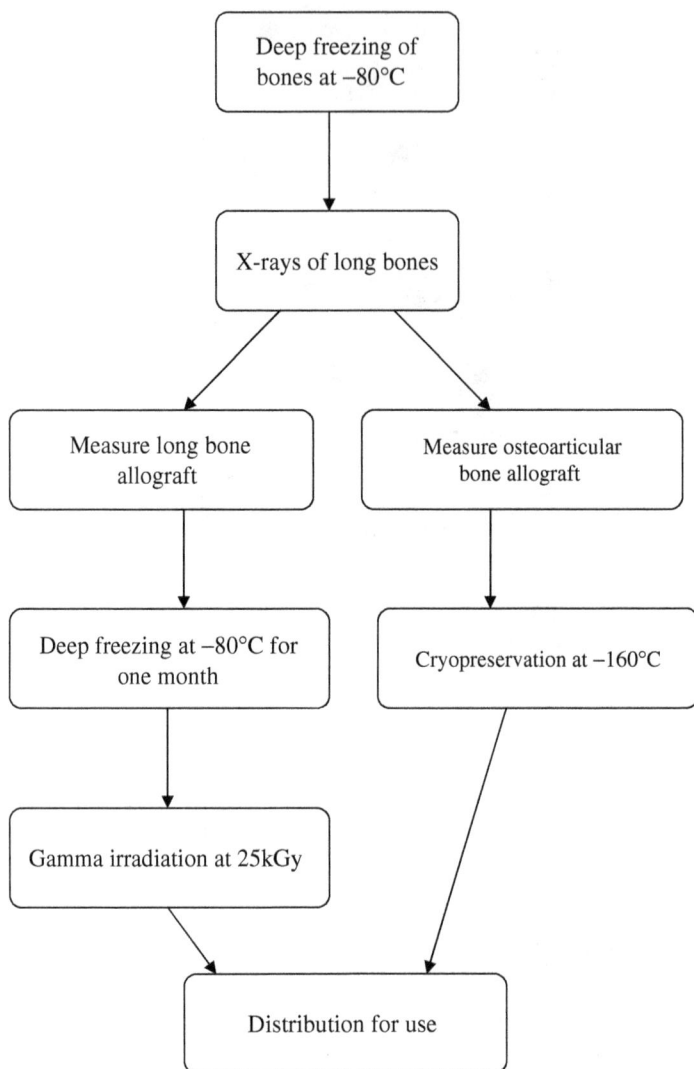

Fig. 6.   Flowchart of deep-freezing process for bones.

at the Yorkshire Regional Tissue Bank has been extensively utilised in the Asia-Pacific region.

Figure 7 demonstrates the stages of the Dexter technique for processing freeze-dried and gamma-irradiated bone allografts.

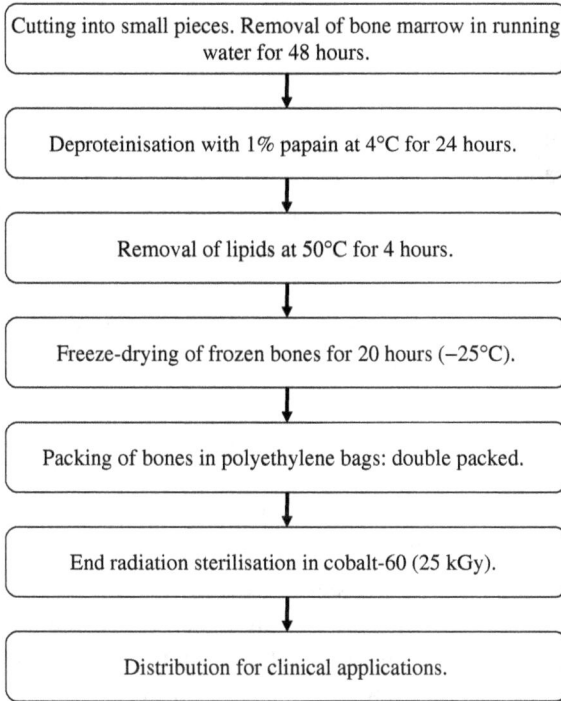

Fig. 7.   Dexter technique for processing freeze-dried bone.

Dexter technique as adopted by the Tata Bone Bank in Tata Memorial Hospital, Bombay, India is as follows:

1. Collect bone in sterile saline in a sterile bin.
2. Place for 1–3 hours in 70% ethanol to destroy HIV and bacteria.
3. Wash in Aqua-Guard filtered water.
4. Remove soft tissues. Cut bone into smaller pieces of desired size and shape. Drill holes/slits into large bones. Label bones (N). Soft tissues must be collected in a separate bag for incineration.
5. Place bone in 2.5% Na_3PO_4 solution. This solution is an anticoagulant and helps to disperse the blood in the bones. The jar containing the bones in Na_3PO_4 is kept in a refrigerator overnight. The low temperature slows down the decay of the tissues.

6. The bone is washed with a powerful water jet to remove as much bone marrow as possible.

7. The bones are placed overnight in 1% papain in the refrigerator. Papain digests proteins and facilitates removal of soft tissues.

8. Bones are placed in fresh 1% papain in flasks and agitated in a water bath shaker at 58°C for 1–3 hours. If surface grease appears, a fresh source of heated papain is introduced, allowing the grease to over-spill. This procedure is repeated until no grease is visible.

9. Residual papain is reduced with a warm-water wash. Rinse finally under filtered tap water.

10. Remove soft tissues with sterile instruments.

11. In case of presence of residual blood or marrow particles, keep the bones overnight in refrigerator in 2.5% Na_3PO_4 solution.

12. Clean with forceful water jet and wash with filter water. At this stage, the grafts should look absolutely white.

13. The grafts are washed thoroughly with distilled water until they are free of Na_3PO_4. Washed samples are tested for residual Na_3PO_4.
    Test: ammonium molybdate + dilute HNO_3 + sample give a yellow-ish white precipitate in the presence of Na_3PO_4.

14. Grafts are placed in freeze-drier trays and stored at −60°C in freezer.

15. Frozen grafts are freeze-dried for 48–72 hours.

16. For bone powder, grind freeze-dried cortical bone and sieve to a particle size of approximately 250–750 microns.

17. Grafts are double packed in polyethylene sleeves and sealed in a laminar flow cabinet. Labels bearing details of the grafts are affixed. Bone powder is double packed in quantities of 0.5 gm.

18. Grafts are sterilised by gamma radiation of 25 kGy (2.5 Mrad).

19. Samples from each batch are sent for microbiological culture.

20. Grafts are quarantined until culture reports are available.

21. Cortical bone required reconstitution in sterile saline/water for 18–24 hours at 4°C.

22. Crushed cortical/cancellous bone does not require prior reconstitu-tion. If desired, antibiotics can be added during reconstitution (e.g. chloromycetin (1 gm) in water (50 cm^3)).

The Dexter procedure produces a bone allograft with the following aims:

- To produce a biological scaffold which is essentially a collagen framework containing hydroxyapatite crystals which promotes osteo-conduction facilitating the ingrowth of new vessels and cells from the adjacent host bone.
- To reduce the antigenicity of the bone allograft.
- To reduce the microbial load of the bone allograft.
- To promote better accessibility to ethylene oxide gas (when this is used for sterilisation of the graft).

Freeze-Drying Process: NUH Tissue Bank Protocol

The Frank Dexter or other freeze-drying process has been shown to significantly reduce the biomechanical strength of freeze-dried bone (Nather *et al.*, 2004).

Currently, the processing of freeze-dried bones is best done in two stages.

1. First stage in a wet-processing laboratory.
2. Second stage in a dry-processing laboratory.

Wet Processing

The first part of freeze-drying process is performed in a wet-processing laboratory (Fig. 8).

The equipment found in the wet-processing laboratory includes:

- electrical deep freezer (−80°C),
- cryofreezer (−160°C to −180°C liquid nitrogen freezer),
- band saw,
- chemical processing (70% ethanol, 0.5% sodium hypochlorite and 0.5 M hydrochloric acid) and
- shaker bath.

The activities in wet processing are shown in Fig. 9.

Fig. 8. Wet-processing laboratory of NUH Tissue Bank.

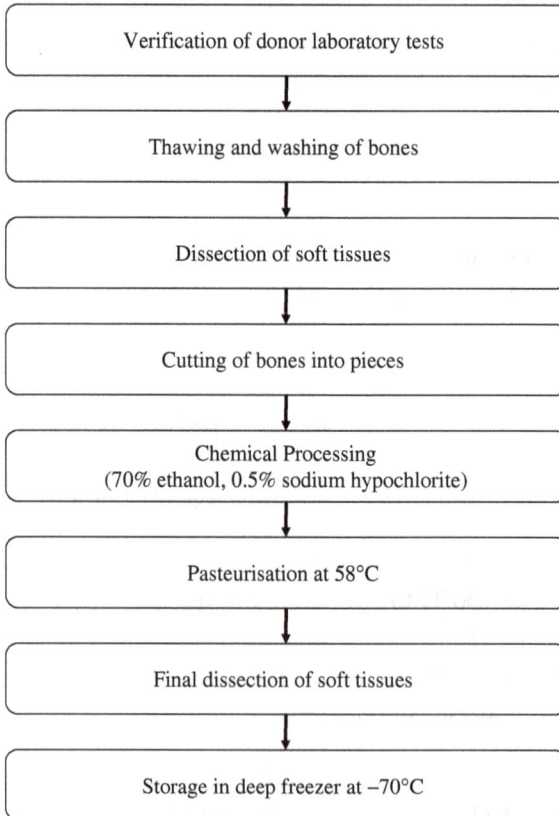

Verification of donor laboratory tests

Thawing and washing of bones

Dissection of soft tissues

Cutting of bones into pieces

Chemical Processing
(70% ethanol, 0.5% sodium hypochlorite)

Pasteurisation at 58°C

Final dissection of soft tissues

Storage in deep freezer at −70°C

Fig. 9. Flowchart of wet processing.

Verification of Donor Bones

The laboratory tests performed for living and deceased donors must first be verified to be negative before processing of bones can begin. These tests include:

- anti-HIV$_1$ and anti- HIV$_2$,
- HBsAg,
- anti-HCV,
- RPR (if positive TPHA) and
- culture and sensitivity for aerobic and anaerobic organisms.

Washing of Bones

The bones from a suitable donor is removed from the deep freezer (−80°C) and allowed to thaw and washed under a tap using a water jet to remove all blood and debris still left behind on the bone from the time it was produced.

Dissection of Soft Tissue

The bone is placed on a clean tray and using a scalpel and tissue forceps, all soft tissues are meticulously dissected off the bone specimen. A periosteal elevator is used to strip off all the periosteum from the bone since these soft tissues are the most antigenic part of the bone allograft. It is likewise important to remove all blood from the bone for the same reason and also because it is through the blood that disease transmission from bone allografts occur in recipients.

Cutting of Bones

Only authorised personnel are allowed to use a band saw (Fig. 10). The operator must wear a disposable gown, cap, mask, protective eye glasses and a pair of gloves (no need sterile) before operating band saw. The gasket of the band saw is first opened and the whole band saw and disposable blade are cleaned and disinfected using 70% ethanol before the saw can be used.

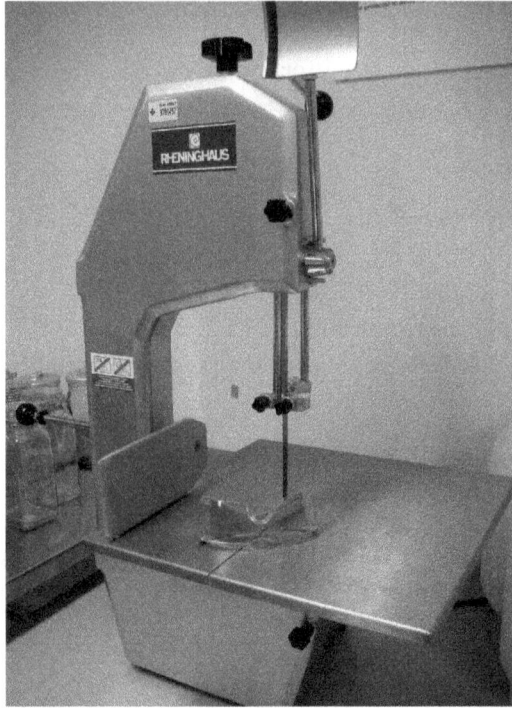

Fig. 10. Band saw.

The band saw is switched on and the bone is cut into appropriate sizes and shapes according to the requirements and collected in a clean glass jar. Waste tissue and debris generated during the cutting are disposed immediately into a biohazard waste bag and the bag tied and removed for incineration by the hospital. Immediately after the cutting process is completed, the band saw is again thoroughly cleaned and disinfected with 70% ethanol.

Washing of Bone Pieces

The bone pieces are washed thoroughly with water using a jet lavage to remove as much blood and other debris present in the bone as possible. The bone in the clean, labelled bottle is kept in the deep freezer until the next step is performed.

Fig. 11. Chemical processing.

Chemical Processing

In the next stage, the bones are placed in a clean bottle containing 70% ethanol (Fig. 11) for at least three hours.

In other tissue banks, chemical processing is performed in the form of treatment with 0.5% sodium hypochlorite for about three hours.

Both 70% ethanol and 0.5% sodium hypochlorite has been shown to be effective in inactivating the HIV.

Upon completion of chemical processing, the chemical solutions are removed and the bones washed with normal saline and stored in a clean, labelled bottle for storage in the deep freezer (−80°C) until the next step is performed.

Pasteurisation

For pasteurisation, the bone pieces are immersed in sterile water (pre-heated to 58°C) in a shaker bath (Fig. 12) containing water pre-heated to 58°C. Pasteurisation is performed for three hours.

The sterile water in the bottle is changed every 50 minutes with pre-heated sterile distilled water (58°C) and the temperature of the water in the shaker bath monitored to maintain temperature of 58°C.

Fig. 12.   Shaker bath.

Final Dissection of Soft Tissues

At the end of three hours, final dissection is performed to remove all soft tissues and periosteum from the pasteurised bone graft pieces. It is found that after pasteurisation, the soft tissues and periosteum stand out from the bone and is easily separated by dissection from the bone. The bone pieces are stored in labelled glass bottles in the deep freezer (−80°C) before dry processing can begin.

Documentation

All the steps are recorded in the wet-processing form, with the date and time performed for each step recorded by the technician, signed and dated.

Dry Processing

The final stage of freeze-drying is performed in a dry-processing laboratory (Fig. 13).

The equipment found in the dry-processing laboratory includes:

- freeze dryer (lyophiliser),
- laminar air-flow cabinet,
- vacuum sealer for sealing polyethylene bags,

Fig. 13. Dry-processing laboratory of NUH Tissue Bank.

- electronic balance and
- oven.

Lyophilisation

A freeze-drying chamber is used for lyophilisation. The chamber is cleaned with 70% ethanol and pre-cooled to −40°C. Note that no pooling of bones from different donors is allowed. Bones from one donor are placed in a tray and recorded diagrammatically. For long specimens, pieces of the specimens are placed in all trays of the chamber and recorded diagrammatically.

Freeze-Drying

The batch number (lyophilisation cycle number, e.g. Batch 306: Cycle on 8.8.08), donor number (tray number: 888 Tray 1; 222 Tray 5, etc.) and graft number (position of graft in each tray, e.g. 12 o'clock Graft No. 1; 9 o'clock Graft No. 4) need to be recorded first. The duration of each freeze-drying cycle varies depends on bone load present. For example,

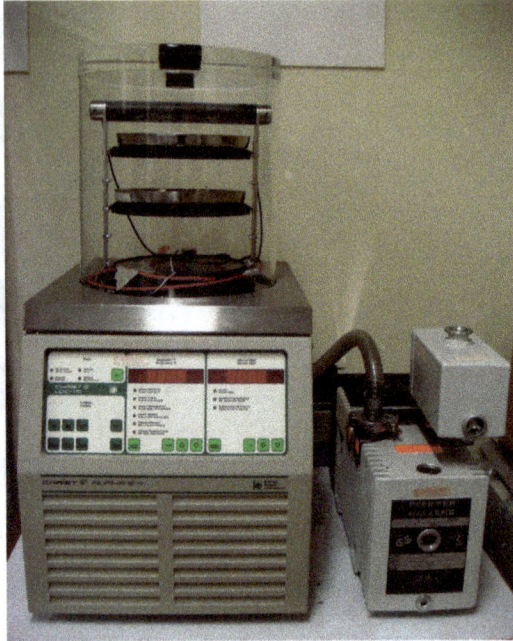

Fig. 14.　Lyophiliser.

a cycle for femoral head pieces takes 1–3 days, while a cycle for one long bone takes 5–7 days to complete.

The chamber of the lyophiliser (Fig. 14) must be disinfected with 70% ethanol after each process.

Quality Control Test

Water content test

Using a small piece of bone from each batch, a water content test is performed for quality control. The test sample must be weighed using an electronic balance (Fig. 15) after lyophilisation (W1). It is then placed in an oven at 60°C for at least three hours or more until constant weight is achieved (W2), i.e. all residual water being removed by the oven.

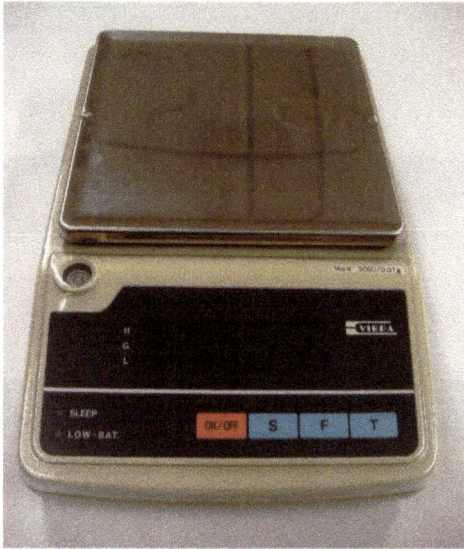

Fig. 15. Electronic balance.

$$\text{Water content} = \frac{\text{W1} - \text{W2}}{\text{W1}} \times 100\%$$

Desired water content is 5–8%.

Packaging, Sealing and Labelling

The laminar air-flow cabinet (Fig. 16) must first be cleaned with 70% ethanol. The blower and UV light are switched on an hour before use. The UV light is then switched off after packaging, sealing and labeling.

The bone graft is inserted into inner polyethylene bag and vacuum sealed (Fig. 17) — 80% vacuum for corticocancellous bones and 90% vacuum for cortical bones. Absolute or 100% vacuum is never used. The bone edges are too sharp for this and breach of polyethylene bag might occur. The label is applied on the outer surface of the inner bag. The Go–No-Go radiation indicator is applied here as well. It should change colour from brown to red. The outer polyethylene bag is then sealed as a final step.

Fig. 16. Laminar air-flow cabinet.

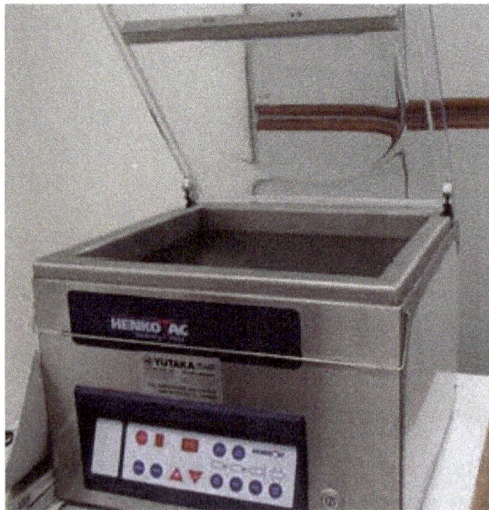

Fig. 17. Vacuum sealer.

Documentation

Each step is recorded in detail in the dry-processing form — date, time and findings.

The activities involved in dry processing are shown in Fig. 18.

```
┌─────────────────────────────────────────────┐
│            Freeze-drying of bones            │
└─────────────────────────────────────────────┘
                       │
                       ▼
┌─────────────────────────────────────────────┐
│        Performing a quality control test     │
└─────────────────────────────────────────────┘
                       │
                       ▼
┌─────────────────────────────────────────────┐
│  Packaging and labelling under laminar air-flow cabinet  │
└─────────────────────────────────────────────┘
                       │
                       ▼
┌─────────────────────────────────────────────┐
│ Vacuum sealing with double packing of polyethylene bags │
└─────────────────────────────────────────────┘
                       │
                       ▼
┌─────────────────────────────────────────────┐
│         Radiation sterilisation of products  │
└─────────────────────────────────────────────┘
                       │
                       ▼
┌─────────────────────────────────────────────┐
│   Documentation and storage at room temperature │
└─────────────────────────────────────────────┘
                       │
                       ▼
┌─────────────────────────────────────────────┐
│       Distribution for clinical applications │
└─────────────────────────────────────────────┘
```

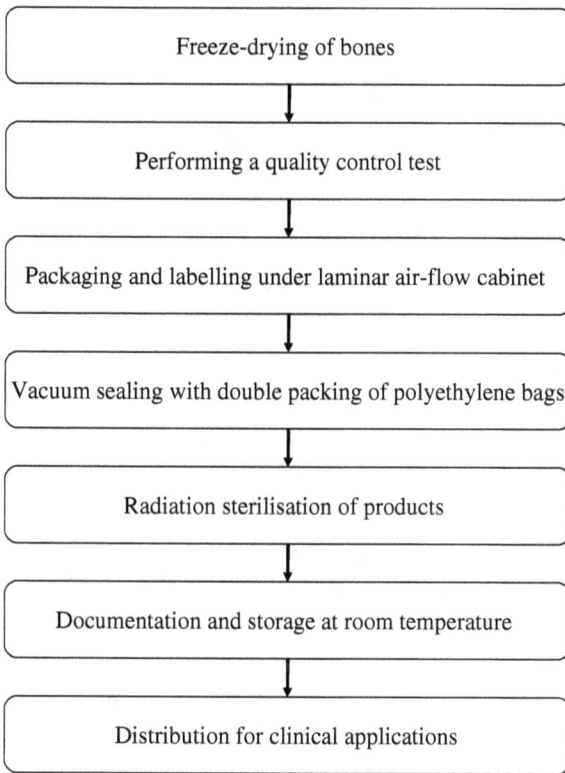

Fig. 18.   Flowchart of dry processing.

Processing of Demineralised Bone

Processing Tools and Equipment

In addition to the usual equipment, the following are required for the frozen bone allograft:

- Bone mill (manual or automatic); it should have the capability of containing liquid nitrogen or water during milling process.
- Bone sieve (standardised); optional for the powder form diameter: 80–300 micron, 300–425 micron, 425–600 micron and 600–1000 micron.
- 0.5 M or 0.6 M HCl.

- PBS (phosphate buffer solution).
- Centrifuge (over 1500 rpm).
- Magnetic stirrer (optional).
- Chloroform and/or methanol (optional).

Steps and Methods of Processing

Bones are washed with distilled water before the dissection of soft tissues. Milling of bones is then carried out with bone mill. The bones are then freeze-dried and sieved to produce different size granules.

Demineralisation takes place with 0.5 M hydrochloric acid (250 ml/5 g for 90 minutes, centrifuge at >1500 rpm for five minutes and repeated three times, pH adjusted to 6.9 using phosphate buffer solution).

During the demineralisation process, the pH of the acid solution has to be controlled because the mineral component of bone (primarily hydroxyapatite) is soluble only at a pH value below 4 (Hilmy *et al.*, 2007).

It is then freeze-dried again, packaged, sealed, labelled and radiation sterilised.

At the end of the process, some bones are taken for validation. The demineralisation process is stopped if the calcium content is 8% or less (AATB, 2002; Basril *et al.*, 2006).

The calcium content is determined via the titrimetric assay method (Hilmy *et al.*, 2007) and the X-ray diffraction method (Basril *et al.*, 2000).

Sieving or Shaping of Each Bone During Processing

When sieving the bones, sieves of various sizes in diameter can be used. When the diameter of bone particle is smaller than 1000–3000 micron, it is called "chip" and if it is larger than 3000 micron, it is called "block". This classification is useful because "chips" and "block" are used in different situations.

Benefits of the Use of Demineralised Bone Allografts

Demineralised bone allograft is claimed by some to induce new bone formation than deep-frozen or freeze-dried allografts. However, this depends on the quality of the demineralised bone allografts produced.

Processing of Musculoskeletal Soft Tissue Allografts

Introduction

Soft tissue allografts are processed only from deceased donors. They include:

- fascia lata,
- patella-tendon ligament tibial tuberosity complex (Fig. 19),
- medial/lateral meniscus,
- tibialis anterior/tibialis posterior tendons, and
- tendo-Achilles calcaneal tuberosity complex.

Deep-Freezing Process

All soft tissues are processed by deep freezing (Fig. 20). Each soft tissue allograft is stored in a sterile double jar which is stored in the electrical freezer at −80°C for at least one month before use.

No Gamma Irradiation

Currently, these grafts are not gamma irradiated before use for fear of weakening the collagen structure of these grafts and thereby significantly reducing the biomechanical strength of tendons and ligaments.

Need for Gamma Irradiation

A 23-year-old man in Minnesota underwent reconstructive knee surgery in November 2001. He died three days after receiving the fresh femoral condyle bone-cartilage allograft. His pre-mortem blood culture revealed the presence of *Clostridium sordellii*.

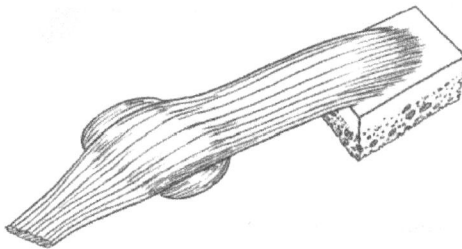

Fig. 19. Patella-tendon ligament tibial tuberosity complex.

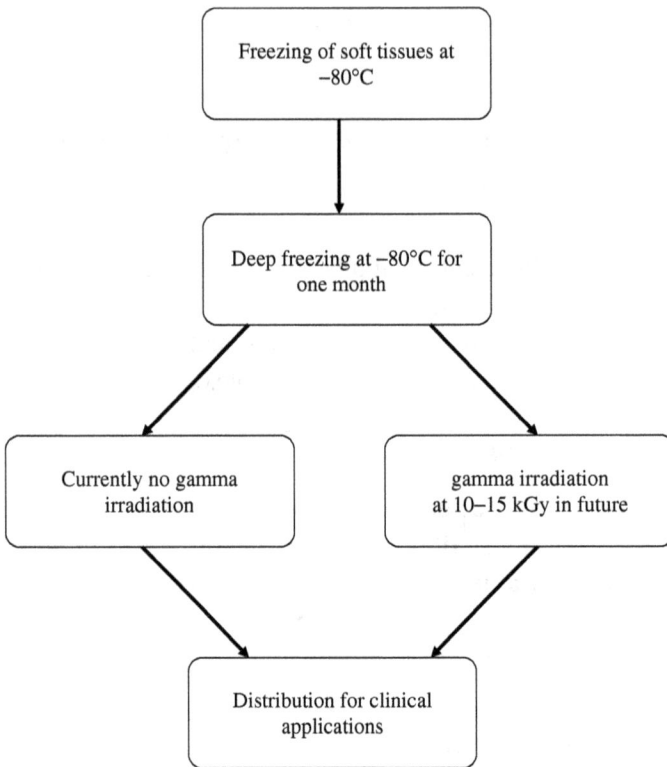

Fig. 20.   Flowchart of deep-freezing process for soft tissue allografts.

Ten tissues from the donor, including the case above, were transplanted into nine patients in eight states. Three resulted in infection. The Centre for Disease Control (CDC) of United States tested 19 non-implanted tissues from the same donor. *Clostridium sordellii* infection occurred in two tissues, one involving fresh femoral condyle and the other a frozen meniscus. This result confirmed that the allografts were the source of infection (Archibald *et al.*, 2002).

The following has been recommended by the CDC (Archibald *et al.*, 2002):

- Tissues should be processed using a method, e.g. gamma irradiation, that can kill bacterial spores. Otherwise, aseptically processed tissues cannot be considered sterile.

- Allograft tissues should be cultured before suspension in anti-microbial solutions. If *Clostridium* or other bowel flora is present, the tissues should not be used.

Australia has started gamma irradiation of all bones and soft tissues at a dose of 10 KGy.

Two questions remain:

1. What is the effect of 10 kGy on biomechanical strength of tendons and ligaments? More research is needed to answer this question. At the moment, the ligaments:

   - turn yellow and
   - have a smoky smell.

   Surgeons may not like to use them and fear that these grafts may not be "good". Professional education is needed to allay fears that grafts are not good and to accept these changes. Even if the grafts are significantly weaker compared to those without irradiation, they should be preferred since they provide better safety than stronger, non-irradiated grafts.

2. Such grafts are not protected from HCV. HCV is killed at 17 kGy (Conrad *et al.*, 1995).

   We then have to rely on donor screening for protection against HCV transmission. The problem is the window period for HCV is 12 days with nucleic acid testing (NAT).

   Likewise where HIV is concerned, our protection is based on donor/screening again with a window period of 12 days with NAT testing.

References

American Association of Tissue Banks (AATB) (2002). *Standards for Tissue Banking*, McLean, VA.

Archibald LK, Jernigan DB, and Kainer MA (Centers for Disease Control and Prevention [CDC]) (2002). Update: allograft-associated bacterial infections — United States, 2002. *MMWR Morb Mortal Wkly Rep* **51**: 207–210.

Basril A, Febrida A, and Nami S (2006). *Validation of Ca Content and Residual Moisture Content of Radiation Sterilization of Demineralized Bone Powder Allografts*, BATAN Research Tissue Bank, BATAN, Jakarta.

Basril A, Febrida A, Surtipanti S, *et al.* (2000). The effects of gamma irradiation and demineralization process on mechanical properties of lyophilized bovine bone. *Proceedings of the 8th APASTB International Conference on Tissue Banking*, Bali, Indonesia, p. 44.

Catto M (1965). A histological study of avascular necrosis of the femoral head after transcervical fracture. *J Bone Joint Surg Br* **47**: 749–776.

Conrad EU, Gretch DR, Obermeyer KR, *et al.* (1995). Transmission of the hepatitis-C virus by tissue transplantation. *J Bone Joint Surg Am* **77**(2): 214–224.

Hilmy N, Abbas B, and Anas F (2007). Validation for processing and irradiation of freeze-dried bone grafts. In Nather A, Yusof N, and Hilmy N (eds.), *Radiation in Tissue Banking*, World Scientific, Singapore, pp. 228–229.

Nather A, Thambyah A, and Goh JC (2004). Biomechanical strength of deep-frozen versus lyophilized large cortical allografts. *Clin Biomech (Bristol, Avon)* **19**(5): 526–533.

http://findarticles.com/p/articles/mi_m0YUG/is_14_14/ai_n17208423h (accessed on 13 January 2009).

Chapter 23

Amnion Processing

Hasim Mohamad*,†

Introduction

The ideal burn wound cover is autologous skin. Yet, there is a shortage of this material for extensive burns. For years, there has been a search for alternative biological dressings (e.g. allograft, cadaveric skin, chorioamnion and xenograft skin) that can mimic autologous skin. However, problems in the availability and immunological rejection of skin allograft have led to further research on artificial skin for burn wound cover. Until permanent repair with autologous skin can be instituted, temporary biological burn wound dressings are life-saving and essential, especially for extensive burns.

Amnion Grafts

Although the application of amnion grafts is referred to as grafting, amnion (being avascular) cannot establish vascular connections between host and graft. On a wound surface, amnion behaves as a biological dressing with all its beneficial properties (Panakova and Koller, 1997).

*School of Medical Science, University of Science, Malaysia.
†Department of Surgery, Hospital Raja Perempuan Zainab II, Kota Bharu, Malaysia.

Chorio-amnion

Air-dried or freeze-dried and gamma-irradiated amnion is used for:

- superficial burns (second-degree burns),
- diabetic ulcers,
- leprosy ulcers,
- post-traumatic wounds,
- vaginoplasty and
- lining for chest wall or abdominal wall defects.

The unlimited amount of amnions allows large wound areas that cannot be dressed by autologous skin to be covered.

The remnants of chorion peeled off from the amnion have been noted to have an interesting property. When the chorionic surface of the amnion is applied to the wound, more granulation is induced when compared to applying the amniotic surface to the wound. This property is useful to promote granulation over wounds with exposed tendons and bone surfaces.

Amniotic Membranes

Amniotic membranes are histologically quite similar to skin. It is made up of two layers: the amnion and the chorion. The amnion is the inner layer, which is smooth and glistening, and is composed of cuboidal cells. Its outer surface consists of mesenchymal connective tissue. The chorion is external to the mesenchymal tissue and is composed of transitional epithelial cells. The amniotic membrane has no blood vessels, nerves, or lymphatic channels. Although it is in intimate contact with the recipient burn wound, there is no occurrence of neovascularisation because there are no blood vessels in the amnion that can be connected with the recipient vessels.

Amniotic membranes are useful alternative to autologous skin for temporary wound dressings on burns, scalds, chronic ulcers, dermal injuries and contaminated wounds. Amniotic membrane is a versatile and effective biological dressing for both superficial and deep superficial burn wounds.

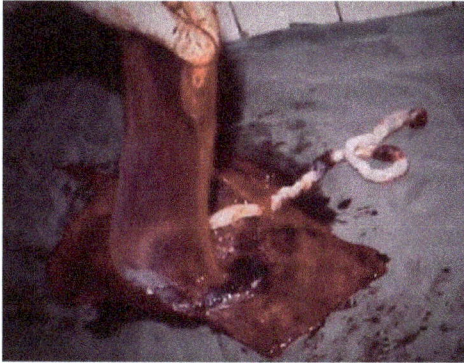

Fig. 1. Harvesting amniotic membrane from placenta.

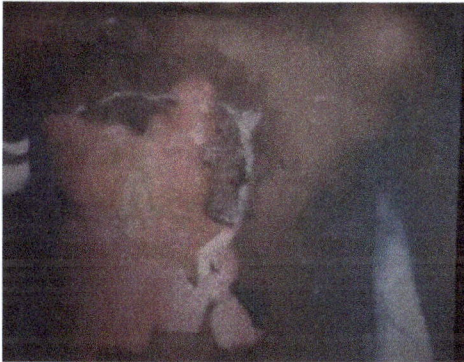

Fig. 2. Amniotic membrane without a layer of gauze.

Amniotic membranes are procured from the placentae of mothers. Before procurement, these mothers have been antenatally screened for communicable or infectious diseases (Figs. 1 and 2). Membranes from placentae with intrapartum complications are discarded. Processing of the membrane involves thorough washing with normal saline, soaking in 0.05% sodium hypochlorite solution for 30 minutes to one hour, shaking several times with normal saline, drying, packing and lastly gamma radiation at 25 kGy or lower according to the bioburden.

The amniotic membrane does not show immunological reaction, as it is non-antigenic. Fresh amniotic membrane is purported to possess angiogenic as well as bacteriostatic effects. Both these properties seem to remain intact in spite of sterilisation by radiation, which helps to prolong the shelf life of the membrane. However, these properties have not been experimentally confirmed.

Amniotic membrane was first reported to be used in the treatment of burn patients by Sabella in 1913. Since then, only a few reports of the use of amniotic membranes for burn wound coverage have been found in the medical literature; but its other applications include treatment of leg ulcers, skin loss in Stevens–Johnson syndrome, pelvic and vaginal surgery, and otolaryngology as well as head and neck surgery.

Preparation of Gamma-Irradiated Amniotic Membrane

Air-Drying Method

Fresh amniotic membranes are obtained from mothers during delivery. Mothers must be seronegative for infectious diseases. They must not have a history of premature rupture of amniotic membranes, endometritis or meconium staining (Table 1).

Under aseptic condition, the membrane is separated from the placenta, placed in a sterile plastic bag containing sterile saline and appropriately labelled. This pack can be frozen and thawed for later processing or can

Table 1. Procurement of amniotic membrane: maternal selection.

Inclusion
    Clean elective cesarean section
    Uncomplicated spontaneous vaginal delivery

Exclusion
    Prolonged rupture of membrane
    Endometritis
    Chorioamnionitis
    Meconium staining
    Drug abuse
    Positive for VDRL/TPHA
    Seropositive for hepatitis/AIDS
    Septicaemia
    Toxemia of pregnancy

be immediately processed. This bag is then stored in a +4°C freezer for later processing.

Processing of the membrane begins by dipping the fresh amniotic membrane in 0.05% sodium hypochlorite solution for 30 minutes to one hour and then washing it repeatedly with tap water until it is completely clear of blood particles and resembles a thin plastic film. After cutting the thin film to the required size (usually 10×10 cm), it is air dried in a laminar flow cabinet or alternatively in a freeze dryer. The procedure is completed by packing the dried membrane in sterile plastic packs inside a laminar flow cabinet and then sealing off.

Figure 3 shows the sequence of events in the preparation of amniotic membrane for packing, after which it is sterilised using cobalt-60 (Co-60) gamma radiation at 25 kGy as soon as possible. The dried and sterilised amniotic membranes are then ready for clinical use.

Freeze-Drying Technique

Fresh amniotic membrane is washed thoroughly with tap water to remove excess blood and then immersed in saline and shaken three times for 15 minutes each to remove amniotic fluid. If necessary, the membrane is spread over a clean flat surface and rubbed gently with sterile cotton gauze.

The washed membrane is then stretched across over sterile cotton gauze and freeze-dried at −40°C. Freeze-dried membrane is similarly packed and sterilised for air-dried membrane.

Clinical Use of Amnions

Radiated amniotic membranes, being thin film dressings, are easy to use clinically. They are also easy to apply onto the burn surfaces because of their good conformability. Radiated amnions adhere firmly to the wound and become dry membrane covering. This protects the wound from contamination and infection. No neovascularisation takes place on the overlying amnions, and the membranes spontaneously peel off after wound healing occurs.

There is no need for regrafting of the recipient wound if it is due to partial-thickness burns or scalds. However, in the case of full-thickness burns, autologous grafting is necessary once granulation tissue have formed.

Wash with clean water

↓

Separate chorion from amnion

↓

Wash with distilled water at pasteurisation temperature of 58°C in shaking water bath for 30 minutes

↓

Treatment with 0.5% sodium hypochlorite for 30 minutes

↓

Clean amnion in distilled water using multi-wrist shaker for 30 minutes, repeated three times

↓

Spread amnion for drying in laminar air-flow cabinet (mount on gauze or plain dryer overnight

or

Freeze-drying in Lyovac machine

↓

Double packed in polyethylene bag

↓

Heat sealed

↓

Labelled

↓

Sterilisation (gamma irradiation)

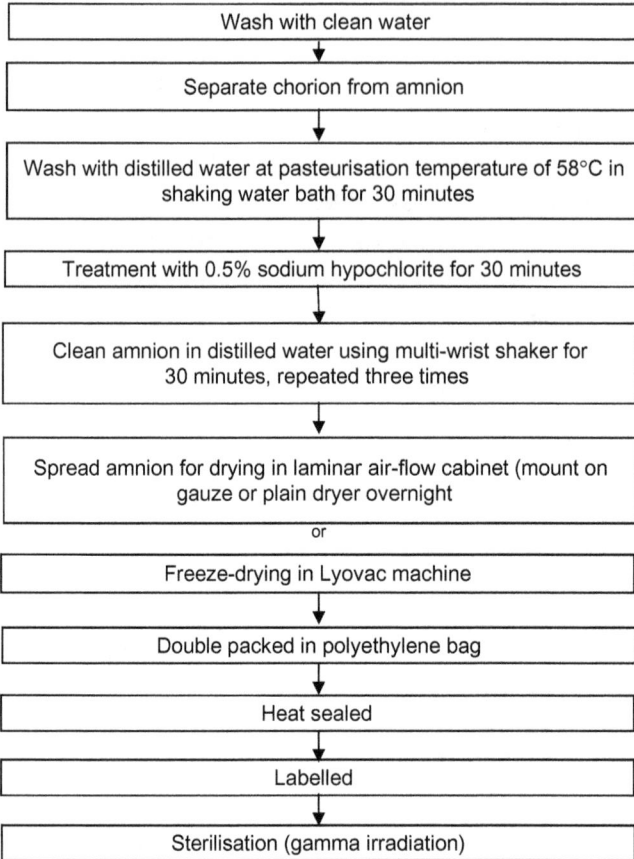

Fig. 3.   Summary of amnion processing.

Clinically, immediate coverage of open burn wounds is necessary to ensure a satisfactory outcome. In fact, it is the most important determining factor for patient recovery. For example, superficial partial-thickness lesions may become contaminated or infected and may progress to full-thickness lesions, resulting in scars or disabilities.

Discussion

Allograft skin satisfies the requirements of a good biological dressing. However, it has the disadvantage of developing rejection. Consequently, frequent dressing changes (about five days from the time of application is necessary) are required — hence there is a need for a constant supply (skin bank) of donors, both living and dead.

To achieve an acceptable clinical outcome, autograft skin should be used; however, the graft available is limited if primary closure of donor site is yet to be achieved. Therefore, amniotic membrane is a good alternative to other biological dressings, as it is easily available at low cost. The successful use of wet and dry amnion membranes has been sporadically reported in medical literature.

In 1972, Robson *et al.* showed in a comparative study that amniotic membranes were found to be equal to isografts and superior to both allograft and xenograft skins in reducing bacterial levels in full-thickness skin defects in skin. Upon application, amniotic membranes alleviate pain. This is particularly useful in the management of paediatric cases. Studies have also shown that they inhibit infection. This antibacterial property is believed to be caused by the lysozyme and progesterone present in the amniotic fluid. Moreover, their adherence to the wound surface prevents the accumulation of fluid and pus on the granulation tissues. Last but not least, radiation does not seem to cause any physical damage or damage to their antibacterial properties.

Carefully procured and radiated amniotic membrane with its long shelf life, plays an important role in burn wound treatment. As mentioned previously, its clinical usage has been proven to be effective. Clinically, its single application without the need for redressing of wounds makes it popular among the nursing staff, unlike the use of silver sulfadiazine, which requires frequent dressing changes.

Reference

Panakova E and Koller J (1997). Utilisation of foetal membranes in the treatment of burns and other skin defects. In Philips GO, Strong DM, von Versen R and Nather A (eds.), *Advances in Tissue Banking*, Vol. 1, World Scientific, Singapore, pp. 165–181.

Part VII

Radiation Sciences

Chapter 24

Principle Concepts of Radiation Sterilisation for Tissue Allografts

Norimah Yusof* and Nazly Hilmy[†]

Introduction

Radiation technology for sterilisation has evolved and is becoming the first choice for thermosensitive materials, the main component of medical disposables. At present, 40–50% of all disposable medical products are radiation sterilised (Chmielewski and Berejka, 2008). Tissue allografts comprised of biological components are also thermosensitive; thus, radiation at appropriate doses is capable of sterilising tissues without causing significant detrimental effects on biological functions of the tissues. The International Atomic Energy Agency (IAEA) has successfully assisted developing member states in establishing the use of radiation technology as a terminal sterilisation process in tissue banking. Almost all tissue banks in the Asia-Pacific region and the Latin America are using radiation to sterilise their processed tissues. More and more tissue banks in the North America are considering the technology. Radiation allows sterilisation of the tissues in their final packaging and the irradiated tissues can be used immediately with no quarantine period and no chemical residues.

---

*Malaysian Nuclear Agency (NM), Bangi 43000 Kajang, Selangor, Malaysia.
[†]BATAN Research Tissue Bank (BRTB), Center for the Application of Isotopes and Radiation Technology, BATAN, Jakarta 12070, Indonesia.

Need for Sterilisation

Sterilisation is the process to eliminate or kill all microorganisms in a product to make the product free from any living organisms. Inactivation of microorganisms by any methods of sterilisation follows exponential law; meaning that there is inevitably always a finite probability of a microorganism that may survive on a product regardless of the extent of the dose delivered. Sterilisation process is therefore required to ensure that no microorganism can proliferate and reduce the microbial count to acceptably low levels, where such levels are defined as the sterility assurance level (SAL), the probability to get a non-sterile product among a population of products. It is the purpose of the sterilisation dose to render the product sterile at a high SAL. There are two levels of SAL being used (IAEA, 1990):

1. SAL 10^{-3} is the probability to get one non-sterile product in a thousand (10^3) of products. This level is applicable to products not intended to come into contact with compromised tissues (i.e. tissue that has lost the protection of the natural body barriers) such as disposable items for laboratory use and packaging materials.
2. SAL 10^{-6} is the probability to get one non-sterile product in a million (10^6) of products. This level is applicable to products intended to come into contact with compromised tissue or opened body part, which are used inside the body such as medical products and tissue grafts.

Microorganisms could be introduced onto the tissue during tissue procurement, processing, drying/preservation, packaging and storage. Radiation can eliminate these microorganisms. However, the risk of infectious disease transmission through tissue allograft is a major concern in tissue banking. The possibility of bacterial, fungal and viral infection of donor origin such as human immunodeficiency virus (HIV), hepatitis B virus (HBV), (hepatitis C virus HCV), cytomegalovirus (CMV), rabies and prion diseases cannot be ignored. High radiation dose is normally required to kill viruses or to inactivate protein. Therefore, the hazard of infectious disease transmission should be tackled through many steps before sterilisation (Eastlund, 2005), as follow:

Careful donor selection using strict selection criteria and proper screening method:

- Donor history screening
  - Medical history: cause of death, current illness, past medical history, risk of HIV, hepatitis, Creutzfeldt–Jakob disease (CJD), malaria, etc.
- Donor physical exam,
- Donor blood tests and
- Autopsy, if one was performed.

Proper tissue procurement:

- Aseptic tissue recovery,
- Minimise contamination and
- Recovery cultures.

Tissue processing steps:

- Disinfection,
- Inactivation of viruses and
- Validated methodologies.

Sterilisation:

- Packaging intact,
- Validated test result and
- Certification.

Tissue safety depends not only on one but several steps where the effectiveness of one step depends on another. Terminal sterilisation will never replace but only to complement these steps to produce disease-free tissue grafts. Sterile procurement of tissues must be practised (Nather, 2001).

Processing procedure must be validated for its effectiveness in cleaning tissue. After sterilisation, package integrity is important as damage to the packaging materials will turn the product unsterile. Proper handling and storage must be given equal attention. Any adverse outcomes must be investigated and any corrective actions are aimed to improve future safety (Eastlund, 2005).

During processing, microbial contamination may come from four main sources, namely environment, personnel, raw material including tissue itself, and machinery. Under tissue banking quality system, some procedures need to be established to minimise contamination, such as the following:

- Air — conduct particulate and bacteria monitoring,
- Processing room surfaces — disinfect tables, clean walls and floor and check microbial count,
- Equipment — calibrate, validate and conduct maintenance schedule,
- Water — check quality and conduct microbial monitoring,
- Supplies — check sterilisation records and certificate,
- Reagents, solutions, packaging materials — check supply and sterilisation batch and
- Personnel — check technique barriers and do routine medical checkup.

Regardless of precaution taken, manufactured/processed products may still carry some microorganisms that we cannot see with naked eyes. Therefore, any processed tissues need sterilisation treatment to ensure for safety use. Comparison of three treatments commonly used for sterilisation in term of their effects on tissue:

1. *Thermal treatment* causes biological and physical damage and cannot penetrate well
2. *Radiation treatment* causes physical damage only at high dose and able to penetrate even final packaging
3. *Ethylene oxide gas (ETO) treatment* leaves toxic residues and has been banned in many countries

Radiation seems to be the best choice if tissue bank can get access to any irradiation facility or having it on site. Dose of 25 kGy has been used for sterilisation and with new developments in radiation research; sterilisation dose can be selected depending on bioburden or initial microbiological count of processed tissues and SAL (IAEA, 2007).

Ionising Radiation

Ionising radiation can cause ionisation of atoms and molecules in the tissue when it interacts with any of them. The interaction will result in cell death either through direct or indirect effect of radiation (Yusof, 2007a). Three types of Ionising radiations commonly used for sterilisation are:

1. Gamma rays are emitted from radioisotopes either cobalt 60 (^{60}Co), which has half life of 5.27 years, or caesium 137 (^{137}Cs), which has half life of 30.1 years. Currently, almost all commercial gamma irradiation facilities use ^{60}Co as the radiation source compared to ^{137}Cs, which is limited in supply. The installed capacity of Co is increasing at the rate of about 6% matching with the growth rate of worldwide use of disposable medical devices (Mehta, 2008). Gamma rays have high penetration capability and can sterilise non-uniform and high-density products in final packaging.
2. Electron beams are generated from accelerators that can be switch on or off when needed. They have low penetration; thus, only thin product can be sterilised unless when high-energy accelerator of 10 MeV is used. However, they have high-dose rate capable of delivering 100 kGy per second compared to gamma rays of approximately 10 kGy/h.
3. X-rays are generated from high-current electron beam accelerators with X-ray converter. They have high penetration comparable to gamma rays.

Major concerns expressed by tissue bankers are whether radiation or any sterilisation method has undesirable effects on tissues and if such method effectively inactivates viruses, especially HIV. Report by LifeNet in 1985 as shown in Table 1 indicated that irradiated tissues did not transmit HIV (Simonds *et al.*, 1992).

Table 1. Recipients with HIV transmitted through organ and tissues (LifeNet, 1985).

	Recipients	HIV+
Fresh organ	4	4
Fresh cornea	2	0
Frozen bone	4	3
Freeze-dried bones, soft tissues and irradiated tissues	47	0

Studies on transmission of HIV in window period donor conducted in the USA from 1999 to 2003 reported that irradiation in sterilising doses can significantly reduce the viral load and in combination with appropriate donor screening and laboratory testing will significantly enhance and improve the safety of tissues being used for transplantation (Strong, 2005). Window period is the period where a donor with viral contamination passed through the system undetected.

Report of the National Center of Infections Disease, USA on the transmission of Hepatitis C virus from donor to recipients from 2000 to 2002 indicated that no infection occurred in 16 recipients of gamma-irradiated bones (CDC, 2003). A rare complication of muskuloskeletal allografts was reported by Centre for Disease Control and Prevention (CDC) in 2002, that 26 cases of infection were found to be caused by contamination of *Clostridium sordellii* and no reports of disease transmission using demineralised bone product as well as radiation-sterilised products (Yusof and Hilmy, 2007) were recorded. Conrad *et al.* (1995) reported that the HCV can be transmitted by bone, ligament and tendon, but no cases of transmission with irradiated bone of 17 kGy.

Direct and Indirect Effects of Radiation

As mentioned earlier, the effects of radiation on cells can be due to either direct action or indirect action of radiation.

Direct Action

The direct action of radiation involves the interaction between the ionising radiation and critical biological molecules, which results in excitation, lesion and scission of polymeric structure (Yusof, 2007a). When passes through a cell, high-energy photon will interact with biological molecules such as deoxyribonucleic acid (DNA) and deposit energy. Due to photon energy deposited on the critical DNA molecule, single and double breaks in the DNA strands occur. Direct action may involve the addition of hydrogen atoms to the opened bonds and damages at the base. Damages in the DNA structure disrupt normal cell functions hence result in cell death. However, this effect is not dominant.

Indirect Action

Ionising radiation interacts with atoms or molecules in or surrounding the cell, in particular, water molecules. The interaction of radiation with the aqueous system results in excitation and ionisation (Yusof, 2007b). Due to the breakdown of water molecules, the radiolysis of water leads to the formation of free radicals and aqueous electron as summarised in the following reaction:

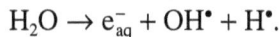

$$H_2O \rightarrow e_{aq}^- + OH^\bullet + H^\bullet.$$

The presence of substantial quantities of water in microorganisms leads to radical formation when radiation photons interact with water molecules, ejecting many electrons at high velocities. Aqueous electron (e_{aq}^-) and the free radicals (hydrogen atoms $H^\bullet$ and hydroxyl radical $OH^\bullet$) are very reactive. They will easily interact further among them, with water molecules, with their own reaction products, or with organic molecules within cell constituents, RH. The organic molecules, RH, also become directly ionised into free radicals $R^\bullet$. These free radicals $R^\bullet$ react with biologically important cellular molecules such as proteins, enzymes, amino acids, metabolites, nucleic material and cause radiobiological damage. These chain reactions, which are called the indirect action of radiation, are generally held responsible for the radiation effects, where these free radicals known as radiolytic products act as intermediaries in the transfer of radiation energy to biological molecules. The indirect effects of radiation responsible for 90% of DNA damage are basically causing biochemical changes within the organism and are mainly associated with the impairment of metabolic reactions. The indirect effect is more prominent.

Much evidence showed that the damages through either direct or indirect effect occur more in DNA molecule compared to other critical sites including membranes and lysosomes of microorganism. The main biological target, DNA, controls the genetic constitution and reproductive process of the cell. DNA is the most vital cell constituent and it presents a relatively large volume in the microbial cell for absorption of radiation and a large surface area for reaction with radiolytic products. Two distinct damages in the DNA have been recognised:

1. breaks in one (single-strand break) or both of the DNA strands (double-strand break) and
2. lesion in the nitrogenous bases.

Some microorganisms are capable of repairing DNA breaks or fractures. It is believed that microorganisms that are sensitive to radiation cannot repair double-strand breaks, whereas radiation resistant species have some capability to repair. The efficiency of repair capability is often reflected by a large increase in the dose of radiation for inactivation.

Radiation Resistance of Microorganisms

The amount of absorbed radiation energy or dose required to inactivate or kill a microorganism in a tissue product depends on the resistance of the microorganism to the radiation. Each microorganism has different degree of resistance and it is influenced by many factors: type or species, cell cycle stage, water content, temperature, presence of oxygen, presence of media or nutrient, presence of radioprotectant/sensitiser and radiation dose rate (Yusof, 2007a; Hamad, 2008).

Radiation resistance varies among species and even strains of the same species. The degree of resistance is expressed as D_{10} value, i.e. amount of radiation dose required to kill 90% of the population of a species of microorganisms. The D_{10} value can be determined from dose-response curve or dose-survival curve for each microorganism when a suspension of the microorganism (at initial number of approximately 10^7–10^8) is irradiated at incremental doses. The number of surviving cell forming colonies after each incremental dose is plotted on log scale

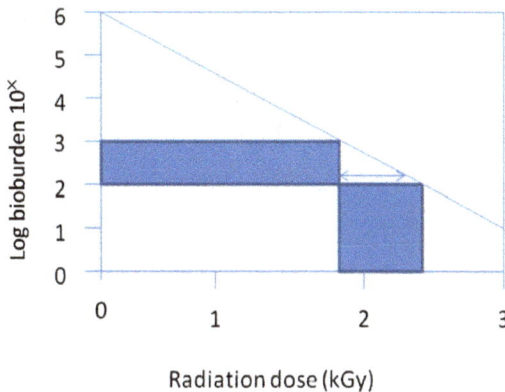

Fig. 1.   Typical dose-response curve for a homogenous microbial population.

against the radiation doses plotted on ordinary scale. The gradient or slope of the curve of the dose-survival curve as shown in Fig. 1 represents the required radiation dose (kGy) to kill 90% of the population or a reduction of one log cycle (10-fold).

The slop of the curve is mostly a straight line; however, certain microorganisms show a "shoulder" at low doses before the linear slope starts (Fig. 2). This "shoulder" might due to multiple targets that need to be inactivated before the killing effect is linearly proportionate to increasing doses, or some repair processes are taking place at low doses. D_{10} values of some common microorganisms have been studied and documented (Yusof, 2007a). In general, D_{10} values for moulds and yeasts are lower than vegetative bacteria, D_{10} values for vegetative bacteria are lower than bacterial spores, while viruses have high D_{10} values. A radiation dose that sufficiently inactivates bacteria will easily eliminate fungal contamination. Conditions in which these microorganisms are irradiated and temperature during irradiation will influence their D_{10} values.

Therefore, the dose required to kill a particular microorganism depends on the initial number of microorganism or bioburden and the type of the microorganism (D_{10} value). The sterilisation dose (D) required to kill a population of microorganism having a known radiation resistance

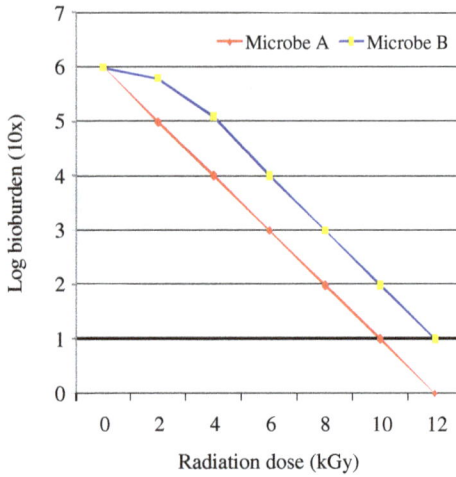

Fig. 2. Dose response curve — microbe A (linear type) and microbe B (shoulder type).

(D_{10}) from initial number/bioburden of N_0 to the final level of N can be calculated using the following formula:

$$D = [(N_0 - N) \times D_{10}] \, kGy.$$

The dose required for sterilisation can be estimated simply by multiplying the total log cycle to be reduced with the D_{10} value. However, bioburden on any tissue product is made of a mixture of various microbial species, each having its own D_{10} value in a specific condition. The final aim in sterilising a product is not just to kill the microorganisms to the level zero or no count but to be able to give an assurance or a probability of the sterility level, i.e. sterility assurance level (SAL) of 10^{-6}. As illustrated in Fig. 3, sterilisation dose varies with bioburden and selected SAL. Bioburden and type of microorganisms are two determining factors in establishing a radiation dose. High dose is required for sterilising product with high bioburden. High dose is also required to kill a radiation-resistant microorganism, which has high resistance level or high D_{10} value compared to radiation-sensitive microbes with low D_{10} value.

Moisture content of a tissue during irradiation is also important. Drying stage by using either freeze-drying or air-drying must be able to

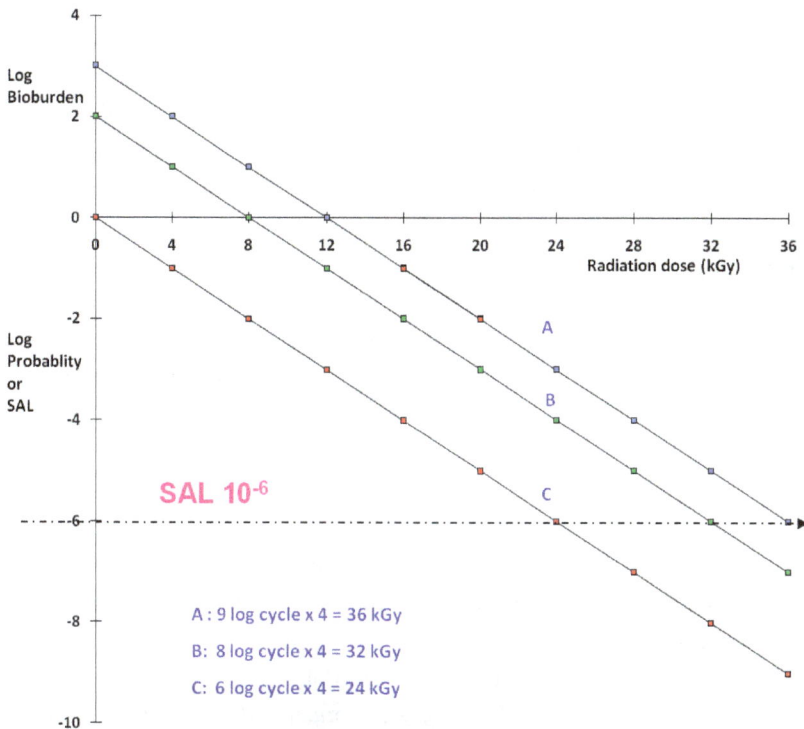

Fig. 3. Sterilisation dose (kGy) at SAL 10^{-6} for a microbe with D_{10} value of 4 kGy, having different bioburden: (A) 10^3 cfu/product, (B) 10^2 cfu/product and (C) 10^0 cfu/product item.

reduce the moisture of the processed tissue less than 10%, preferable between 4% and 7%. More damage to tissue will be observed when the moisture content is high, due to active movement of free radicals resulting in more interactions thus more damage. However, for frozen tissue, this will not happen as the water molecules are stagnant in freezing condition, less movement means less interaction thus resulting in less damage.

Normally radiation dose selected for sterilisation is 25 kGy. With recent developments in disease detection methods at molecular level, new processes to inactive viruses and microbes and increasing demand to minimise radiation damage to soft tissues, it is now a trend to go for lower doses. However, the chosen sterilisation dose must be validated to ensure the irradiated tissues are safe for clinical use.

Conclusion

Radiation provides a convenient and effective method to sterilise tissues in their final packaging. Radiation dose for tissue sterilisation is determined by initial number of microorganisms on the product, i.e. bioburden and the type or the resistance of the microorganisms to radiation, i.e. D_{10} value. Clean product with low bioburden can be easily sterilised by radiation to achieve SAL of 10^{-6}. Radiation or any sterilisation technique will complement good tissue banking practices in producing a clinically safe tissue. Therefore, tissue must be properly screened, procured under controlled environment and processed properly in a clean environment in order to produce high-quality tissue with low bioburden before sterilisation.

References

Chmielewski AG and Berejka AJ (2008). Radiation sterilisation centres worldwide. In *Trends in Radiation Sterilization of Health Care Products*, IAEA, Vienna, pp. 49–61.

Center for Diseases Control and Prevention (CDC) (2003). Hepatitis C virus transmission from an antibody-negative organ and tissue donor — United States, 2000–2002. *MMWR Morb Mortal Wkly Rep* **52**(13). 273–274, 276.

Conrad EU, Gretch DR, Obermeyer KR, Moogk MS, Sayers M, Wilson JJ, and Strong DM (1995). Transmission of the hepatitis-C virus by tissue transplantation, *J Bone Joint Surg Am* **77**: 214–224.

Eastlund T (2005). Viral infections transmitted through tissue transplantation. In Kennedy JF, Phillips GO and Williams PA (eds.), *Sterilisation of Tissues Using Ionising Radiations*, CRC Woodhead Publishing, England, pp. 255–278.

Hamad AA (2008). Microbiological aspects of radiation sterilization. In *Trends in Radiation Sterilization of Health Care Products*, IAEA, Vienna, pp. 119–128.

International Atomic Energy Agency (IAEA) (1990). *Guidelines for Industrial Radiation Sterilisation of Disposable Medical Products (Cobalt-60 Gamma Irradiation)*. IAEA-TECDOC Series No. 539, IAEA, Vienna.

International Atomic Energy Agency (IAEA) (2007). *Radiation Sterilisation of Tissue Allografts: Requirements for Validation and Routine Control — A Code of Practice*. IAEA, Vienna.

Mehta K (2008). Gamma irradiators for radiation sterilization. In *Trends in Radiation Sterilization of Health Care Products*, IAEA, Vienna, pp. 5–25.

Nather A (2001). Sterile procurement of bones and ligaments. In *Advances in Tissue Banking*, Philips GO and Nather A (eds.), World Scientific Publishing, Singapore, Vol. 5, pp. 265–306.

Simonds RJ, Holmberg SD, Hurwitz RL, Coleman TR, Bottenfield S, Conley LJ, Kohlenberg SH, Castro KG, Dahan BA, Schable CA, Rayfield M, and Rogers M (1992). Transmission of human immunodeficiency virus type 1 from a seronegative organ and tissue donors. *N Engl J Med* **326**(11): 726–732.

Strong DM (2005). Effects of radiation on the integrity and functionality of soft tissue. In Kennedy JF, Phillips GO and Williams PA (eds.), *Sterilisation of Tissues Using Ionising Radiations*, CRC Woodhead Publishing, England, pp. 163–172.

Yusof, N and Hilmy N (2007). Need for radiation sterilization of tissue grafts. In Nather A, Yusof N and Hilmy N (eds.), *Radiation in Tissue Banking: Basic Science and Clinical Applications of Irradiated Tissue Allografts*, World Scientific Publishing, Singapore, pp. 11–21.

Yusof N (2007a). Radiation killing effects on bacteria and fungi. In Nather A, Yusof N and Hilmy N (eds.), *Radiation in Tissue Banking: Basic Science and Clinical Applications of Irradiated Tissue Allografts*, World Scientific Publishing, Singapore, pp. 121–132.

Yusof N (2007b). Interaction of radiation with tissues. In Nather A, Yusof N and Hilmy N (eds.), *Radiation in Tissue Banking: Basic Science and Clinical Applications of Irradiated Tissue Allografts,* World Scientific Publishing, Singapore, pp. 99–107.

Chapter 25

Good Radiation Practices
for Sterilisation of Allografts

Norimah Yusof* and Noriah Mod Ali*

Introduction

Radiation sterilisation for commercial application was first started in the United Kingdom and Australia since more than 50 years ago, followed by the United States and Scandinavian countries. Sutures were the first radiation-sterilised medical item provided by Johnson & Johnson in 1954. Following that a wide range of medical products, pharmaceuticals and cosmetics have been irradiated routinely. Since 1990, the number of tissue grafts sterilised by radiation is increasing. At present, over 160 gamma irradiators for sterilisation are operating in 47 countries; more than 100 of these plants are located in Europe and North America (Chmielewski and Berejka, 2008). Besides gamma irradiators, an increasing number of electron accelerators is in operation for sterilisation service. Although both types of irradiation, gamma rays and electron beams, can be used in this process; however, the sterilisation efficiency of irradiation will depend on its high effectiveness in inactivation of microorganisms without causing significant changes in tissue structure and function. Gamma ray has the advantage due to its high penetration capability, thus it can easily irradiate non-uniform and high-density tissue products. Electron beam on the other hand, although limited by product thickness that can be irradiated, its exposure time is shorter than gamma ray, thus enabling high throughput for a

*Malaysian Nuclear Agency (NM), Bangi 43000 Kajang, Selangor, Malaysia.

commercial plant. Irradiation of tissues mainly using gamma rays has been expanding in parallel with the popularity of bone transplantation in North America (Tomford, 2005) while in the Asia-Pacific and the Latin America regions, it was due to the great success of the IAEA programme in promoting radiation sterilisation for tissue banking.

Radiation Sterilisation

Radiation is a form of energy emitted from a source, which travels through tissues either as particles or as waves (Yusof *et al.*, 2007). They lose energy along the way mainly through ionisation and excitation. Advantages in using ionising radiation for tissue sterilisation are:

- Radiation is a cool process. With no significant increase in temperature to cause any changes, tissues retain their biological properties.
- High penetration of gamma rays can sterilise hard and soft tissues in the final packaging even in boxes or in containers filled with dry ice.
- No toxic residues are left after irradiation; therefore, tissue grafts can be used without delay as there is no quarantine period.
- There is only one parameter, i.e. exposure time for precise process control.

Most of gamma irradiators are using gamma emitted from radioactive cobalt-60. As shown in Fig. 1, gamma rays in the electromagnetic spectrum have short wavelength. The shorter the wavelength, the higher energy the radiation has to cause ionisations and excitations of atoms and molecules in a matter it passes through.

Parameters for Radiation Sterilisation

In considering radiation as a sterilisation process, several parameters have to be given attention.

Radiation Dose

The minimum radiation dose commonly used to sterilise products is 25 kGy, whereby 1 Gray (Gy) is equivalent to 1 Joule kg^{-1}. However,

THE ELECTROMAGNETIC SPECTRUM

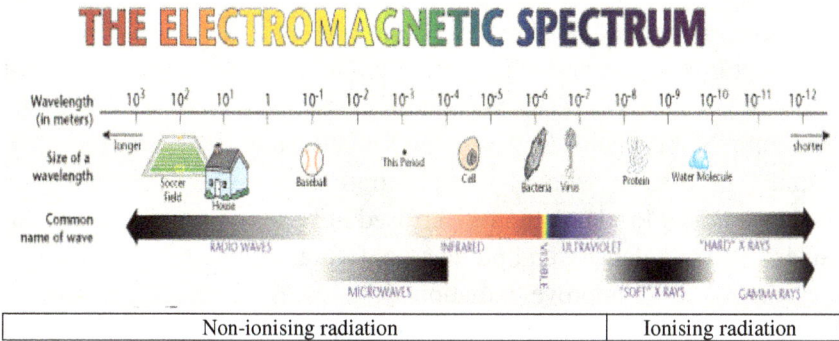

Non-ionising radiation	Ionising radiation

Fig. 1. Gamma rays in the electromagnetic spectrum.

tissue banks in Poland have used 35 kGy with over 40 years of experience (Dziedzic-Goclawska *et al.*, 2008), while in some other countries such as the North America and Australia, there is a tendency to lower the dose to 10–15 kGy to cater for soft tissues. Actually, sterilisation dose can be selected and determined based on bioburden, i.e. the microbial count on the tissue product prior to sterilisation and types of the microbes. Dose can be selected according to ISO document ISO 11137:2006 (ISO, 2006), which will be elaborated in Chapter 26 of this book. In doing the dose validation, pre-sterilisation doses must be delivered as accurate as possible within $\pm 10\%$ variations. Therefore, a radiation facility is necessary not only for routine service but also during process and dose validation for a particular tissue product. A tissue banker is responsible to establish radiation dose; however, the responsibility to achieve a sterile product is shared between the tissue banker and the sterilisation centre.

Temperature During Irradiation

Temperature and condition during irradiation will influence radiation effects on tissues. For air- and freeze-dried tissues, irradiation is done at room temperature. Frozen bones must be irradiated together with dry ice to ensure that the bones are frozen throughout the irradiation process. Chilled skin is packed with enough ice before irradiation.

Dose Uniformity

It is very important to ensure that radiation dose is evenly distributed in a product box or container. In a box of tissue products, all products must get doses not lower than the selected sterilisation dose, while the maximum dose cannot be too high to cause any undesirable effects. Gamma irradiation with higher penetration compared to electron is the best choice. Products can be sterilised in boxes or in ice-packed containers. Product density and product configuration in a box or container are important to improve radiation dose distribution. For any given irradiation condition, it is necessary to specify the absorbed dose in the particular tissue product.

Product Integrity

When radiation sterilisation is being considered for tissues, the effects of radiation on tissues as well as on the packaging materials must be taken into account. The tissue banker must make sure that tissues maintain their functional properties after irradiation. The minimum dose given must be able to sterilise the tissue whilst the maximum dose will not result in detrimental effects on tissue components. Packaging materials used must also be of radiation compatible and not easily damaged by radiation and not degraded during storage. Tissue bankers, as the product manufacturers, are responsible to demonstrate the integrity of the irradiated tissue product in relation to safety and quality.

Good Radiation Practice (GRP)

Radiation sterilisation is an integral part of the total manufacturing process whereby all relevant elements in quality assurance must be considered. Good Manufacturing Practice (GMP) and Good Radiation Practice (GRP) are the main components of the quality assurance for processing and irradiation, respectively. Tissues to be sterilised must be processed under conditions that fully comply with the requirements of GMP. GRPs are the actions and procedures which must be followed during the irradiation facility to ensure that the tissues are exposed to the required radiation doses. Even though GRP is more applicable to irradiation personnel, staff of tissue banks must also be aware of GRP, as radiation is considered as part of tissue processing.

GRP in an irradiation facility with respect to tissue allografts covers the following aspects (Yusof *et al.*, 2007).

- *Irradiator* — work on irradiator system towards irradiation plant commissioning.
- *Dosimeters* — primary and secondary (for routine irradiation) dosimeters to measure absorbed doses and traceable to international body on dosimetry.
- *Dose mapping* — dose distribution in a product box with specific product load to determine position of minimum dose, position of maximum dose and dose uniformity ratio (DUR).
- *Material compatibility* — stability of tissue products and packaging materials after radiation exposure and storage.
- *Product control* — control of product movement, before and after irradiation.
- *Routine process control* — establish process parameters for routine irradiation and controls for product release and certification.

Irradiator

An irradiation facility comprises four basic components.

1. *Radiation source* — radioactive source for gamma rays or accelerator machine for electron beams. For gamma rays, the radioactive source is stored in a water pool and it will be lifted up when in use.
2. *Irradiation room or bunker* — the area where the radiation source is placed must have concrete biological shield, i.e. concrete wall of approximately two-feet thick.
3. *Carrier/conveyor system* — automated system is used to carry products from non-irradiated area into the irradiation room and then carry the product out into irradiated area.
4. *Control and safety system* — control panel is where exposure time or cycle time is set, depending on the dose requested to sterilise

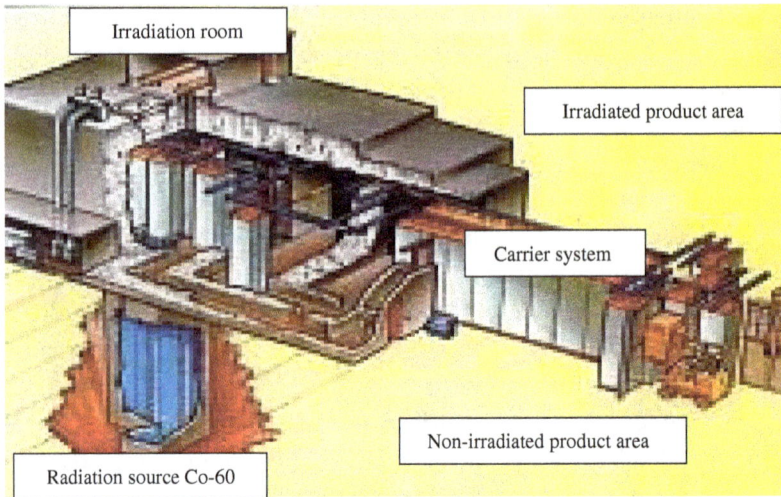

Fig. 2.   Layout of a gamma irradiation facility with radioactive cobalt-60 as a source.

tissues and the dose rate of the irradiator or the strength of cobalt-60 in the case of gamma rays. Plant safety system is routinely monitored and maintained. In the event of critical component failure during operation, the radiation source must be in safe position.

Figure 2 shows the layout of a gamma irradiation facility with radioactive cobalt-60 as a source stored in the water pool when the irradiator is not in operation. Figure 3 shows the layout of an accelerator for electron beams with products on the conveyor system moving slowly underneath the scan horn during irradiation process.

During the commissioning of irradiation facility, it is extremely important to characterise the magnitude, distribution and reproducibility of absorbed dose in products of certain density and relate these parameters with operating conditions. All systems of the facility must be calibrated, well maintained and functioning correctly.

For tissue banks equipped with gamma cell, they need to work with a local nuclear authority or nuclear research institute to assist in the commissioning of the irradiation equipment. The institute will also be

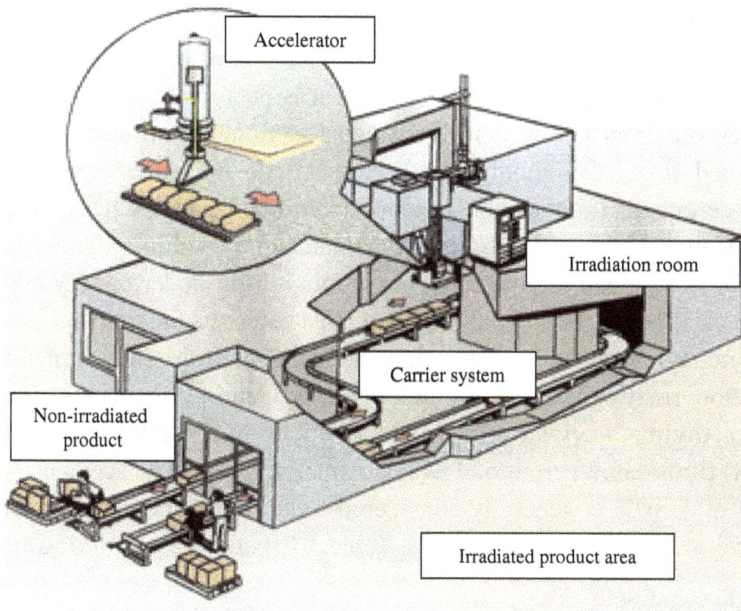

Fig. 3. Layout of an accelerator for generating electron beams.

able to suggest the dosimeters to be used for commissioning, calibration and routine irradiation.

Dosimeter

Dosimeters are devices that are capable of providing a quantitative and reproducible measurement of dose through a change in one or more of the physical properties of the dosimeters in response to the exposure to ionising radiation energy (Mod Ali, 2007). A dosimetry system consists of dosimeters, measurement instruments and their associated reference standards, and procedures for using the system. The measuring instrument must be well characterised, so that it gives reproducible and accurate results.

The selection of an appropriate dosimetry system will depend on a variety of factors, including the dose range needed to achieve a particular technological objective, cost, availability and ease of use. A variety of dosimetry systems are available. Reference standard dosimeters are dosimeters of high metrological quality that can be used to calibrate other dosimeters. In turn, they need to be calibrated against a primary standard. They must have a radiation response that is accurately measurable, and the response must have a well-defined functional relationship with the absorbed dose. Commonly used reference dosimeters include Fricke, ceric-cerous, dichromate, ethanol monochlorobenzene (ECB) and alanine dosimeters. Routine (or working) dosimeters are used in radiation facilities for dose mapping and for process monitoring under quality control. They must be frequently calibrated against reference dosimeters, as they may not be sufficiently stable and independent from environmental or radiation field conditions. Commonly used routine dosimeters include polymethylmethacrylate (PMMA), radiochromic, cellulose triacetate (CTA) films, ceric-cerous and ECB dosimeter. Dosimetric accuracy of the order of 5–10% is generally considered to be necessary for the effective control of the sterilisation process. Table 1 lists the classes of dosimeter.

Table 1. Useful dose ranges for various routine and reference standard dosimeters. (Classes: R: routine; S: reference standard)

Dosimeter	Useful dose range (Gy) $10^{-5}\ 10^{-4}\ 10^{-3}\ 10^{-2}\ 10^{-1}\ 1\ 10^1\ 10^2\ 10^3\ 10^4\ 10^5$	Class
Calorimeter		S
Alanine		SR
Cellulose triacetate		R
Ceric-cerous		SR
Clear PMMA[b]		R
Dyed PMMA[b]		R
Ethanol chlorobenzene[a]		SR
Ferrous cupric sulphate[a]		R
Ferrous sulphate[a]		S

[a] Aqueous solution.
[b] PMMA stands for polymethylmethacrylate.

For gamma irradiation, we use Fricke or ferrous sulphate solution as reference dosimeter and ceric-cerous as routine dosimeters (Yusof, 2007). For electron irradiation, graphite calorimeter is the primary or reference dosimeter while CTA film is widely used as routine dosimeter.

For irradiation, it is advisable to pool only the same type of tissue with the same density in the same box. For instance, a box of femoral heads may contain a collection of femoral heads procured from several donors; they are individually packed to avoid cross-contamination, and then they are irradiated in frozen form. However, it is possible to mix different types of tissue of almost the same density to be irradiated at the same temperature, for example, we can pool packages of freeze-dried amnion with packages of freeze-dried bone (small pieces or in morsalised/powder form) in a box to be irradiated at room temperature. It is also possible to put together packages of freeze-dried bones of different sizes and shape. A box of frozen bones may contain different types and shapes of bones from one donor. Product configuration (arrangement) in a box must be kept constant after the dose mapping of a particular box size with a particular density has been conducted. In the following irradiation batches, routine check on the weight of the box should be carried out as part of process control.

Dose Mapping

For any given irradiation condition, it is necessary to specify the absorbed dose in the particular material of interest because different materials have different radiation absorption properties (Mod Ali, 2007). Dose mapping is the measurement of dose distributions throughout a reference product package in its complete passage. It is also known as plant commissioning for a new facility or product validation for a new product.

Dose mapping of actual products is an integral part of performance qualification according to ISO11137:2006. For dose mapping work, a product box is filled with tissue products in their final packets/packages. Routine dosimeters are placed in between the products inside the box at specified points. Dosimeters are also placed outside the box as in Fig. 4. It is important to determine the minimum absorbed dose received for every batch because the minimum dose received by the product must be higher than the requested dose for sterilisation. It is also important to

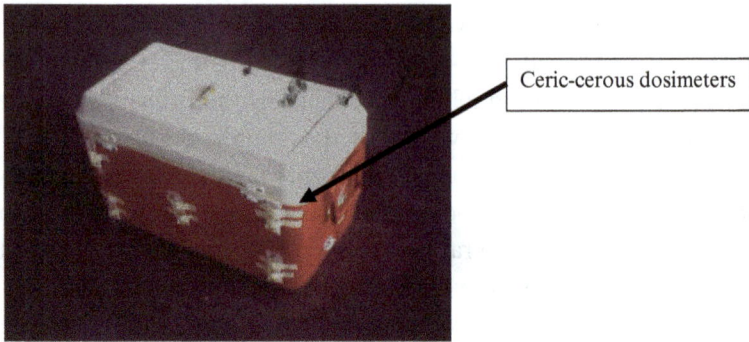

Ceric-cerous dosimeters

Fig. 4.   Dosimeters are placed outside the product box.

determine the maximum absorbed dose received by each batch because we do not want the tissues to be overexposed.

As the dose mapping will be repeated several times, it is possible to use a "dummy" material to substitute the real tissues. The dummy material to be selected must have almost similar density to human tissues such as wood chips or wood stick and some polymers. The materials must be in the same size and packed exactly like the real tissues. The use of dummy materials is very useful especially for replacing frozen tissues where the box is also filled up with dry ice (solid carbon dioxide) to maintain low temperature throughout the irradiation time. Dosimeters such as ceric-cerous and Perspex cannot be used in frozen state. To simulate the frozen condition, spaces between the dummy bones are filled up with other materials, which are equivalent to dry ice in density and preferably easy to obtain and are cheap. You need to be creative in choosing the replacement materials, some may use plastic beads and some may consider animal feed, rubber/plastic beads or many other possible materials.

Different facility will give different positions of minimum and maximum doses in an irradiated box, as shown in Fig. 5. These findings must be documented as part of quality record. In routine irradiation, dosimeters will only be placed at these two positions. The dose uniformity ratio (DUR) can then be determined as follow:

$$DUR = \frac{\text{maximum dose } (D_{max})}{\text{minimum dose } (D_{min})}$$

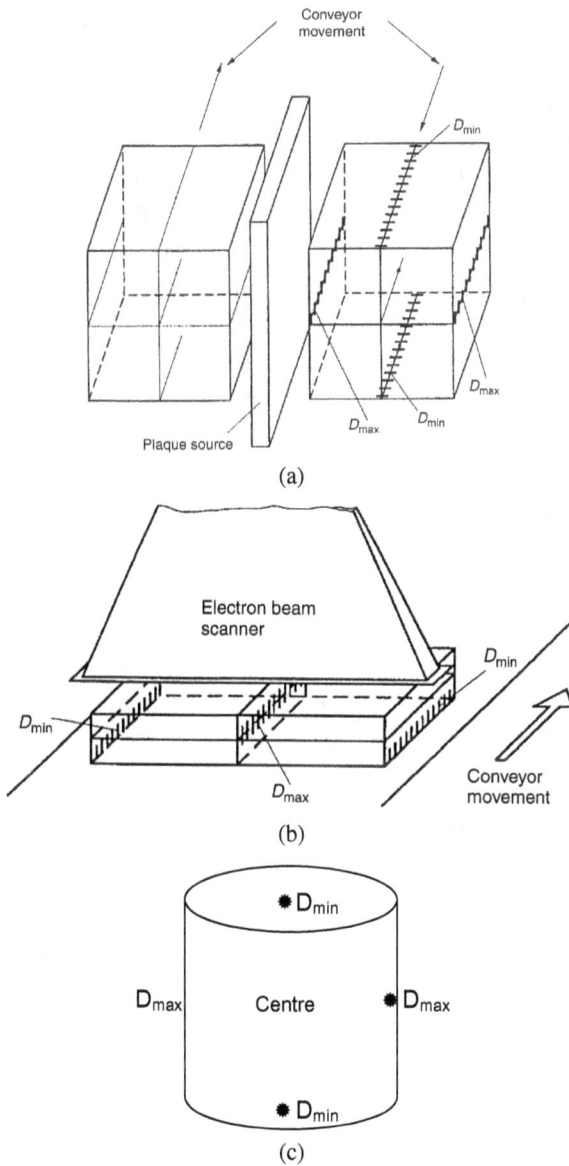

Fig. 5. Typical positions of minimum dose (D_{min}) and maximum dose (D_{max}) of box/package irradiated in (A) commercial gamma irradiator, (B) electron beam accelerator and (C) gamma cell.

DUR is a useful indicator to verify that the dose distribution in the irradiated box is uniform. Based on our experience, the best DUR obtained for a box of air-dried amnion with density of 0.18 g/cm^2 is 1.06 (Yusof, 2007). The DUR recommended by AAMI Guidelines (AAMI, 1992) ranges from 1.0 to 1.05. Such low DUR can be achieved by reducing the box/container size and/or rotating the box/container during irradiation or by doing more passes in the irradiation room.

Details in conducting the dose mapping has been described earlier (Yusof, 2007; Mod Ali, 2007). Figure 6 illustrates how to prepare a box containing long bone allografts for routine irradiation in frozen state. Bones, each packed in three layers (plastic, linen and plastic), are arranged in an insulated box and the space present in between the bones is filled up by pieces of dry ice in equivalent amount. At irradiation facility, the box will not be allowed to open. Dosimeters are placed outside the box at different positions including the positions of the minimum and maximum doses obtained from the dose mapping exercise. After irradiation, the reading of the absorbed dose at the minimum dose outside the box must be multiplied by a correction factor to determine the actual minimum absorbed dose inside the box. The correction factor should be determined during the dose mapping exercise when the dummy material was used, as follow.

$$\text{Correction factor} = \frac{D_{mix} \text{ inside the box}}{D_{min} \text{ outside the box}}$$

It is recommended to irradiate long or massive bones in a frozen state at 25 kGy to minimise radiation effects on bone physical properties. The correction factor obtained from our dose mapping exercise for frozen bones was 1.03 (Yusof, 2007).

Material Compatibility

As part of product validation, tissue bankers must conduct material compatibility tests after irradiation, particularly to evaluate:

- the tissue stability by conducting physical and chemical analyses and
- the packaging materials and the sealing condition, as the sterility can only be guaranteed when the packaging is intact or not damaged.

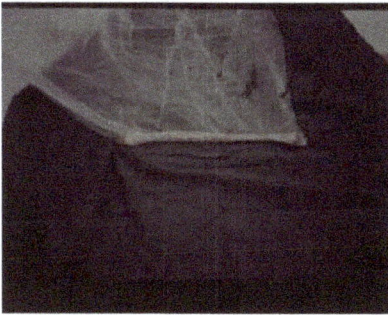

(a)

(b)

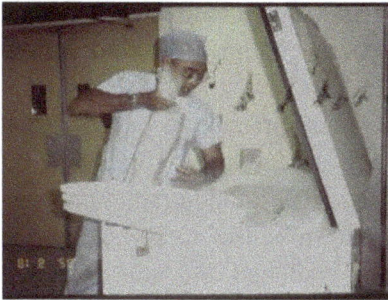

(c)

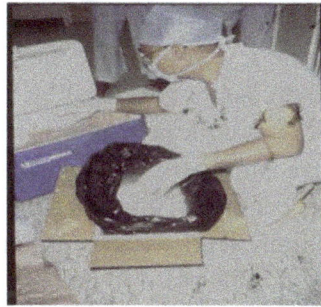

(d)

(e)

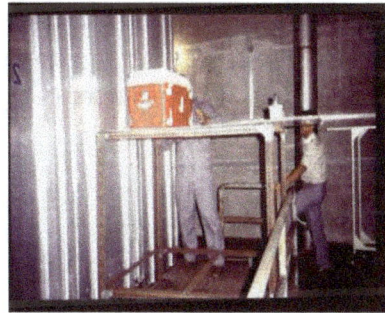

(f)

Fig. 6. Preparation of a box of bone allografts for irradiation in frozen state: (a) pack each bone in linen-plastic-linen, (b) packed bones are frozen at −80°C, (c) transfer frozen bones from the freezer into an insulated box, (d) fill in space between bones with dry ice, (e) place dosimeters on the box including at D_{min} and D_{max} positions and (f) position the box in the irradiation room.

Packaging materials that are radiation compatible must be selected for packing tissue grafts. A general guide on packaging material is given in ISO11137 (ISO, 2006), of which several radiation stable thermoplastic materials are listed. Due to rapid development in polymer science, it is recommended that tissue bankers must refer to plastic manufacturers for new radiation-stable packaging materials.

Product Control

Irradiation facilities must ensure that irradiated or sterile products are segregated from non-irradiated products (Mehta and Abdel-Fattah, 2008). In a well-designed facility, the area of untreated product and treated product is separated by tall fence. This will simplify the product inventory control procedures. It is a common practice to put radiation sensitive indicators sometimes referred to as "go-no go" indicator on product package. This indicator will change colour upon radiation, generally from yellow to red or sometimes from yellow to green. This is merely a guide to differentiate between irradiated and non-irradiated products. Use of the "go-no go" indicator does not replace the routine product dosimeters, as they are only qualitative indicators. In addition, the colour change is not stable after irradiation and will fade away during storage or when expose to light or heat. However, the indicators are useful for facility operators as a guide to differentiate the products before release.

Routine Process Control

Under process control, the process parameters must be carefully established, monitored and controlled. All measurements are performed with the best precision and accuracy attainable. All measurement systems and instrumentation must be calibrated with traceability to national or international standards. All recorded values of the process parameters are examined to verify their compliance with specifications, taking into account the uncertainty of the measurement system. Results from the dose mapping and the sterilisation dose validation exercises will be useful in preparing the standard procedure for irradiation of tissue products. Routine irradiation cannot deviate from the written procedure. The sterilisation centre has full responsibility to apply the validated irradiation procedure to the product during routine sterilisation and certificate will normally be issued indicating minimum and maximum absorbed doses received by each irradiation batch. Radiation centres are responsible to define the frequency at which dosimeters are used in routine processing. Any significant changes in the product configuration or bioburden will warrant new dose mapping and dose validation exercises.

Conclusion

Good Radiation Practices will ensure that tissue allografts will receive the requested sterilisation dose as accurate as possible and within an acceptable range. Allografts handled and processed according to Good Manufacturing Practices may have very low bioburden that will allow the possibility of using sterilisation doses lower than 25 kGy. Tissue bankers, as product manufacturers, are responsible for dose establishment in achieving the selected sterility assurance level for their products. Dose validation exercise is conducted according to ISO11137:2006 and records generated will be part of quality documents of the tissue bank. Each time a modification is performed the effect on the whole system must be evaluated and revalidated.

References

Association for the Advancement of Medical Instrumentation (AAMI) (1992). *AAMI Guidelines for Radiation Sterilization*, Arlington, VA.

Chmielewski AG and Berejka AJ (2008). Radiation sterilization centres worldwide. In *Trends in Radiation Sterilization of Health Care Products*, IAEA, Vienna, pp. 49–61.

Dziedzic-Goclawska A, Kaminski A, Uhrynowska-Tyszkiewicz I, Michalik J, and Stachowicz W (2008). Radiation sterilization of human tissue grafts. In *Trends in Radiation Sterilization of Health Care Products*, IAEA, Vienna, pp. 231–260.

International Organization for Standardization (ISO) (2006). *Sterilization of Health Care Products — Radiation*, ISO11137, Switzerland.

Mehta K and Abdel-Fattah AA (2008). Dosimetry and the radiation sterilization process. In *Trends in Radiation Sterilization of Health Care Products*, IAEA, Vienna, pp. 91–118.

Mod Ali N (2007). Dosimetry and requirements for process qualification. In Nather A, Yusof N and Hilmy N (eds.), *Radiation in Tissue Banking: Basic Science and Clinical Applications of Irradiated Tissue Allografts*, World Scientific Publishing, Singapore, pp. 171–186.

Tomford WW (2005). Effects of gamma irradiation on bone — clinical experience. In Kennedy JF, Phillips GO and Williams PA (eds.), *Sterilisation of Tissues Using Ionising Radiations*. CRC Woodhead Publishing, England, pp. 133–140.

Yusof N, Mod Ali N, and Hilmy N (2007). Type of radiation and irradiation facilities for sterilization of tissues grafts. In Nather A, Yusof N and Hilmy N (eds.), *Radiation in Tissue Banking: Basic Science and Clinical Applications of Irradiated Tissue Allografts*, World Scientific Publishing, Singapore, pp. 109–119.

Yusof N (2007). Validation of radiation dose distribution in boxes for frozen and nonfrozen tissue grafts. In Nather A, Yusof N and Hilmy N (eds.), *Radiation in Tissue Banking: Basic Science and Clinical Applications of Irradiated Tissue Allografts*, World Scientific Publishing, Singapore, pp. 187–199.

Chapter 26

Establishing Radiation Dose for Sterilisation of Tissue Allografts: 25 kGy and Lower

Norimah Yusof* and Nazly Hilmy[†]

Introduction

Radiation dose of 25 kGy has been widely used to sterilise medical products and it becomes so synonymous with sterilisation process. Many manufacturers consider it as a "magic" number, the dose that must be used for sterilisation regardless of what they produce or manufacture. However, this magic number is hard to use when their products are sensitive to radiation or not radiation compatible. As there are many new materials coming into the market for manufacturing new types of medical devices and new biomaterials, some of these materials may not be stable after radiation sterilised at 25 kGy. Similarly in tissue banking, 25 kGy has been reported to be rather too high for some soft tissues such as fascia lata, dura mater, heart valve and tendon, especially when the tissues are irradiated at room temperature. The dose has been found to enhance damage caused by freeze-drying in bones and skin (Yusof, 2000).

At present, radiation sterilisation of medical products is regulated by ISO standard, ISO 11137:2006 (ISO, 2006). The ISO 11137 provides clear guides on how to validate a radiation dose that can effectively

---

*Malaysian Nuclear Agency (NM), Bangi 43000 Kajang, Selangor, Malaysia.
†BATAN Research Tissue Bank (BRTB), Center for the Application of Isotopes and Radiation Technology, BATAN, Jakarta 12070, Indonesia.

sterilise medical products to attain the sterility assurance level (SAL) of 10^{-6}. The ISO standard comprised of three parts establishes the requirements for the development, validation and routine control of sterilisation process in the quality management system of an irradiation centre as well as the sterilisation process itself. This ISO is established for the radiation sterilisation of health care products, which include medical devices, medicinal products (pharmaceuticals and biological agents) and *in vitro* diagnostic products. In some countries, the ISO is applicable to tissue grafts where tissue products are considered a manufactured or a processed item. However, tissues are produced in a limited number and with physical and biological characteristics that are very different from medical products. In addition, tissue allografts are not products of commercial activity as they are not uniform in size, shape and density. Realising these differences, the International Atomic Energy Agency (IAEA) has established a Code of Practice for the radiation sterilisation of tissue allografts and its requirements for validation and routine of the sterilisation of tissues (IAEA, 2007). The IAEA Code has been prepared based on several ISO and AAMI documents. This chapter will guide you on how to use the ISO document and also the IAEA Code of Practice when you intend to select radiation doses for your tissue allografts and more importantly how to validate the dose. However, the approach is adopted only for tissues that have been screened for viral contamination and subjected to proper handling and processes that inactivate viruses. Even tissues contaminated with Gram-negative bacteria should be used.

Selection of a New Dose Other than 25 kGy

The International Organization for Standardization (ISO) is a worldwide federation of national standard bodies. The ISO document ISO 11137:2006 Part 2 (ISO, 2006) describes two methods for the establishment of the sterilisation dose. These methods are based on a probability model for the inactivation of microbial populations, as described in Chapter 24 of this book. The probability model, as applied to bioburden made up of a mixture of various microbial species, assumes that each species has its own unique D_{10} value. The probability that an item will possess a surviving microorganism after exposure to a given dose of radiation is defined in terms of:

- the initial number of microorganisms present on the item prior to irradiation (*bioburden*) and
- the D_{10} values of the microorganisms (*type of microorganisms*).

The ISO 11137-2 for establishing the sterilisation dose, offers:

- Methods 1 and 2 for dose setting to obtain a product specific dose and
- Method VD_{max} Approach for dose substantiation to verify a preselected dose of either 25 kGy or 15 kGy.

These methods involve performance of tests of sterility on product items that have received doses of radiation lower than the sterilisation dose. The test results are then used to predict the dose needed to achieve a predetermined sterility assurance level (SAL).

Method 1 of the ISO 11137:2006

This method depends on experimental verification that the radiation resistance of the bioburden is less than or equal to the resistance of a microbial population having the standard distribution of resistance (SDR). The SDR as in Table 1 specifies resistance of microorganisms in terms of D_{10} values and the probability of occurrence of values in the total population. The individual doses required to each SAL is established using computational methods. Method 1 requires a total of 130 product items, i.e. 10 items from each of three independent production batches for bioburden determination, and then select 100 of production items from a single batch of product, preferably one of the three batches for which a bioburden determination

Table 1. Standard distribution of resistance (SDR) microorganisms, used in Method 1 of ISO 11137:2006.

D_{10} (kGy)	1.0	1.5	2.0	2.5	2.8	3.1	3.4	3.7	4.0	4.2
Probability (%)	65.487	22.493	6.302	3.179	1.213	0.786	0.350	0.111	0.072	0.007

was carried out. These 100 items are exposed to a verification dose as tabulated in the document. Looking at the number of samples required for the test, this method is impossible to be used for tissue products.

Method 2 of the ISO 11137:2006

This method is used on products with microbial population not having the SDR. The method uses the results of tests of sterility conducted on product items that have been exposed to a series of incremental doses to estimate the dose at which one in 100 items is expected to be non-sterile. The D_{10} value estimated from this result is used to determine sterilisation dose by extrapolating to SAL values below 10^{-2}. Method 2 requires even more product items, i.e. 280 items from each of three production batches for incremental dose experiment and 100 for verification dose experiment. Method 2A is the method that has been generally applied for medical items, whereas Method 2B has been developed for products with a consistent and very low bioburden.

Method A of IAEA Code of Practice

As the number of samples for both Methods 1 and 2 are high, it is very unlikely for tissue banks to be able to establish a sterilisation dose based on these two methods. The IAEA Code of Practice offers Method A, which has similar approach as Method 1 of ISO 11137 but requires less sample size. Method A of the Code allows the use of ten samples to determine the verification dose. In this modification, the dose needed for an SAL value of 10^{-1} is used to establish the dose required for an SAL value of 10^{-6}. The total number of samples required is 20, ten for bioburden estimation and ten for the verification dose exercise.

There are two methods offered under Method A.

Method A

This method is for products having microbial population of SDR. It can be used for tissue allografts.

Method A(i)

This method is for products when the distribution of microbial radiation resistance is known and is different from the SDR.

Work examples are provided in Tables 2 and 3 for Method A and A(i), respectively. These simple work examples describe how to conduct the exercises systematically and at the same time hoping to guide the tissue bankers in understanding the Code requirements. At least 20 samples should be used, ten for bioburden test and ten for the verification dose experiment. In the work example, we are using Table III in Annex III of the Code to get the verification dose. The verification dose ±10% must be delivered. In this case, the gamma cell (self-shielded gamma irradiator) is preferable to deliver such an accurate dose compared to gamma irradiation facility. For both Methods A and A(i), the IAEA Code reminded us that low bioburden levels combined with low sample numbers will give rise to an increased probability of failure of the verification dose experiments. In the event of failure, it is suggested to choose 25 kGy and substantiate it as sterilisation dose and this may decrease the risk of failure. Hilmy *et al.* (2007) managed to use Method A to establish sterilisation doses of 19.4 kGy and 15.2 kGy for amnion and bone grafts, respectively.

Sampling According to IAEA Code of Practice

If samples can be prepared from tissues that are reasonably reproducible in shape, size and composition and also in sufficient numbers for statistical purposes, then the usual sampling procedures apply. However, if allograft products are few in number (less than ten) and cannot be considered as identical products, then it may be necessary to take multiple sample item portions (SIPs) of a single tissue allograft product for both bioburden analysis prior to sterilisation and for the purpose of establishing a sterilisation dose.

Assuming that the distribution of microorganisms is the same throughout the sample, the SIP shall validly represent the microbial challenge presented to the sterilisation process. In fact, SIPs may be used to verify that microorganisms are distributed evenly. It is important to ascertain that the samples to be used or items of the SIPs are reasonably uniform not only in shape, size and composition but also in bioburden.

Table 2. Work example in establishing sterilisation dose using Method A of the IAEA Code of Practice for products having microbial population of SDR. Sample: Amnion from screened mothers, treated with sodium hypochlorite for HIV inactivation and preserved by air-drying.

Stage	Value	Comments
Stage 1		
Production batch size	30	10×10 cm amnion samples, obtained from one processing batch of three amnions (sample sizes between ten and 100)
Test sample size for bioburden determination	10	Bioburden test on ten samples
Test sample size for the verification dose exercise	10	Verification dose required for SAL 10^{-1} ($=1/10$)
Stage 2		
Obtain samples	20	Ten samples for bioburden and ten samples for verification dose experiment
Stage 3		
SIP	1	The entire amnion is used
Average bioburden	7.6 cfu	Average bioburden of ten samples: 10, 9, 13, 6, 11, 3, 7, 5, 4 and 8 cfu/amnion. The average bioburden is less than 1000 cfu and therefore Method A may be used.
		Note: If an SIP < 1 was taken, then the bioburden of the whole amnion should be corrected and should be less than 1000 cfu per amnion for this method to be valid.

(Continued)

Table 2. *(Continued)*

Stage	Value	Comments
Stage 4		
Verification dose calculation	2.8 kGy	Assuming that the microbial population of amnion is of SDR, the verification dose is obtained from Table III-4 of the IAEA Code: for bioburden 8.0 and SAL 10^{-1} or 1/10, the verification dose is 2.8 kGy.
Stage 5		
Verification dose experiment	2.9 kGy delivered	The actual dose delivered to ten samples must be within 10% variation, i.e. 2.52–3.08 kGy. The experiment is accepted.
Sterility test	No positive	The verification exercise is accepted.
Stage 6		
Sterilisation dose	17.2 kGy	Refer to Table III-8, radiation dose required to achieve SAL of 10^{-6} at bioburden 8.0 is 17.2 kGy.

Note: The exercise used figures obtained directly from the tables as indicated in the IAEA Code and we did no calculate as suggested. The use of tables based on the bioburden estimation is simpler.

Table 3. Work example on establishing sterilisation dose using Method A(i) of the IAEA Code of Practice for products when the distribution of microbial radiation resistance is known and is different from the SDR. Sample: Amnion from screened mothers, treated with sodium hypochlorite for HIV inactivation and preserved by air-drying.

Stage	Value	Comments
Stage 1		
Production batch size	30 pieces/ batch	10×10 cm amnion samples, obtained from one processing batch of three amnions (minimum batch sizes of 20–79)
Test sample size for bioburden determination	10	Bioburden test on 19 samples
Test sample size for the verification dose exercise	10	Verification dose required for SAL 10^{-1} (=1/10)
Stage 2		
Obtain samples	20	Ten samples for bioburden and ten samples for verification dose experiment
Stage 3		
SIP	1	The entire amnion is used
Average bioburden	7.63 cfu	Average bioburden of ten samples: 10, 9, 13, 6, 11, 3, 7, 5, 4 and 8 cfu/amnion. The average bioburden is less than 1000 cfu and therefore Method A(i) may be used. Note: If an SIP < 1 was taken, then the bioburden of the whole amnion should be corrected and should be less than 1000 cfu per amnion for this method to be valid.

(*Continued*)

Table 3. (*Continued*)

Stage	Value	Comments
Stage 4		
Verification dose calculation $N_{tot} = 7.63 \times 10-(D/1.8)$ $N = N0 \times 10-(VD/D10)$ $VD = 1.8 \times (\log 7.63 - \log 10^{-1})$ $VD = 1.8 \times (0.883 + 1)$	3.4kGy	The microbial population of the amnion is known and differs from the SDR. The survival equation in Annex I is used to calculate the verification dose (VD) for an SAL value of 0.1 (N_{tot}), i.e. the reciprocal of the number of samples used, the total initial number of microorganisms (N_o) per amnion (SIP = 1) is 6.7 cfu and the D_{10} value of the known microbe is 1.8. The verification dose is calculated: $N_{tot} = N_o \times 10-(VD/D10)$ or $VD = D_{10} [\log N_o - \log N_{tot}]$ $= 1.8 [\log 7.63 - \log 10^{-1}]$ $= 3.39 \, kGy$
Stage 5		
Verification dose experiment	3.5kGy delivered	The actual dose delivered to ten samples must be within 10% variation, i.e. 3.06–3.74 kGy. The experiment is accepted.
Sterility test	No positive	The verification exercise is accepted.
Stage 6		
Sterilisation dose	12.4kGy	Sterilisation dose is calculated to achieve SAL of 10^{-6} at bioburden 7.63: $SD = D_{10} [\log N_o - \log N]$ $= 1.8 [\log 7.63 - \log 10^{-6}]$ $= 12.39 \, kGy$

Note: Sterilisation dose obtained is lower than in Work Example 1 even though the bioburden is the same.

The dose to be selected and validated can be lower than 25 kGy depending on the bioburden of the tissue products. By looking at Table III-8 of the Code and Table 5 of the ISO, this can be achieved when bioburden not exceeding 1000 colonies forming unit (cfu) per item. Tissues when processed properly could have bioburden less than 100 for amnion and in some cases bones could have bioburden less than ten (Hilmy *et al.*, 2007). However, skin is a naturally highly contaminated tissue, which has high bioburden even after radiation, with bioburden ranging from >10 to 162.7×10^3 cfu/100 cm^3 (Kairiyama *et al.*, 2009). When cleaning process is conducted in a clean room, bioburden less than one can be obtained (Baker *et al.*, 2005). Product with low bioburden is preferred to avoid the presence of any endotoxin. In practice, only clean products can be converted into sterile products because dirty products require high radiation dose to kill high microbial counts. Radiation can only clean dirty products but not sterilise them. Therefore, it is strongly recommended that products must be properly processed according good manufacturing practice (GMP) or good hygienic practice before considering any method of sterilisation. This sterilisation dose to be established is based solely on the number and the types of microorganisms on and in the tissues prior to irradiation (Yusof, 2005).

Validation of 15 kGy and 25 kGy Using VD$_{max}$ Approach

ISO 11137:2006 offers another method to establish sterilisation dose. The Method VD$_{max}$ allows the substantiation of two doses, i.e. 25 kGy or 15 kGy as the sterilisation dose. In carrying out the substantiation, the method verifies that bioburden present on product prior to sterilisation is less resistant to radiation than a microbial population of maximal resistance consistent. The method can be used to substantiate that the selected dose can achieve the sterility assurance level of 10^{-6}. In the exercise, it also requires a determination of bioburden and the performance of a verification dose.

When using VD$_{max}$ approach, sterilisation dose is already fixed. The method is generated from AAMI TIR 27 (2001). Tissue bankers are expected to know the performance or bioburden of their finished products before deciding on the dose. The bioburden from several production batches should be fairly constant and the tissue banker can assure that the bioburden is kept consistent by having GMP in the bank.

Method VD_{max}^{25} to Substantiate 25 kGy

The VD_{max}^{25} is applicable to products having an average bioburden less than or equal to 1000 cfu for multiple production batches. At least ten product items are selected from each of three independent production batches for average bioburden determination. Dose verification is conducted at an SAL of 10^{-1} using ten product items from a single batch of product. Table 9 of the ISO is referred to get the verification dose or VD_{max}^{25}. Use of SIP for less than one item is not permitted for product with an average bioburden less than 0.9 cfu per item. Items are not irradiated for VD_{max}^{25} of 0.0 kGy or bioburden 0. The dose 25 kGy is substantiated or accepted as the sterilisation dose when no more than one positive after sterility test. For a single production batch, ten samples for bioburden are taken from a single batch.

Method VD_{max}^{15} to Substantiate 15 kGy

This is applicable to product having an average bioburden less than 1.5 cfu per item. At least ten product items are selected from each of the three independent production batches for average bioburden determination. Dose verification is conducted at an SAL of 10^{-1} using ten product items. Use of SIP for less than one item is not permitted. Table 10 of the ISO is referred to get the verification dose or VD_{max}^{15}. Items are not irradiated for VD_{max}^{15} of 0.0 kGy. The 15 kGy dose is substantiated or accepted as the sterilisation dose when there is no more than one positive after sterility test.

Work examples as in Tables 4 and 5 are presented to give better understanding in using VD_{max}^{15} and VD_{max}^{25}, respectively.

The method allows tissues to be sterilised at 15 kGy; thus, the damage due to radiation can be definitely minimised, especially for soft tissues whereby radiation doses higher than 25 kGy have been reported to affect the physical properties of the tissues (Tomford, 2005; Koller, 2005; Yusof, 2000).

Figure 1 illustrates the steps involved in conducting dose selection and dose validation exercises in establishing sterilisation dose for air-dried amnion.

Table 4. Work example on substantiating 25 kGy as sterilisation dose using VD_{max}^{25} Approach of the ISO 11137:2006 (SIP = 1.0). Sample: Amnion from screened mothers, treated with sodium hypochlorite for HIV inactivation and preserved by air-drying.

Term	Value	Comments
Stage 1		
SAL	10^{-6}	This method substantiates 25 kGy as a sterilisation dose to achieve maximally an SAL of 10^{-6}.
SIP	1.0	The entire amnion was used for testing.
Number of product items	40	Ten from each of three batches for bioburden determination and ten for the verification dose experiment.
Stage 2		
Overall average bioburden	7.63 cfu	Overall average bioburden of three processing batches was 7.63 cfu:
		Batch 1: 9.3 cfu/amnion,
		Batch 2: 6.1 cfu/amnion and
		Batch 3: 7.5 cfu/amnion.
		None of the individual batch average bioburdens was twice the overall average bioburden, therefore 7.63 is used to calculate the verification dose. As the bioburden is <1000 cfu, the method can be used.
Stage 3		
Verification dose	6.9 kGy	Use Table 9 to obtain the verification dose. A bioburden of 7.63 cfu is not listed in the table, so the next higher bioburden of 8.0 is used. The VD_{max}^{25} dose for an SIP of one is 6.9.
Stage 4		
Results of tests of sterility	0 positives	The highest dose to any items was 7.0 kGy and the mean was 6.5 kGy. The highest dose to the product items did not exceed VD_{max}^{25} by more than 10% and the arithmetic mean of the highest and the lowest doses delivered to the product items is >90% of the VD_{max}^{25}.
Sterilisation dose	25 kGy	The tests of sterility results were acceptable as the allowed limit is one positive test. Therefore, 25 kGy as a sterilisation dose for SAL 10^{-6} is substantiated.

Note: This example is using VD_{max}^{25} for multiple production batches. For tissues seldom processed, VD_{max}^{25} for a single production batch can be considered. Ten samples for bioburden determination and ten samples for verification experiment are taken from the single batch.

Table 5. Work example on substantiating 15 kGy as sterilisation dose using VD_{max}^{15} Approach of the ISO 11137:2006 (SIP = 1.0). Sample: Amnion from screened mothers, treated with sodium hypochlorite for HIV inactivation and preserved by air-drying.

Stage	Value	Comments
Stage 1		
SAL	10^{-6}	This method substantiates 15 kGy as a sterilisation dose to achieve maximally an SAL of 10^{-6}.
SIP	1.0	In applying the VD_{max}^{15}, the entire product shall be used. Therefore, the entire amnion was used for testing.
Number of product items	40	Ten from each of three batches for bioburden determination and ten for the verification dose experiment.
Stage 2		
Overall average bioburden	0.73 cfu	Overall average bioburden of three processing batches:
		Batch 1: 0.8 cfu/amnion.
		Batch 2: 0.7 cfu/amnion and
		Batch 3: 0.7 cfu/amnion.
		As the bioburden is ≤1.5 cfu, the method can be used. None of the individual batch average bioburdens was twice the overall average bioburden; therefore, 0.73 is used to calculate the verification dose.
Stage 3		
Verification dose	2.3 kGy	Use Table 10 to obtain the verification dose. A bioburden of 0.73 cfu is not listed in the table, so the next higher bioburden of 0.8 is used.
Stage 4		
Results of tests of sterility	0 positives	The highest dose to any items was 2.5 kGy and the arithmetic mean was 2.3 kGy. The dose delivered to the product items was within the specified range.
Sterilisation dose	15 kGy	The results of tests of sterility were acceptable as the allowed limit is one positive test. Therefore, 15 kGy as a sterilisation dose (SAL 10^{-6}) is substantiated.

Note: This example is using VD_{max}^{15} for multiple production batches. For tissues seldom processed, VD_{max}^{15} for a single production batch can be considered. Ten samples for bioburden determination and ten samples for verification experiment are taken from the single batch.

1. Procure amnion from placenta.

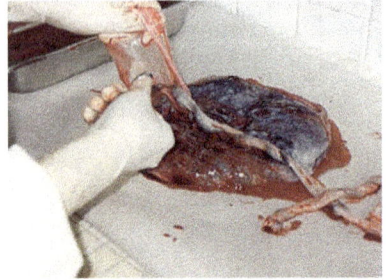

2. Process amnion.

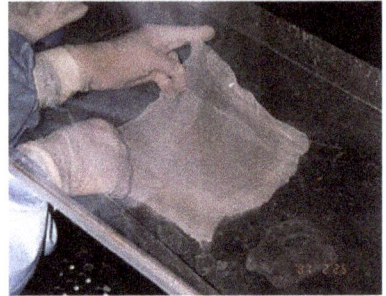

3. Air-dry amnion.

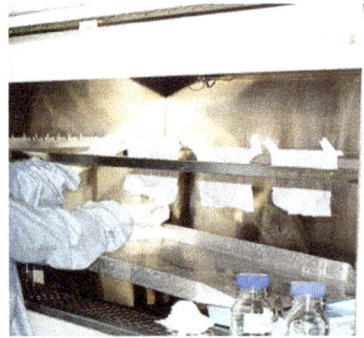

4. Collect ten samples from each of the three batches for bioburden determination and ten sample for the verification dose experiment.

   SIP = 1 when entire amnion is tested

Fig. 1.   Processing and establishing sterilisation dose for air-dried amnion.

5. Determine average bioburden for each of the three batches and then determine the overall average bioburden.

6. Expose ten samples to verification dose.

7. Conduct test for sterility.

8. Send product for sterilisation when the verification exercise is accepted.

Fig. 1. (*Continued*)

Acceptance of the Verification Dose Experiment

The verification experiment is valid and the verification dose is accepted when the tests of sterility are conformed to the total positives as stated by each method. Do not accept the verification if the positives are exceeding the allowed number. The verification dose experiment can be repeated following implementation of corrective action if only the outcome can be ascribed to incorrect performance of the bioburden test, incorrect performance of sterility test and incorrect delivery of the verification dose. If the outcome cannot be ascribed to a cause addressed by corrective action, i.e. any of the three mentioned above, an alternative method shall be used. It is advisable to review the procedures and the practices in the tissue bank before embarking on another verification exercise or considering an alternative method.

Routine Irradiation

It is advisable that the validation exercise on radiation sterilisation dose is conducted only when tissues produced are of consistent quality. The tissues must have been processed by trained staff and according to the validated procedures under the GMP. The sterilisation dose for routine irradiation can be any doses other than 25 kGy after it has been determined during the verification dose experiment based on the bioburden performance according to Method A of the IAEA Code. Alternatively, the sterilisation dose will be the preselected dose, either 15 or 25 kGy, which has been substantiated using VD_{max} approach. The established sterilisation dose can now be used as a validated dose for routine production, provided that there is no significant deviation in the processing and the tissue bank practices that can influence the product bioburden.

Auditing Sterilisation Dose

According to ISO 11137:2006, once the sterilisation dose has been established, periodic audits shall be carried out to confirm the continued appropriateness of the sterilisation dose. The audits shall be performed every three months to determine the continued validity of the established dose.

However, the method requires 110 samples, ten samples for bioburden determination and 100 to be irradiated at the dose originally used to establish the SAL of 10^{-2} and then tested for sterility. Verification is accepted when there are no more than two positive tests of sterility from the irradiated 100 samples. This method is very difficult to adopt for tissue allografts. At this juncture, it is advisable for tissue banks to monitor closely the performance of processed tissues for every batch. Bioburden should be made as one of the quality control parameters of the processed tissues prior to sterilisation by radiation. Re-establishment of the sterilisation dose must be conducted when there are changes in the tissue processing, environment or tissue components.

Conclusion

Tissue bankers are responsible to establish sterilisation dose other than 25 kGy using ISO 11137 document or IAEA Code of Practice. Generally, by lowering bioburden, radiation dose can also be lowered. Therefore, radiation damage in tissue, especially soft tissue, can be minimised or avoided, hence restoring the functional roles of the tissue. When 25 kGy is still the choice, dose validation, as required under the quality system, can be conducted using Method VD_{max} of ISO 11137.

References

American Association of Medical Instrument (AAMI) (2001). *Sterilization of Health Care Products — Radiation Sterilization — Substantiation of 25 kGy as a Sterilization Dose*. AAMI TIR 27, Arlington VA.

Baker TF, Ronholdt CJ, and Bogdansky S (2005). Validating a low dose gamma irradiation process for sterilizing allografts using ISO 11137 method 2B. *Cell Tissue Bank* **6**: 271–275.

Hilmy N, Febrida A, and Basril A (2007). Experiences using IAEA Code of practice for radiation sterilisation of tissue allografts: Validation and routine control. *Radiat Phys Chem Oxf Engl* **76**: 1751–1755.

International Atomic Energy Agency (IAEA) (2007). *Radiation Sterilisation of Tissue Allografts: Requirements for Validation and Routine Control — A Code of Practice*, IAEA, Vienna.

International Organization for Standardization (ISO) (2006). *Sterilization of Health Care Products — Radiation — Part 2: Establishing the Sterilization Dose*, ISO 11137, Switzerland.

Kairiyama E, Horak C, Spinosa M, Pachado J, and Schwint O (2009). Radiation sterilisation of skin allograft. *J Radiat Phys Chem* **78**: 445–448.

Koller J (2005). Effects of radiation on the integrity and functionality of amnion and skin grafts. In Kennedy JF, Phillips GO, and Williams PA (eds.), *Sterilisation of Tissues Using Ionising Radiations,* CRC Woodhead Publishing, England, pp. 197–220.

Tomford WW (2005). Effects of gamma irradiation on bone — clinical experience. In Kennedy JF, Phillips GO, and Williams PA (eds.), *Sterilization of Tissues Using Ionising Radiations*, CRC Woodhead Publishing, England, pp. 133–140.

Yusof N (2000). Gamma irradiation for sterilising tissue grafts and for viral inactivation. *Malays J Nucl Sci* **18**(1): 23–35.

Yusof N (2005). Is the irradiation dose of 25 kGy enough to sterilise tissue grafts? In Nather A (ed.), *Bone Grafts and Bone Substitutes — Basic Science and Clinical Applications*, World Scientific, Singapore, pp. 189–212.

Chapter 27

Bioburden Estimation in Tissue Banking

Norimah Yusof*, Asnah Hassan* and Nazly Hilmy[†]

Introduction

A sterile tissue product must be free from viable microorganisms. In producing a sterile product, the tissue must be processed under clean environment and all practical means to minimise contamination are applied. Nevertheless, even product processed under strictly hygienic condition and good manufacturing practices in accordance with the requirements of quality systems for tissue banking may, prior to sterilisation, still have microorganisms on it. Therefore, such product is non-sterile and terminal sterilisation process is required to inactivate the microbiological contaminants. Sterilisation, using either physical or chemical method, is considered as a special process because its process efficiency cannot be verified by inspection and testing of the treated product. For this particular reason, sterilisation processes have to be validated before use; their performances have to be monitored routinely and the equipment need to be properly calibrated and maintained. Three international standards specifying procedures for validation and routine control of the commonly used sterilisation processes are available: ISO 11134:1994 for moist heat, ISO 11135:1994 for ethylene oxide and ISO 11137:1995 for radiation sterilisation (ISO, 1995). However, the provision of assurance that a product is

*Malaysian Nuclear Agency (NM), Bangi 43000 Kajang, Selangor, Malaysia.
[†]BATAN Research Tissue Bank (BRTB), Center for the Application of Isotopes and Radiation Technology, BATAN, Jakarta 12070, Indonesia.

sterile is not only determined by the validated and accurately controlled sterilisation processes, it is very important to determine the microbiological challenge of the product that is presented to any sterilisation process, in terms of number and type. According to standard document ISO 11737-1:1995 (ISO, 1995), the presterilisation microbiological contamination is commonly termed as *bioburden,* which is defined as the quantity that describes the population of viable microorganisms present on a product or item.

Bioburden Estimation

Bioburden reflects the microbiological contamination level of a product. The contaminants may originate from various sources: the environment, in which the product is processed or manufactured, the personnel who process/manufacture it, the machinery and tools used during processing and the raw materials or components used. It is not possible to determine bioburden exactly. In practice, a viable count of microorganisms is determined using a defined technique, which requires validation exercises to be performed to relate the viable count to a bioburden estimate on a product by the application of a correction factor. In ISO 11737-1:1995, bioburden estimation is defined as a value established for the number of microorganisms comprising the bioburden by applying to a viable presterilisation count a factor compensating for the recovery efficiency. Determination of a correction factor will be described in the later part of this chapter.

Bioburden estimation is performed in practice not only for products prior to sterilisation process, where bioburden is used as product quality control, but also for other aspects of tissue banking, as follow:

- validation and routine control of tissue processing procedures,
- validation and revalidation of a sterilisation process,
- validation and routine control of a manufacturing process,
- overall environmental monitoring programme,
- assessment of the efficiency of a cleaning process in removing microorganisms,
- monitoring products that require microbiological cleanliness and
- monitoring of incoming materials: tissues, components (water, solutions, reagents and media) and packaging materials.

Bioburden estimation is performed under controlled condition in a microbiology laboratory. If located on the site of a tissue bank, the management and the operation of the facility should be within the tissue bank's quality system. Generally, tissue banks of any hospital usually rely on a common microbiology laboratory of the hospital. Unfortunately most of the hospital microbiology laboratories may conduct only microbiological growth test in their routine work. In that case, the tissue bank must specify the requirement for the laboratory to conduct bioburden estimation. Alternatively, tissue bank may outsource the microbiological analyses to independent laboratories.

Requirements for Bioburden Estimation

The ISO 11737-1:1995 has described in detail the components involved in conducting bioburden estimation. The following components will be discussed in view of their practicality in tissue banking:

- personnel,
- equipment,
- microbiological media and eluents,
- selection of product samples for test,
- selection of technique,
- validation of techniques and
- calculation of bioburden estimate.

Personnel

Specific personnel must be assigned for bioburden estimation work if the tissue bank has a microbiology laboratory in its setup. Training should be provided/conducted in accordance with documented procedures. Under quality system, records of the relevant qualifications, training and experience of the technical personnel shall be maintained.

Equipment

Basic equipment required in a microbiology laboratory is listed in Table 1. Scheduled maintenance must be conducted in accordance with documented

Table 1. Laboratory equipment and tools for bioburden estimation work.

Equipment	Tools	Others
Laminar air-flow cabinet	Membrane filtration system and pumps	Surgical gloves
Incubators	Anaerobic jars	Head covers
Autoclave	Forceps	Shoe covers
Water purification system	Scissors	Laboratory coats
Water bath	Spreaders	Face masks
Microscopes	Inoculating loops	Disinfectants
Dry steriliser (oven)	Thermometers	Hand soaps and creams
Refrigerators	Petri dishes	Staining kit
Vortex mixer	Glass slides and covers	Media
Magnetic stirrer hotplates	Glassware (bottles, beakers, flasks, test tubes, pipettes, etc.)	Chemicals
Ultrasonic cleaner	Automatic pipettes and tips	Identification kit
Sealing machine	Bunsen burner and gas tank	
Electronic balances		
pH meter		

Note: This list is not exhaustive.

procedures and records of maintenance shall be retained. Calibration must be carried out on equipment having measurement or control function.

Any equipment or parts of the equipment that comes into contact with tissue, eluent, media, etc. during processing or testing should be sterile.

Microbiological Media and Eluents

All microbiological media and eluents used to remove microorganisms from product should be prepared sterile. The selected media should be able to support growth of microorganisms. A growth promotion test is done on each batch of medium using an inoculum of low numbers between 10 and 100 colony forming units (cfu) of a selected microorganisms. Examples of media and incubation conditions as suggested in Table 2 can be used as a guide in choosing media. All non-selective anaerobic culture methods may also permit the growth of facultative anaerobic organisms.

Table 2. Examples of media and incubation conditions (ISO, 1995).

Types of microorganism	Solid media	Liquid media	Incubation[a] conditions
Non-selective aerobic bacteria	(Soya bean casein digest agar) Tryptone soya agar Nutrient agar Blood agar Glucose tryptone agar	(Soya bean casein digest broth) Tryptone soya broth Nutrient broth	30°C to 35°C For two days to five days
Yeast and moulds	Sabouraud dextrose agar Malt extract agar Rose Bengal Chloramphenicol agar (Soya bean casein digest agar) Tryptone soya agar	Sabouraud dextrose broth Malt extract broth (Soya bean casein digest both) Tryptone soya broth	20°C to 25°C For two days to five days
Anaerobic bacteria	Reinforced clostridial agar[b] Schaedler agar[b] Pre-reduced blood agar[b] Fastidious anaerobe agar[b] Wilken–Chalgren agar[b]	Robertson's cooked meat broth Fluid thioglycollate broth	30°C to 35°C For two days to five days

Note: This list is not exhaustive.

[a] The incubation conditions listed are those which are commonly used for the types of microorganisms listed.

[b] Cultured under anaerobic conditions.

During bioburden estimation, eluents are used to remove microorganisms from the product and transfer removed microorganisms for enumeration, while diluents may be employed to obtain suspensions containing microorganisms in countable numbers. Compositions of selected eluents and diluents should not cause any proliferation or inactivation of microorganisms. The use of proper eluents and diluents can influence the overall efficiency of the bioburden method used. Commonly used eluents and diluents as listed in Table 3 can be considered for use.

When a liquid is used for the removal of microorganisms from solid surfaces, the addition of a surfactant such as polysorbate into the

Table 3. Examples of eluents and diluents (ISO 11737-1:1995).

Solution	Concentration in water	Applications
Ringer	¼ Strength	General
Peptone water	0.1%–1.0%	General
Buffered peptone water	0.067 M phosphate 0.43% Sodium chloride 0.1% Peptone	General
Phosphate-buffered saline	0.02 M phosphate 0.9% Sodium chloride	General
Sodium chloride	0.25%–0.9%	General
Calgon Ringer	¼ Strength	Dissolution of calcium alginate swabs
Thiosulfate Ringer water	¼ Strength	Neutralisation of residual chloride Dilution of aqueous samples Preparation of isotonic solutions of soluble materials prior to counting

Note: This list is not exhaustive.

selected eluent may be considered. Recovery of microorganisms from product surfaces may be enhanced by the presence of surface-active agent in the eluent and can be improved further by subjecting the sample to a physical treatment while in eluent, e.g. stomaching, ultra-sonicating and shaking. The eluents and diluents to be used in routine work must be validated to show their efficiency in giving the highest bioburden count. Documentation of the validation exercises and the rationales for decisions taken will assist in subsequent review of procedures.

Selection of Product Samples for Test

In sampling product prior to sterilisation for bioburden estimation, samples in their standard finished packaging are normally taken at random. Due to the limited availability of tissue products for sampling, it is possible to take products that are not suitable for clinical use but have undergone all processing steps including cleaning and packaging, preferably treated in the same processing batch, i.e. using the same batch of washing solutions and chemical, if use any, and carried out by the same personnel (Yusof, 2001). During sampling and handling of the samples, any introduction of advertent contamination and significant alterations to the numbers and nature of microorganisms in the sample must be avoided. Sampling procedures should be consistent; thus, comparisons of data can be made over a period of time.

The bioburden estimation should utilise the whole product whenever possible. If not feasible, such as when dealing with large bone, a portion can be used. The selected portion must be able to reflect as closely as possible the bioburden of the whole products. Ideally, bioburden should be estimated for each tissue type on a regular basis. However, in tissue banking there is a wide range of products, from amniotic membrane to soft tissues and bones, often produced in small batches and quantities. In such circumstances, products may be grouped on the basis of:

- product type, e.g. amnions, femoral heads or skin,
- processing method, e.g. freeze dried or freezing and
- production batch.

The rationale for grouping such products should be documented and should ensure that the bioburden are representative of all products in the group.

Sampling may be performed at a frequency based on time, e.g. monthly or based on production batch, e.g. every batch, alternate batches, or every three to five batches. However, in order to establish baseline levels, it is advisable to perform bioburden estimations for every production batch during the initial production of a new tissue product. The frequency can be reduced as the knowledge of the bioburden develops or the bioburden shows consistency. The frequency of bioburden estimations should allow detection of changes in bioburden due to changes in processing/manufacturing, changes in materials, changes of personnel or even environment. Bioburden estimation is therefore a useful indicator for monitoring the production.

Selection of Technique

The sequence of main stages of the technique for bioburden estimation of microbial contamination is illustrated in Fig. 1 (Yusof, 2001).

After product sampling, samples are subjected to treatment. Swabbing is recommended instead of taking samples for large and irregularly shaped tissue such as long bones. The swab moist with buffer or liquid medium is used to wipe a predetermined sampling surface area. The tissue sample or the swab is immersed in a known volume of an eluent. If surfactant is used to enhance recovery, the concentration is practically low enough because at high concentration it may be inhibitory to microorganisms. Any physical treatment used should be reproducible and should avoid conditions that are likely to affect the viability of microorganisms, such as excessive shear forces, temperature rise or osmotic shock. The time of treatment must be specified for optimum removal of microorganisms released from the samples. The time of contact between the sample and the eluent, the rate of flushing and the volume of fluid should be specified.

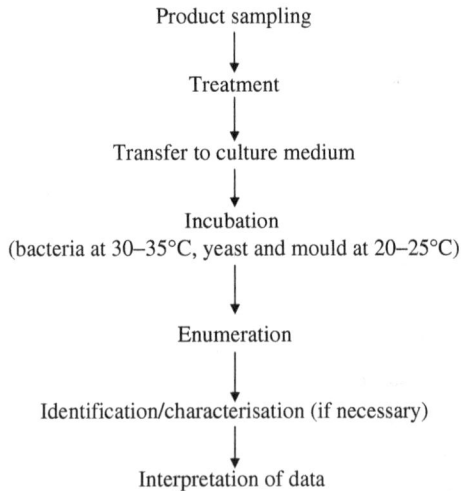

Product sampling
↓
Treatment
↓
Transfer to culture medium
↓
Incubation
(bacteria at 30–35°C, yeast and mould at 20–25°C)
↓
Enumeration
↓
Identification/characterisation (if necessary)
↓
Interpretation of data

Fig. 1. The sequence of main stages of the technique for bioburden estimation of microbial contamination.

The transfer of culture medium is carried out after the microorganisms are well separated into single cell suspension in the eluent. There are several enumeration methods for colony counts where each viable microorganism should be able to express itself as a visible colony on medium agar. Three most commonly used methods are described here.

Membrane Filtration

Cell suspension is filtered through a membrane filter followed by incubation of the filter on appropriate growth medium to produce visible colonies. This method is an effective means to detect low bioburden where the membrane filter having pore size of 0.45 μm is generally used to capture the microorganisms that are passing through. The membrane is washed by filtering sterile water and then is placed on agar surface or on an absorbent pad soaked in nutrient medium for incubation. Colonies will appear on the surface of the membrane filter after five to seven days. They are counted and isolated for characterisation and identification. The count is expressed as cfu per product unit.

Pour Plating

Cell suspension is mixed with molten agar at a temperature not exceeding 45°C, which is then poured into a petri dish. The mixture is swirl gently to disperse the microorganisms evenly before allowing the gel to solidify. Colonies grown within and on the agar surface are counted and isolated. Pour plating is used for products of high bioburden and after a serial dilution, only small volume of cell suspension is used.

Spread Plates

Cell suspension is spread on agar surface using a spreader. The cell suspension has to be absorbed into the agar so that discrete colonies can develop. The volume of the suspension taken for the test is rather small.

Agar plates are incubated invertedly at 30°C to 35°C for bacterial cultures and at 20°C to 25°C for yeast and mould cultures. Incubation period as recommended in ISO 11737-1:1995 varied from two to seven days as mentioned in Table 3.

Validation of Techniques

The validation of bioburden estimation techniques largely depends upon:

- the ability of the selected media to support growth (ensuring recovery of the microorganisms comprising the bioburden) and
- the ability of the selected temperature and incubation time to support the growth in the selected medium.

The validation of the bioburden estimation techniques ultimately should lead to an insight into the microflora existing on a product. The ISO 11737-1:1995 requests all the techniques used be validated and the recovery efficiency determined. Two approaches available for validation of the efficiency of removal of microorganisms from tissues are:

- repetitive treatment of a sample product and
- product inoculation with known levels of microorganisms.

The first approach has the advantage of using the naturally occurring microbiological contamination but requires a relatively high initial bioburden. The second approach creates a model system for testing purposes suitable for low levels of natural contamination.

Repetitive Recovery Method

This approach uses the bioburden as it occurs naturally on product, some-times referred to as an "exhaustive recovery". The bioburden estimation method should be repeated until there is no significant increase in the accumulated number of microorganisms recovered. After each repetition, the eluent is totally recovered from the product. Results accumulated from the consecutive recoveries are compared.

Product Inoculation Method

An artificial bioburden is created by inoculating known numbers of selected microorganisms onto product in order to establish recovery efficiencies.

Aerobic bacterial spores are most commonly used. A ratio of the recovered microbes to initial inoculums establishes the recovery efficiency for the particular method and product.

Calculation of Bioburden Estimate

Enumeration of microbial colonies grown on agar plates is the sum of the total of bacterial colonies (aerobic and anaerobic) and the total colonies of yeast and mould. The bioburden estimation of a product is obtained by applying a correction factor to the enumeration, compensating for the recovery efficiency of the observed colonies. It is not necessary to isolate and identify each type of contaminants. From years of experience in tissue banking, we suggest that identification of the most commonly found contaminants is sufficient and preferably if the most resistant colonies to the selected sterilisation treatment can be isolated. In the case of radiation as the sterilisation treatment, some species of *Bacillus* are radiation resistant.

When bioburden estimates are being used to establish the extent of the sterilisation treatment/process such as determining radiation dose, it is important that an accurate bioburden estimate is obtained. It is important to employ a precise method for estimating bioburden to detect changes before a level is reached at which the validity of the sterilisation process is affected. The bioburden is required to confirm the adequacy of an established sterilisation process.

Calculation of Correction Factors

Two examples as described in ISO 11737-1:1995 are presented here to illustrate how the ca. correction factor is determined.

Example A: Using repetitive recovery method

Bioburden estimates of five replicates of a tissue product are shown in Table 4. In each replicate, bioburden estimate is repeated four times with fresh eluent for each treatment. At the final stage, agar overlay is conducted by coating the surfaces of the product with a molten agar culture medium (at a maximum temperature of 45°C) and incubated to produce

Table 4. Colony counts determined from repetitive treatment for five replicates of a product and calculation of correction factor (ISO 11737-1:1995).

Treatment	Replicate count					Mean colony count
	1	2	3	4	5	
1	60	50	70	55	45	56.0
2	10	12	5	2	3	6.4
3	1	0	2	0	0	0.6
4	0	1	0	0	1	0.4
Agar overlay	10	5	7	4	2	5.6
Total colony count	81	68	84	61	51	69.0
% Removal*	74	74	83	90	88	

Average recovery = 81.8% (range = 74%–90%)
Correction factor = 100/81.8 = 1.22

*% Removal = treatment 1/total colony count.

visible colonies. Percentage of recovery is calculated by dividing colony count in the first treatment with total colony count for each replicate. The average recovery from five replicates is then obtained. The correction factor is determined as follow:

Correction factor = 100/average recovery = 100/81.8 = 1.22.

In routine bioburden estimation, the cfu of the product are multiplied with the correction factor to get the bioburden estimation, as follow:

Bioburden estimate = [colony count at presterilisation × 1.22] cfu.

Example B: Using product inoculation method

An aqueous suspension of *Bacillus subtilis* var. *niger* is prepared and the viable count of the suspension is determined. A dilution of the suspension is prepared such that 0.1-ml aliquots contain 100 spores. A selected portion of the sample or the whole product is inoculated with 0.1 ml of this diluted suspension and allowed to dry under laminar air flow. The inoculated product is subjected to the chosen removal technique and the mean number of the *B. subtilis* spores removed is obtained. Therefore, the correction factor for recovery efficiency is calculated as below:

Number of *B. subtilis* spore inoculated = 100 cfu

Number of *B. subtilis* spore removed
= 35 (with the range from 25 to 40).

Therefore,

Correction factor = 100/35 = 2.9

Bioburden estimate = colony counts × 2.9.

Table 5. Bioburden estimations of tissues and correction factor for recovery efficiency (ISO 11737-1:1995).

Tissues	Bioburden estimates	Correction factor	References
Skin	>10 to 1.62×10^5 cfu/100 cm^2	NA	Kairiyama *et al.* (2009)
Cancellous bone chips	1.95 cfu/packet	2.5	Hilmy *et al.*
Demineralised bone powder	1.90 cfu/packet		(2007)
	22.0 cfu/packet		
Femoral heads (frozen)	32.5 cfu/packet		
Amnion			
Femoral heads	0.0 cfu/packet	3.5	Nguyen *et al.*
Milled bone	0.0 cfu/packet	1.65	(2008)
Structural bone	0.0 cfu/packet	1.4	
Incoming allograft tissues (Achilles tendon, femur, fibula, hemi-pelvis, iliac crest, tibia, humerus)	<20 to >1.6×10^6 cfu/tissue	3.5	Ronholdt and Bogdansky (2005)
Cortical, cortical/cancellous and soft tissue	7 to 99 cfu/article	NA	Ronholdt *et al.* (2005)
Amnion	14.5 cfu/100 cm^2		Yusof *et al.* (2005)

NA: Not available/not mentioned in the references.

In routine bioburden estimation, the cfu of the product are multiplied with the correction factor to get the bioburden estimation, as follow:

Bioburden estimate = [colony count at presterilisation × 2.9] cfu.

Some bioburden estimates of allografts reported by several tissue banks are summarised in Table 5.

Conclusion

Bioburden estimates of tissues elaborated in this chapter are obtained by adopting the methodology specified in the ISO document ISO 11737-1:1995. All media and eluents to be used in routine bioburden estimation must be validated. Enumeration method chosen must be able to capture almost all the viable counts. Apply correction factor to the presterilisation colony counts in order to obtain the bioburden estimates. Accurate bioburden estimate is required when the estimate is used to establish the extent of sterilisation process. As for radiation treatment, bioburden is used to select and verify the sterilisation dose. Bioburden estimate can be used as a sensitive tool for detecting changes in manufacturing processing or environment.

References

Hilmy N, Febrida A, and Basril A (2007). Experiences using IAEA Code of Practice for radiation sterilization of tissue allografts: validation and routine control. *Radiat Phys Chem* **76**: 1751–1755.

International Organization for Standardization (ISO) (1995). *Sterilization of Medical Devices — Microbiological Methods — Part 1: Estimation of Population of Microorganisms on Products*. ISO 11737-1:1995, Geneva.

Kairiyama E, Horak C, Spinosa M, Pachado J, and Schwint O (2009). Radiation sterilization of skin allograft. *Radiat Phys Chem* **78**: 445–448.

Nguyen H, Morgan DAF, Sly LI, Benkovich M, Cull S, and Forwood MR (2008). Validation of 15 kGy as a radiation sterilisation dose for bone allografts manufactured at the Queensland Bone Bank: application of the VDmax 15 method. *Cell Tissue Bank* **9**: 139–147.

Ronholdt CJ and Bogdansky S (2005). Determination of microbial bioburden levels on pre-processed allograft tissues. In Kennedy JF, Phillips GO and William PA (eds.), *Sterilisation of Tissues Using Ionising Radiation*, Woodhead Publishing Limited, pp. 311–318.

Ronholdt CJ, Bogdansky S, and Baker TF (2005). Establishing an appropriate terminal sterilisation dose based upon post-processing bioburden levels on allograft tissues. In Kennedy JF, Phillips GO and William PA (eds.), *Sterilisation of Tissues Using Ionising Radiation*, Woodhead Publishing Limited, pp. 303–309.

Yusof N (2001). Bioburden estimation in relation to sterilisation. In Phillips GO, Strong DM, von Versen R and Nather A (eds.), *Advances in Tissue Banking*, Vol. 5, World Scientific, Singapore, pp. 200–211.

Yusof N, Shamsudin AR, Mohamad H, Hassan A, Yong AC, and Rahman MNF (2005). Bioburden estimation in relation to tissue product quality and radiation dose validation. In Kennedy JF, Phillips GO and William PA (eds.), *Sterilisation of Tissues Using Ionising Radiation*, Woodhead Publishing Limited, pp. 319–329.

Chapter 28

Elimination of Viruses, Prion and Emerging Infectious Viral Agent in Allografts

Nazly Hilmy* and Paramita Pandansari*

Introduction

Studies of the possibility of disease transmission through transplantation of contaminated allografts by viral agents, such as HIV 1/2, hepatitis B/C virus (HBV and (HCV), cytomegalovirus (CMV), human T-lymphotropic virus 1 (HTLV-1), as well as by prion, have been published by several authors around the world (CDC, 1987; Estlund T, 2005).

Allografts and the viral agents that can be transmitted were reported by CDC (2003) and Hilmy *et al.* (2007) as follows:

Frozen Bone: HIV-1 Hepatitis (B, unspecific), human T-cell leukaemia virus
Frozen tendon: Hepatitis C, HIV
Cornea (viable tissues): Hepatitis C, rabies, herpes simplex virus
Heart valve (viable tissues): Hepatitis B
Saphenous vein: Hepatitis C

A prion has been defined as a *small proteinaceous infectious particle* which resists inactivation by modifying nucleic acids, and it has been reviewed in many publications (Prusiner, 1996). The risk of disease

*Batan Research Tissue Bank (BRTB), Center for the Application of Isotopes and Radiation Technology, BATAN, Jakarta 12070, Indonesia.

transmission in musculoskeletal tissue allografts depends upon the type of graft used and how it is processed. In general, the risk can be eliminated through well-screened donor, processed tissues (by removing blood and bone marrow), pasteurisation (at 58–60°C and wash with chemical reagents), freezing, freeze-drying and then followed by radiation sterilisation (using gamma rays or particle electrons) of the finished product (Dziedzic-Goclawska A, 1997; Thomford WW, 1993; Conrad EU *et al.*, 1993; Strong DM *et al.*, 1993; Eastlund T, 2005; Hilmy N *et al.*, 2006; Paramita *et al.*, 2008). Transfer of prions (Creutzfeld–Jakob Disease; CJD) through dura mater transplant have been reported; however, up to now no cases of an infection of this agent through allogenic bone transplant nor blood transfusion have been reported. Rabies have been transmitted through corneal transplant and dura mater implant; CJD and kuru disease had been shown to be transmissible by injecting extracts of the diseased brains into the brains of healthy animals (Gajdusek DC *et al.*, 1972; Prusiner SB, 1991; Brown P *et al.*, 1992).

Emerging infectious disease caused by agents such as virus, including outbreaks of previously unknown disease or known disease whose incidence in human beings is low, has increased significantly in the past two decades. Re-emerging diseases are known diseases that have reappeared after a significant decline in incidence. New emerging and re-emerging infectious diseases caused by viruses and prions are breaking out around the world. Transmission of those diseases through contaminated blood, faeces and other body liquid from unscreened donor, as well as food/feed products, will increase the outbreak.

Viruses are very small microbes, have DNA or RNA, are resistant to radiation but only to a certain degree as they can be eliminated by radiation. Research on elimination of their contamination on products by irradiation or by chemical reagent is limited. Prions and several viruses are responsible for human epidemics/pandemics; they have made a transition from animal host to human host and now several of them are transmitted from human being to human being. Human immunodeficiency virus (HIV) is responsible for the AIDS epidemic; severe acute respiratory syndrome (SARS) is suspected to be caused by corona virus; bird flu or avian influenza (H5N1) is caused by virus of the family

Orthomyxoviridae; swine flu is caused by virus AHINI; mad cow diseases and CJD are caused by prions. These are examples of emerging infectious diseases (CDC, 2003; WHO, 2003; Pruss *et al.*, 2005; WHO, 2009).

New infectious diseases continue to evolve and emerge. Changes in human demography are changing the transmission dynamics by bringing people into closer and more frequent contact with pathogens, human behavior and land use are factors that are contributing to new disease emergences. This may involve exposure to animal or arthropod carriers of those diseases. Zoonotic pathogens are more likely to be associated with emerging diseases than non-emerging ones (Murphy, 1998). Increasing trade in exotic animals as pets and food sources has contributed to the rise in opportunity for pathogens to jump from animal reservoirs to human beings. For example, close contact with exotic rodents imported to the United States as pets was found to be the origin of the recent US outbreak of monkey pox, and use of exotic civet cats for meat in China was found to be the route by which the SARS coronavirus made the transition from animal to human hosts, while migration of birds was found to be the route for outbreak of avian flu type H5N1 (CDC, 2003; NIAID, 2005; NCID, 2005; WHO, 2006). Radiation technology has been used to sterilise medical devices, as well as allografts, xenografts and implanted devices to be used in surgery without risk of infection. This chapter describes briefly infectious viruses and prion as well as new emerging agents, possibility of them contaminating the allografts and xenografts, and finally the possibility of eliminating the contaminated agents, viruses and prions from tissue bank products by irradiation or combined treatment between irradiation with other methods such as pasteurisation or soaking in certain chemical reagent for safe utilisation.

Viruses, Prion and Emerging Infectious Agent/Diseases and How Far it can Affect the Allografts and Xenografts (Hilmy *et al.*, 2006)

Emerging infectious agents/diseases caused by viruses and prions can be defined as infections that have newly appeared in a population or have existed but are rapidly increasing in incidence or geographic range.

Most of the viruses come from animals (zoonoses) and jump to human beings (host jumping) such as coronavirus (SARS), HIV, bird flu/avian flu H5N1, hepatitis viruses, dengue fever virus, West Nile virus (WNV), hantavirus, Marburg haemorrhagic fever virus, Hendra virus, Nipah virus and the latest one is virus H1N1 which caused outbreak of swine flu in Mexico in 2009. Transmission of those agents through transplantation of contaminated tissue allograft or xenografts to recipient can happened if the donor was not screened properly. The donor can be screened from some of those viruses such as HIV, hepatitis viruses and WNV. Processing of tissue allografts by pasteurisation, washing and soaking in H_2O_2 and soap can eliminate contaminated viruses to a certain amount, and these findings have been reported by several authors (Hilmy *et al.*, 2006).

Although viruses are generally resistant to radiation but to a certain degree they can be well eliminated by radiation. Their D_{10} values vary from 4 kGy to 13 kGy. This chapter discribes briefly the possibility of using combined treatment of processing, lyophilisation and sterilisation by radiation to overcome problems of non-screened donor carrying contaminated viruses.

Viral Diseases (See Table 1)

SARS coronaviruses (SARS Co-V) emerged in several Asian countries such as Hongkong, China, Taiwan, and Singapore in 2003 (CDC, 2003; Eastlund, 2005). West Nile virus (WNV) was first isolated in Uganda in 1937. Today it is most commonly found in Africa, West Asia, Europe, and the Middle East. In 1999, it was found in the Western Hemisphere for the first time in New York City. In early spring 2000, it appeared again in birds and mosquitoes and then spread to other parts of the eastern United States. By 2004, the virus had been found in birds and mosquitoes in every state except Alaska and Hawaii. WNV or summer virus infected through mosquitoes, emerged in USA and Canada in 1999, nearly 4000 human cases were identified and 254 deaths were recorded. It can be transmitted through blood transfusion, organ transplantation and needlestick (CDC, 2003). Between January 1, 2004, and January 11, 2005, WNV caused 2470 cases of disease, including 88 deaths (NCID, 2005). In 2008, a new agent of diseases called H1N1 virus broke out in Mexico and then

Table 1.   Emerging and re-emerging diseases caused by viruses.

Virus	Diseases	Countries affected
West Nile virus (Flavivirus)	Encephalitis, meningitis (neuroinvasive disease)	United States (1999–2005)
Coronaviruse	SARS (Severe Acute Respiratory Syndrome)	China, Hong Kong, Taiwan, Singapore, Canada, US (2003), Africa (1978,1993)
Bunyaviridae family	Rift Valley fever	Madagaskar (1991) Saudi Arabia (2000)
Monkeypox virus (Orthopoxvirus group)	Monkey pox	Central and West Africa (1978) US (2003), Germany, Yugoslavia
Nipah virus	Encephalitis and respiratory illness	Malaysia, Singapore (1999)
Human Imuno-deficiency Virus	AIDS (Acquired immunodeficiency syndrome)	Worldwide
Hendravirus	Respiratory and neurologic disease	Brisbane, Australia (1994)
Hantavirus	Hantavirus Pulmonary Syndrome	Korea (1950), Finland (1980), New Mexico (1993)
Ebola virus (Filovirus)	Ebola haemorrhagic fever	Zaire, Sudan, Uganda, Gabon (1976–2001)
Bird flu virus	Avian influenza (H5N1)	Hong Kong (1997), Vietnam, Indonesia (2004–2006)
Swine flu virus	Virus H1N1	Mexico (April 2009) and then around the world (2009)

the pandemic continue in several Asian countries such as Japan, China, Singapore, and Indonesia (2009). The latest outbreak is swine flu virus, AH1N1, which started in Mexico, April 2009 and then spread to the rest of the world as well as to Asian countries in 2009. In Indonesia, August 2009, 900 cases of AH1N1 infection have been detected and six of them died.

Ebola virus of the *Filoviridae* family is an RNA virus that caused haemorrhagic fever. It has emerged in several African countries such as Zaire, Sudan, Gabon, and South Africa. Marburg virus from the *Filoviridae* family is an RNA virus that emerged in Africa in 1967 and re-emerged in 2005 in Angola-Africa. HIV and hepatitis B/C viruses are still emerging in Asia, Africa and elsewhere, and they can be transmitted through blood transfusion and organ or allografts transplantation (Conrad *et al.*, 1995; O'brien, 1996; Chisari *et al.*, 1997; Pruss, 2005; Eastlund, 2005).

Nipah virus from the *Paramyxoviridae* family has a single-stranded DNA. It emerged in Malaysia and Singapore between 1998 and 1999. Belonging to the same family, Hendra virus emerged in 1994 in Brisbane, Australia. Hantaan virus caused a Korean haemorrhagic fever in 1950 to 1960, a different species, Puumala virus, emerged in Finland in 1980, followed by another species, Sin Nombre virus which emerged in 1993 in New Mexico.

Avian influenza (AI) or bird flu virus type H5N1 is caused by an RNA virus (family *Orthomyxoviridae*). It emerged in Hongkong in 1997, Vietnam and Indonesia in 2004 to 2006, Turkey in 2005, Macedonia and Romania in 2005, Nigeria, Germany and Austria in February, 2006. This virus consists of two vital proteins, i.e. hemaglutinin(H) and neuraminidase(N). Virulent type H5N1 virus that is able to jump from bird/animal to human being makes the disease become fatal. Cumulative number of confirmed human cases of avian influenza A/H5N1, up to February 2006, are as follows: 88 peoples have been killed, among of them 18 in Indonesia, 42 in Vietnam, 14 in Thailand, 4 in Turkey, and 4 in China. Millions of birds have been killed and the endemic will continue year by year, following bird's migration (WHO, 2006). In 2009, another flu virus break out in Mexico, i.e. swine flu H1N1 and subsequently in several Asian countries as well as in Europe (WHO, 2008).

If the tissues obtained from screened donors, who are only free from HIV and hepatitis B/C viruses as stated in several Tissue Bank Standards, but are infected by one of those new emerging viruses, without any process to eliminate them, those diseases might be transferred to the recipient through transplantation of the contaminated allografts or xenografts.

Prions Diseases (*See Table 2*)

Prion diseases are caused by infection agents which do not have a nucleic acid genome. It seems that the protein alone is the infectious agent. The infectious agent has been called a prion. It has been defined as a "*small proteinaceous infectious particle* which resists inactivation by procedures that modify nucleic acids" (Prusiner, 1996).

Prion diseases are often called spongioform encephalopathies because of the post-mortem appearance of the infected brains with large vacuoles in the cortex and cerebellum. Specific examples of those diseases in animals are as follows: Scrapie (sheep), TME (transmissible mink encephalopathy) on mink, CWD (chronic wasting disease) on muledeer, elk, and BSE (bovine spongiform encephalopathy) on cows.

Humans are also susceptible to several prion diseases, they are: CJD (Creutzfeld–Jakob disease), GSS (Gerstmann–Sträussler–Scheinker syndrome), FFI (fatal familial insomnia) and BSE (bovine spongiform encephalopathy) that jump host from cows to human beings (Prusiner, 1995; Prusiner, 1996; WHO, 2003; Pattison, 1998; Pruss, 2005).

Specific factors precipitating disease emergence can be identified in virtually all cases. These include ecological, environmental and demographic factors that place people at increased contact with a previously unfamiliar microbes or their natural host. Changing human demographic

Table 2.   Prions disease in animals and human.

Diseases
Animals
Scrapie
Transmissible Mink Encephalopathy (TME)
Chronic Wasting Disease (CWD)
Bovine Spongiform Encephalopathy (BSE)
Human
Creutzfeld–Jakob Disease (CJD), variant CJD
Gertsmann–Sträussler–Scheinker Syndrome (GSS)
Fatal Familial Insomnia (FFI)
Bovine Spongiform Encephalopathy (BSE)

and human behaviour, increase accessibility of international travel, advances in technology, microbial adaptation and change, and breakdown in public health measures are also factors that contribute to disease emergence (Murphy, 1998; Woolhouse *et al.*, 2005). If all of those infectious agents could not be eliminated from screened grafts properly (which screened only for HIV, hepatitis B and C), then the agents could be transmitted to recipient through the implanted grafts.

Zoonotic Virus

Many members of the *Flaviviridae* and *Bunyaviridae* families appear to be zoonotic viruses that can also infect humans from animal host origin. HIV is one of the zoonotic viruses that can also transfer between human hosts. The source of HIV is the green monkey from Africa.

   Zoonoses (diseases caused by zoonotic pathogens) can be broken down into two basic groups: those spread by direct contact with the infected animal and those spread via an intermediate vector. It can infect human beings through many vectors; for example, members of the genus Hantavirus are spread by rats, West Nile virus by mosquitoes, avian flu by pigs or direct contact with birds. Virus, protozoa, bacteria and prion can all be transmitted either directly or via vector. All zoonotic members of the familia *Flaviviridae* and *Togaviridae* are transmitted via an intermediate vector, while all zoonotic members of the families *Arenaviridae, Paramyxoviridae*, and *Filoviridae* are transmitted through direct contact. Bird flu in Asia was first detected in Hong Kong in 1997, it then spread to other Asian countries, Europe and Africa, and affects the poultry industry severely. Influenza avian A1 type H5N1 is a bird virus that infects humans by jumping from birds via pigs to human beings or through direct contact with birds, while Swine flu H1N1 is a virus that jumps from pigs to human beings.

   The migration of birds and rapid movement of people around the globe today presents such newly evolved viruses with unparalleled opportunities to spread to the human species, perhaps with catastrophic consequences. The Sin Nombre hantavirus that caused an outbreak in New Mexico in 1993 is an example of such zoonotic virus. In this case, deer mice are thought to be the natural host and pass the virus through their urine and faeces. The good growing conditions following an especially wet spring in 1993 allowed an increase in the deer mouse population, which in turn resulted in more human beings coming into contact with the infected animals. In 1999, the West Nile flavivirus, an African virus as its name indicates, emerged in the United States. The normal host of the virus is birds but human beings may become infected when mosquito feeds first on an infected bird and then a person. The virus spreads rapidly because it appears that numerous species of birds can serve as hosts and that the virus can be transmitted by a wide variety of mosquitos. Some zoonotic viruses can transmit from human being to human being under special circumstances. The filoviruses that cause Ebola fever are zoonotic viruses whose natural host and mode of transfer to human beings have not yet been identified. Like other zoonotic viruses, Ebola virus causes a severe disease with a very high mortality rate. Unlike most other zoonotic viruses, however, Ebola virus can be transmitted via blood, tissues, or other body fluids from one person to another. Hendra virus is a member of the family *Paramyxoviridae* that was first isolated in 1994 from specimen obtained during an outbreak of respiratory and neurologic disease in horse and human beings in Hendra, a suburb of Brisbane, Australia. The human infections were due to direct exposure to tissues and secretions from infected horses. Nipah virus, also a member of the family *Paramyxoviridae,* is related but not identical to Hendra virus. Nipah virus was initially isolated in 1999 upon examining samples from an outbreak of encephalitis and respiratory illness among adult men in Malaysia and Singapore. This virus was transmitted to human beings, cats, and dogs through close contact with infected pigs.

Classification of viruses make possible predictions about details of replication, pathogenesis and transmission. This is particularly important when a new virus is identified. Without a classification scheme, each newly discovered virus would be like a black box. The current classification scheme allows most newly described viruses to be placed in a box with a label. In the best cases, much can be assumed about the biology of the virus. Even in the worse case a framework for investigation would be suggested, because there are so few virus discoveries now being made which do not fit into the existing classification scheme. We can state with a degree of confidence that most of the major groupings of viruses infecting human beings and domesticated animals have been identified (Bruce, 2002).

For a virus to multiply, it must obviously infect a cell. Viruses usually have a restricted host range, i.e. animal and cell type in which this multiplication is possible. A comprehensive review by Murphy (1998) has identified zoonotic status as one of the strongest risk factors for a disease emergence. Roughly 75% of emerging pathogen are zoonotic; and zoonoses are twice as likely to be considered emerging as non-zoonoses. The virus classification is as follows:

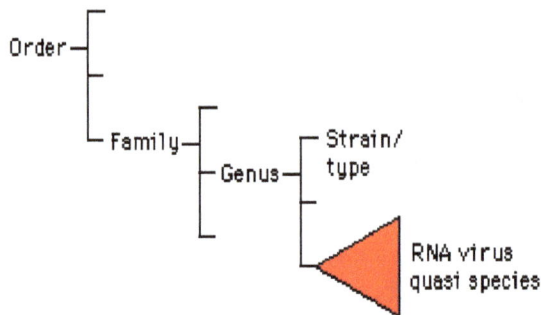

Viruses can be subdivided by genome type as follows:

- double-stranded (ds) DNA, such as *Herpesviridae* and *Picornaviridae*
- single-stranded (ss) DNA such as *Parvoviridae*
- dsRNA such as *Reoviridae*
- ssRNA, divided into positive-sense RNA such as *Picornaviridae* and negative-sense RNA such as *Paramyxoviridae* (Bruce, 2002)

Humans might be infected by virus and prion in several ways:

Animal to human being: Those diseases can be transferred from animal to human being by direct contact with animal's products as well as close contact with animals and pets through faeces, saliva and environment.

Human being to human being: Diseases can be transmitted through human organs and tissues transplantation such as kidney, liver, cornea, bone, soft tissues, as well as through blood transfusion, bone marrow, and platelets. The diseases also can be transmitted through close contact to infected humans such as in hospitals or at home.

The Possibility of Radiation to Reduce the Outbreak

Increasing international trade in animals and animal products has contributed to the rise in the opportunity for pathogens to jump from animal
reservoirs to human beings and also from human being to human being.
Those products which could be contaminated by virus or prion are meat
and meat products, poultry product, dairy products, bone meal, eggs and
egg powder, dried blood plasma and other blood products, xenografts,
allografts, feathers (from birds), animal skin, etc. They pose a risk of
spreading the disease to human beings (Prusiner, 1995; IAEA Code, 2004;
Eastlund, 2005; WHO, 2006).

To minimise and to eliminate the risk of disease transmission from
animal products to human beings, proper processing of products such as
washing with chemical reagent, pasteurisation for 3 hours at 58–60°C,
drying, packaging and then followed by irradiation at room temperature
or in frozen state should be carried out. In case of allografts, strict donor
screening combined by processing and irradiation have been conducted
by several tissue banks around the world (Pruss *et al.*, 2005; Eastlund,
2005; AATB, 2002; Hilmy *et al.*, 2001; Fideler *et al.*, 1994; Hilmy *et al.*,
2006). Allografts without terminal sterilisation can be affected by virus in
window period (Table 3). Window period is the period between infection
and when the virus is detectable by screening tests.

Effects of Radiation on Viruses and Prions

It has been known that high energy radiation of gamma rays and electron
beams has the ability to generate reactive species during interaction with
matter. The process involves ionisation and excitation. Ionising radiation

Table 3.   Window period (WP) of several virus.

Virus	HIV	HCV	HBV
WP using FDA	22 days (anti-HIV)	70 days (anti-HCV)	56 days (HBsAg)
WP using NAT*	7–12 days	10–29 days	41–50 days

*Nucleic Acid Test.
Source: Busch MP and Kleinman SH (2000). *Transfusion* **40**: 143–146.

can affect the materials in two ways, i.e. direct and indirect. Direct effect usually means that the radiation causes ionisation or excitation of the molecules and then followed by damage in the molecules. Indirect effect usually refers to the damage done to molecules by radiolytic products of irradiated water, oxygen or other materials in the medium.

Virus is a subcellular organism with a parasitic intracellular life cycle and has no metabolic activity outside the host cell. Therefore, it cannot actually stay alive outside the host cell. The antibiotics and chemotherapeutic agents which inactivate bacteria are generally ineffective against viruses. Viruses are as a rule considerably more resistant to radiation than either bacteria or bacterial spores.

The size of viruses range mostly from 20 nm to almost 14000 nm. Genomes (DNA and RNA) are the major targets for the biological effects of ionising radiation in killing the microbes. Depending on the size and type of genome, virus can be very resistant or sensitive to irradiation. The main cause of virus inactivation is protein damage. The dose of radiation required to inactive an infectious virus or its nucleic acid is much greater under direct condition than indirect condition. The damage of the viral nucleic acid appears to be almost solely responsible for the loss of infectivity. Sensitivity of the targets depends very much on their sizes. A large target is more sensitive to radiation compared to a smaller one. In general, a cell with a large nucleus and DNA is more vulnerable than one with a small genome. Compared to genomes of bacteria, yeasts and moulds, viral genomes are very small. This is one of the reasons why viruses are more resistant to radiation compared to bacteria and yeast. The radiation resistance of different virus groups show considerable differences. Viruses with single-stranded genomes are about ten times more sensitive than viruses with double-stranded genomes, although the genomes are smaller. Viruses with large genomes may be five times more sensitive than viruses with small genomes. However, their resistance may vary as much as ten-fold, depending on a number of factors, particularly the concentration of oxygen (O_2) and water, the organic materials in the suspending substrate, the temperature (frozen and room temperature), the pH (which is unfavourable to microbes) during irradiation. Irradiation at the frozen state (-70 to $-80°C$) where water is immobilised, as well as in organic substrate, such as in a high concentration of protein and alcohol, will increase the radiation resistance of viruses. In contrast, radiation in wet condition at room temperature will increase the sensitivity of viruses compared to radiation in dry state. Although strand breakage has been reported as an important cause for radiation inactivation of single-stranded nucleic acid, the combination of base damage and intrastrand cross link formation is also important. The alteration is lethal because the viruses cannot reproduce after their nucleic acids are damaged. Many D_{10} values of viruses exceed 5 kGy, and some of them, such foot-and-mouth disease virus, have D_{10} values of 13 kGy when irradiated in frozen state. The effects of radiation on microorganisms, including viruses, are exponential. The radiation sterilisation dose used depends on the viral bioburden (Fideler *et al.*, 1994; Hilmy *et al.*, 2001; Pruss *et al.*, 2005).

Table 4 shows several D_{10} values of viruses that can cause emerging diseases (Pruss *et al.*, 2005). It can be seen that radiation will and does kill viruses. The amount of radiation dose required to accomplish a log reduction of viruses (D_{10} value) is higher than those of bacteria and mold/fungi, although some bacteria spores are very resistant to radiation, such as spores of anaerobic *Clostridium sordellii*, the bacteria that led to the death of one tissue recipient (Grieb TA *et al.*, 2005).

Table 4. D_{10} values of enveloped, non-enveloped viruses and bacteria.*

Virus	D_{10} value (kGy)[†]
HIV 1–2[‡]	4–7.09
BPV (bovine parvovirus)[§]	7.27
PV-1 (polio virus)[§]	7.13
HAV (hepatitis A virus)[§]	5.31
PRV (pseudorabies virus)[‡]	5.29
BDBV (bovine viral diarrhoea virus)[‡]	<3.0
Bacteria, mold and yeast	<2.0

*Hilmy *et al.*, 2001; Pruss A *et al.*, 2005.
[‡]Enveloped virus.
[§]Non-enveloped virus.
[†]The value was within the range of variation determined by different procedures.

Infectious risks of tissue and organ transplantation have often been identified after first being recognised as blood transfusion-transmitted infection. Although the susceptibility of those viruses to gamma irradiation or other sterilants is unknown wholly, the routine use of sterilisation may provide some protection from transmission of diseases through tissue transplantation (Eastlund, 2005). The same was also observed by Ernest U. Conrad *et al.* (1995), who reported that the hepatitis C virus can be transmitted by processed unirradiated bone, ligament and tendon, but no infection cases exist with irradiated bone of 17 kGy. Studies on transmission of HIV in window period donor have been done in USA from 1999 to 2003, and the results showed that irradiation in sterilising doses can significantly reduce the viral load and in combination with appropriate donor screening and laboratory testing, these will significantly enhance and improve the safety of tissues being used for transplantation (Strong MD, 2005).

In April 2006, a new version of ISO 11137 has been established to replace the ISO 11137:1995, which consist of three parts. The ISO 11137 part 2 (establishing the sterilisation dose) consists of (i) dose setting to obtain a product-specific dose and (ii) dose substantiation to verify a preselected dose of 25 kGy (VD max^{25}) and 15 kGy (VD max^{15}). Method of VD max^{25} (adopted from AAMI TIR 27:2001) is used for samples with bioburden of 1000 cells per piece or less and VD max^{15} for those with the bioburden of 1.5 cells per piece or less. The samples needed for the verification/VD experiment at the selected Sterility Assurance Level (SAL) per production batch are 20, i.e. ten for bioburden determination and the remaining ten for sterilisation test. The two VD max methods can be used for the establishment of RSD for most of the tissue allografts. If the VD experiment pass, the RSD for VD max^{25} is 25 kGy and 15 kGy for VD max^{15}. However contamination by viruses and prions are excluded from this ISO.

Unlike virus, a prion is an infectious protein, produced by modifying the structure of a specific prion protein (PrP) which is found on nerve cell membranes. The most common form of prion disease in human being is sporadic CJD and variant CJD. Less than 1% of CJD cases are infectious and most of those appear to be iatrogenic. Between 10% and 15% of prion disease cases are inherited, the remaining are sporadic. PrP^{CJD} has been found in the brains of most patients who died of prion disease CJD. The term PrP^{CJD} is preferred by some investigators when referring to the abnormal isoform of human PrP in human brain. PrP^{Sc} is always used after human CJD prions passage into an experimental animal. At present prions have gained wide recognition as extraordinary protein agents that cause a number of infectious, genetic and spontaneous disorders (Prusiner SB, 1995). Prions are very resistant to endogenous protease that would normally destroy the protein, temperatures above 100°C, formalin, extremes of pH, non-polar organic solvents, and even years of burying will not eradicate them. They are able to pass through 0.1 μm filters as well. Infectivity of prions can be destroyed by 0.1N NaOH for one hour at room temperature, or one hour in 0.5% solution of sodium hypochlorite as well as 30–60 minutes at 130°C. Since prion is a very small modified protein without any nucleic acid component which can be degraded by ionising radiations, it is resistant to ultraviolet and gamma radiation which break down the nucleic acid. Up to now no detailed findings have ever been reported on their D$_{10}$ values. Irradiation dose of 25 kGy could not eliminate prions from lyophilised dura mater (CDC, 1987; Hilmy *et al.*, 2001; Pruss *et al.*, 2005).

Several Methods to Eliminate Undetected Viral Agent in Blood Plasma and Tissue Allografts (WHO, 2003, 2004; NYBC, 2004)

- Pasteurisation for one to three hours (for both enveloped and non-enveloped viruses)
- Terminal dry heat (80°C) of dry/lyophilised products (for both enveloped and non-enveloped viruses)
- Vapor heated (for both enveloped and non-enveloped viruses)
- Solvent detergent treatment (for enveloped viruses)
- Low pH (for both enveloped and non-enveloped viruses)
- UVC Light irradiation (for both enveloped and non-enveloped viruses)
- Gamma irradiation at −80°C or at room temperature (for both enveloped and non-enveloped viruses)
- Pasteurisation, defatting, washing and followed by soaking in H$_2$O$_2$ with solvent detergent (DePaula CA *et al.* (MTF), 2002;

Table 5. Viral clearance in cortical bone allografts processing in detergent and an hour in H_2O_2 and ethanolo.*

Virus	RNA/DNA	Model for	Virus log reduction
Envelope			
BVDV- Bovine Viral Diarrhea	RNA	HCV	➤ 10.62
HIV	RNA	HIV	➤ 15.22
PrV/Pseudo Rabies	DNA	CMV	➤ 12.23
Non-Envelope			
HAV	RNA	HAV	➤ 6.46
Polio	RNA	POLIO	➤ 10.96

*Source: DePaula *et al.*, 2002.

HILMY *et al.*, 2003; Febrida *et al.*, 2008; Hilmy *et al.*, 2008 (for both enveloped and non-enveloped viruses). This process can eliminate viruses up to 15 log step for HIV (see Table 5).

Suggestions and Future Prospect of Eliminating Undetected/ Unknown Virus Agents and Virus at Window Period from Allografts and Xenografts: A Combined Treatment Processed and Irradiation at 15 kGy (Hilmy *et al.*, 2008; ISO 11137, 2006)

Since new emerging diseases caused by viruses and prions will be recurring and the possibility of those diseases jumping from animals to human beings is increasing so will the risk of infection from using products of animals and human origin such as allografts and xenografts. If there is no effort to eliminate those virus and prions from those products, we will face severe problems in health, global trade and economy. One way to overcome those problems is by combining radiation with other technologies, such as washing, lyophilisation, freezing or pasteurisation, which will provide the opportunity to eliminate those contaminations by excluding prions from allografts:

- Processing: Pasteurisation at 60°C, washing, defatting, soaking in detergent and H_2O_2, followed by radiation sterilisation at 15 kGy (VD_{max}^{15})

- Processing: Pasteurisation/moist heat at 80°C, Washing, defatting, lyophilised, followed by radiation sterilisation at 15 kGy (VD_{max}^{15})

The sterilisation dose of 15 kGy can be done if the bioburden of the allografts are less than 1.5 cells per product (ISO 11137, 2006). By combined treatment the bioburden can be reduced to lower than one cell per product (Febrida *et al.*, 2008). The current inactivation methods are only efficient to a limited extent and most of the procedure cannot be recommended for bone and soft tissues due to their effects on the biological properties of tissues. All of the combined treatments are also subjected to further validation.

References

Bruce AV (2002). In *The Biology of Viruses*, Second Edition, McGraw Hill, Singapore, p. 341.

Busch MP and Kleinman SH (2000), *Transfusion* **40**: 143–146.

CDC, Center for Disease Control and Prevention (1987). Update: Creutzfeldt Jacob Disease in a patient receiving a cadaveric dura mater graft. *JAMA* **258**: 309–310.

CDC, Center for Disease Control and Prevention (2002). Provisional surveillance summary of the West Nile virus epidemic — US, January–November 2002. *MMWR* **51**: 1129–1133.

CDC, Center for Disease Control (2003). Update: Outbreak of severe acute respiratory syndrome — worldwide 2003. *MMWR* **52**: 241–248.

CDC, Center for Diseases Control and Prevention (2003). Hepatitis C virus transmission from antibody-negative organ and tissue donor. *MMWR* **52**(13): 273–276.

Chisari FV and Ferrari C (1997). Viral hepatitis. In Nathanson N, Ahmed R, Scarano FG, Griffin DE, Holmes KV, Murphy FA and Robinson HL (eds.) *Viral Pathogenesis*, Lippincott-Raven Publisher, Philadelphia, pp. 745–747.

Conrad EU, Gretch D, Obermeyer K, Moogk M, Sayers M, Wilson J, and Strong DM (1995). The transmission of hepatitis C virus through tissue transplantation, *J Bone Joint Surg* **77-A**: 214–224.

DePaula CA *et al.* (2002). Muskuloskeletal Transplant Foundation (MTF), USA.

Eastlund T (2005). Viral infections transmitted through tissue transplantation. In Kennedy JF, Phillips GO and Williams PA (eds.), *Sterilisation of Tissues Using Ionizing Radiation*, Wood Head Publishing Limited, Cambridge, 255–278.

Febrida A, Basril A, Suryani N, and Hilmy N (2008). Microbes clearance from biological tissues by soaking in a mixture of non-ionic detergent and hydrogen peroxide, 5th World Congress on Tissue Banking and 12th International Conference of Asia Pacific Association of Surgical Tissue Banks, Kuala Lumpur, June 2008.

Fideler BM, Vangsness CT, Moore T, Li Z, and Rasheed S (1994). Effects of gamma irradiation on human immunodeficiency virus, *J Bone Surgery* **76–A**: 1032–1035.

Grieb TA, Forng R-Y, Lin J, Wolfinbarger L, Melgarej JS, Sharp C, Drohan WN, and Burgess WH (2005). In Kennedy JF, Phillips GO and Williams PA (eds.), *Sterilisation of Tissues Using Ionising Radiations*, CRC Woodhead Publishing, England, 285–302.

Hilmy N and Lina M (2001). Effect of ionizing radiation on viruses, proteins and prions, *Advances in Tissue Banking*, World Scientific, Singapore, p. 358.

Hilmy N and Pandansari P (2007). New emerging diseases caused by viruses and prion, In Nather A, Yusof N, Hilmy N (eds.), *Radiation in Tissue Banking*, World Scientific, Singapore, pp. 133–146.

Hilmy N and Pandansari P (2008). Possibility of using combined treatment of processing and ionizing radiation to eliminate contaminated viruses from non-screened donor of tissue allografts, 5th World Congress on Tissue Banking and 12th International Conference of Asia Pacific Association of Surgical Tissue Banks, Kuala Lumpur, June 2008.

IAEA-International Atomic Energy Agency (2004). Code of Practice for the Radiation Sterilization of Tissue Allografts-Requirements for Validation and Routine Control, IAEA, Vienna.

ISO 11137-2 (2006). Sterilization of Health Care Products-Radiation Part 2. Establishing Radiation Sterilization Dose, International Organization for Standardization, Geneva.

Murphy FA (1998). *Emerging Zoonoses, Emerging Infectious Disease*, Vol. 4 No. 3, University of California, USA.

NCID, National Centre for Infectious Diseases; Centre for Disease Control and Prevention (2005). Infectious Disease Information: Emerging Infectious Diseases. Available at http://www.ncid.cdc.com

NIAID National Institute of Allergy and Infectious Diseases (2005). Emerging Infectious Diseases, Division of Microbiology and Infectious Diseases, Maryland, USA. Available at http://www.niaid.nih.gov/dmid/eid/#top

O'brien WA and Pomerantz RJ (1996). HIV infection and associated disease. In Nathanson N, Ahmed R, Scarano FG, Griffin DE, Holmes KV, Murphy FA and Robinson HL (eds.), *Viral Pathogenesis*, Lippincott-Raven Publisher, Philadelphia, pp. 815–836.

Pattison J (1998). The emergence of bovine spongiform encephalopathy and related diseases; Emerging in infectious disease. Available at http://www. cdc.gov/ncidod/eid/pattison.htm

Prusiner SB (1995). Prion disease. *Sci Am* **272**: 48–57.

Prusiner SB (1996). Prion. In Fields BN and Knipe DM (eds.), *Fields Virology*, Lippincott-Raven, Philadelphia, pp. 2901–2949.

Pruss A, Von Versen R, and Pauli G (2005). Viruses and their relevance for gamma irradiation sterilisation of allogenic tissue transplant. In Kennedy JF, Phillips GO and Williams PA (eds.), *Sterilization of Tissues Using Ionizing Radiation*, Woodhead Publishing Limited, Cambridge, 235–254.

Strong DM, (2005). In Kennedy JF, Phillips GO and Williams PA (eds.), *Sterilisation of Tissues Using Ionising Radiations*, CRC Woodhead Publishing, England, 163–172.

WHO, World Health Organization (2003). Health Technology and Pharmaceutical Cluster, Guidelines on Transmissible Spongiform Encephalopathies in relation to Biological and Pharmaceutical Products, WHO/BCT/QSD/13.01, Geneva.

WHO, World Health Organization (2006). Epidemic and Pandemic Alert and Response (EPR), Disease Outbreak News, WHO, Geneva. Available at http://www.who.int/

Woolhouse EJM and Sonya GS (2005). Host Range and Emerging and Reemerging Pathogens, Emerging Infectious Diseases, Centre for Infectious Disease, University of Edinburg, UK. Available at http://www.cdc.gov/eid

Chapter 29

Interaction of Radiation with Tissues

Norimah Yusof* and Awang Hazmi Awang Junaidi†

Introduction

Sterilisation is required to sterilise tissue products and it can be an important safety measure to avoid disease transmission in addition to strict donor screening and sterile retrieval. Heating or autoclaving is a common sterilisation method, but heating reduces the osteoinductive activities of bone, while autoclaving denatures the organic matrix. Ethylene oxide sterilisation is an efficient method to inactivate almost every bacterial spore and virus in medical devices and implants. Unfortunately, the effect of ethylene oxide sterilisation depends on many parameters, such as time, temperature, product size, product configuration, water and fat content, that need to be controlled. Ethylene oxide itself is toxic when remains in soft tissues where defatting and dehydration processing are incomplete. Therefore, ethylene oxide sterilisation is not safe to use on biological materials. The best and most efficient sterilisation method for tissue products is irradiation. Gamma sterilisation has been widely used for sterilising medical products and has been reported safe without significant influence on the biological properties.

Radiation sterilisation of musculoskeletal tissue grafts was initiated more than 50 years ago, started by using cathode rays in 1955 to gamma irradiation in the recent years. Long bones and osteoarticular grafts which

*Malaysian Nuclear Agency (NM), Bangi 43000 Kajang, Selangor, Malaysia.
†Faculty of Veterinary Medicine, Universiti Putra Malaysia, 43400 UPM Serdang, Selangor, Malaysia.

by virtue of their massive size are easily sterilised by high penetration of gamma radiation (Tomford, 2005). Use of irradiated tissues in clinical applications has increased tremendously in the past 20 years. Like any other methods, radiation induces physical and chemical changes in tissues. This chapter will discuss the changes in the physicochemical properties of tissues, as reported in literatures, and also the mechanism behind the changes.

Ionisation and Excitation

When radiation passes through tissue components, energy deposited causes ionisation and excitation. The depth of penetration of the radiation depends on energy level of the radiation, density and atomic number of the atoms of the tissues. Due to the interactions, the radiation is progressively weaker as it goes deeper into the tissues. *Direct effect* of radiation on critical biological molecules may cause lesion, scission and cross-link of biopolymeric structures. For viable cells such as microbes and viruses, damage occurred in the cell at critical components, mainly DNA molecules, will lead to the killing of cell due to failure to reproduce/proliferate. Other critical components in the cell are believed to be membranes and lysosomes. In fresh tissues containing high amount of water, *indirect effect* of radiation will take place (Yusof, 2000). Water molecules when interact with radiation will result in the formation of free radicals. The free radicals, $H^{\bullet}$ and $OH^{\bullet}$, and aqueous electrons (e_{aq}^{-}) will interact further among them and also with water molecules, their own reactive products, and organic molecules. The free radicals react with biological molecules such as protein, enzyme, amino acid, metabolites and nucleic acid, generating chain reactions which are held responsible for radiation-induced defects. These chain reactions have been elaborated in Chapter 24 of this book and in our earlier publications (Yusof, 2007). In tissues containing high water content, the presence of oxygen may enhance indirect effects of radiation. Oxygen reacts with free radicals to produce hydroperoxy radicals (HO_2) and hydrogen peroxide (H_2O_2) and with organic free radicals to produce organic peroxy radicals (RO_2). These products will generate more radiation damage. Thus, in practice, irradiation of bone in frozen state will minimise radiation damage. Freeze-dried bones and amnions with moisture content less than 10%, preferably in the range of 4–7%, will lessen indirect effects of radiation, hence minimising damage.

Interaction with Tissue Constituents

Tissue is a complex matter comprising of biological molecules and various components which can give rise to various effects. The main components of tissues are:

- organic — such as collagen, fibres, protein and enzyme,
- inorganic — such as salt, minerals and hydroxyapatite and
- water molecules.

For instance, bone has an organic matrix of collagen that is impregnated with calcium salts, mainly in the form of calcium phosphates in crystalline hydroxyapatite. Amnion is a thin membrane of five layers with epithelium and fibroblast cells with the basement containing collagen, enzymes and growth factors.

Collagen and Fibril

Earlier work on native collagen and elastin showed that most changes were observed at very high radiation doses (Dziedzic-Goclawska, 1978). Tensile strength of hydrated collagen is reduced to one-third of its original strength after a very high dose of 460 kGy (Braams, 1963), while fibril structure also changed at high dose of more than 100 kGy (Bailey and Tromans, 1964). In general, irradiation of collagen resulted in extensive changes in its physical properties and relatively small changes in the chemical composition. Changes might be due to disorganised secondary structure (Bailey and Tromans, 1964). Radiation at 10–50 kGy induced minor cleavage of alpha chains of the collagen triple helix molecule (Tomford, 2005). Intramolecular and intermolecular hydrogen bonds are broken at 10 kGy, mainly due to scission and structure breakdown. Cross-link at 50–100 kGy is suppressed below 0°C, as movement of reactive free radical was limited in frozen state. Amino acids composition in collagen did not alter even at high doses of 50–100 kGy (Yusof, 2000).

The effect differed markedly depending on whether collagen was irradiated in dry or wet form and temperature during irradiation.

Tendon

Tensile strength of dry tendon is reduced by two-third of its original value when irradiated at 180 kGy (Braams, 1963). Irradiation at 25 kGy and freeze drying did not change histological pattern of the tendon, and it is still comparable to the fresh at six months after implantation, while 20 kGy and 40 kGy showed no effect on biomechanical properties at six months after surgery.

Therefore, most effects on tissue constituents were observed at doses above 25 kGy, which is the generally recommended dose for sterilisation (Yusof, 2000).

Interaction with Tissues

Effects of radiation on non-viable tissues are not summation of effects of all tissue constituents (Dziedzic-Goclawska, 1978). Desirable effects of radiation may include decrease in immunogenicity and increase in resorbability, while undesirable effects may include reduction in biomechanical properties and decrease in osteoinductive capacity. These effects are influenced by the type of tissue, type of radiation sources and conditions during irradiation (temperature and water content).

Radiation is not used solely but is also applied after tissues have been subjected to washing and preservation, including freezing, demineralisation, air-drying and freeze-drying/lyophilisation for prolonging storage. Therefore, changes in the biomechanical properties are the effects of combined treatments of radiation with physical and chemical processes carried out prior to sterilisation.

Gamma rays and electron beams have been used with doses ranging from 17 to 80 kGy to evaluate effects of radiation on various soft tissues, including cartilage, heart valves, dura mater, skin, fascia, sclera, amnion, meniscus, and tendons (Strong, 2005). Damage generally occurred with increasing doses of irradiation. At low doses (17–20 kGy), effects on strength and modulus were not consistently significant but at doses higher than 25 kGy, biochemical, biophysical, mechanical, and material properties were more significantly altered. Only tissue grafts that are not required to maintain structural integrity but rather are used as coverings such as amnion or in non-weight-bearing reconstruction such as cartilage may have more clinical success.

Soft Tissues

The solubility of dried collagen-derived membrane, used for wound dressing, increased by 50% after irradiation at 35 kGy in dry frozen state. This was due to the direct effect of scission of polypeptide chains hence decreased tensile strength (Sliwowski and Dziedzic-Goclawska, 1976). When collagen is irradiated in the presence of water, the solubility decreased due to covalent intermolecular cross-links that resulted from random reactions of free radicals on the polypeptide chains structure. At 50–100 kGy, amino acid composition in collagen was not altered.

Even though freeze-drying is used to reduce water content of tissues in order to minimise indirect effect of radiation, unfortunately the process itself has detrimental effects on tissue structure (Yusof, 2000). In amnion, irradiation up to 30 kGy did not have significant effect on biomechanical characteristics of freeze-dried amnion; however, reduction by 50% in tensile strength and elongation was observed after one year of storage (Hilmy *et al.*, 1992). When using air-drying method, physical properties of amnion irradiated at 17–25 kGy were not affected and were even better than freeze-dried amnions. No change in physical properties was observed even after six years of storage (Yusof and Hilmy, 2007). Although radiation was claimed by some authors to cause epithelial cell degeneration and destruction of the compact and fibroblast layer (von Versen-Hoeynck *et al.*, 2008), there was no change in the chemical structures evaluated by means of infrared spectroscopy for air-dried amnions irradiated at 25, 36 and 50 kGy (Singh *et al.*, 2007a). There were no significant differences in the water absorption capacity and water vapour transmission rate of the irradiated amnions. At 25 kGy, there was no adverse effect of gamma-irradiated amnion on the treatment of second degree burns, and sterilised and air-dried amnion was easy to apply on wound compared to glycerol-preserved amnion even though both amnions favoured epithelisation in patients (Singh *et al.*, 2007b).

Skin is the largest organ of our body. The tensile strength of freeze-dried and then rehydrated skin was reduced by 80% but with radiation before rehydration, tensile strength is reduced by only 25% (Kwarecki *et al.*, 1977). However, the solubility of collagen increased by 5–15% after 33 kGy. Permeability of freeze-dried dermoepidermal grafts decreased by 50% after 25 kGy (Klen and Pascal, 1975). Since skin serves mainly as a wound covering, effects of radiation on physical and bio-mechanical properties may not affect its clinical usage.

Frozen heart valves irradiated at 29 and 32 kGy maintained the bio-mechanical strength of aortic wall; however, the tensile strength decreased when the freeze-dried heart valves were irradiated at 25 and 33 kGy (Dziedzic-Goclawska, 1978). The freeze-dried tissues were crack and brittle, which might due to damage in the structure (Little, 1973).

Rosomoff and Malinin (1976) could not observe any damage after freeze-dried dura mater was irradiated at 20–25 kGy. Before that Triantafyllou and Karatzas (1975) reported changes in biomechanical properties and decrease in permeability in freeze-dried dura mater irradiated at 25 kGy. They concluded that freeze-drying instead of radiation significantly contributed to the overall changes.

Freeze-dried fascia lata irradiated at 25 kGy decreased in permeability and needed longer time to reach steady state by two to six times compared to that of the unirradiated controls (Klen and Pascal, 1975). Since fascia lata is used as a supporting material, radiation sterilisation may not be used if it affects the biomechanical properties.

Grafted tendon, after freeze dried and irradiated at 25 kGy, did not change in the histological pattern at six months after implantation and is still comparable to fresh one (Feher *et al.*, 1975). Irradiation in the presence of water decreased the solubility and the tendon became more resistant to collagenase digest. When irradiated in dry state, the solubility increased resulting in faster resorption thus time was insufficient to stimulate mesenchymal cells to differentiate. Based on this finding, it is suggested that tendon is irradiated in frozen state instead of freeze drying. Rasmussen *et al.* (1994) reported that radiation at 40 kGy significantly weakened the patellar tendon by 26% but in their later study, 20 and 40 kGy did not show any significant effects on biomechanic properties at six months after surgery. Shiba *et al.* (1999) reported that when Achilles tendon collagen was irradiated under hydrated (wet) condition, the degradation was insignificantly lower than that of the dried samples. Collagen chain of type I was still intact even at 50 kGy. The hydration state seemed to enhance cross-linking of collagen during irradiation and strengthen it.

Bone

Doses greater than 30 kGy produced irreparable damage to the collagen; therefore, bones may be sterilised at doses lower than 30 kGy, preferably in a frozen state where free radicals are inactive, to prevent an extensive inflammatory response to the irradiated collagen. Frozen massive allografts irradiated at 25 kGy using gamma rays gave encouraging clinical results with no infection after three years (Kang *et al.*, 2005). The result was confirmed when both deep-frozen and freeze-dried allografts irradiated at 15–25 kGy, used in reconstruction for malignant bone tumours, were as good as non-irradiated. However, rapid thawing may be the major cause of cellular damage and delayed rupture in cryopreserved arterial allografts; therefore, use of slow thawing is recommended.

Effects of radiation on bone are not only on collagen that sustains the biomechanical properties of bone but also on the osteoinductive factors or the so-called bone growth factors. Radiation at 25–30 kGy has a minimal adverse effect on bone biomechanical properties and healing; however, effects on osteoinductivity need further research (Yusof, 2007).

Combined treatment of radiation and freezing has no effect on the elastic and strength properties of collagen. There was no significant effect on compressive stress of deep-frozen cancellous bone subsequently irradiated at 10 and 30 kGy, slight change at 50 kGy and failure only observed at >60 kGy (Anderson *et al.*, 1992). Bone rings irradiated at −78°C were less brittle and had less collagen damage compared to when irradiated at room temperature (Hamer *et al.*, 1999). Irradiation is believed to cause destruction of collagen alpha chains, probably mediated by free radicals generated when radiation interacts with water molecules. However, in frozen state, the mobility of water molecules was reduced and therefore the production of free radicals was decreased. Low temperature during irradiation may protect the osteoinductive properties.

Freeze-drying (lyophilisation) causes deterioration of biomechanical characteristics of bones and the effect exaggerated after irradiation (Yusof, 2000). Freeze-drying and lipid extraction of femoral heads reduced compressive strength by 20% and irradiation with 25 kGy further reduced to 42.5% (Tomford, 2005). Deep freeze when combined with 25 kGy neither changed the SEM structure nor reduced elasticity of bone, but freeze dry with radiation caused microcracks and reduced elasticity (Voggenreiter *et al.*, 1994). Decrease in elasticity was also observed in freeze-dried cancello-cortical and cortical bones irradiated at 25 kGy (Dziedzic-Goclawska, 1978). Based on these findings, irradiated and freeze-dried bones are suitable for small grafts or morselised particles, which will be used as filler and not for weight bearing.

Remodelling of Bone

Bone remodelling comprised of graft resorption followed by substitution by host's own tissues and then bone corporation (Czitrom, 1994). The newly formed bone subsequently substitutes the resorbed graft. The formation of new bones is important to achieve union between the bone graft and the host side. The influence of radiation on remodelling process is still unclear. Some reported that radiation delayed corporation while some found that irradiated and freeze-dried bone stimulated new bone formation (Dziedzic-Goclawska, 1978). Prewett *et al.* (1990) found that gamma irradiation at 10 and 15 kGy prior to processing did not diminish the osteoinductive property but irradiation after grinding and freeze-drying rendered the bone matrix susceptible to radiation damage. Irradiation of bone matrix gelatine at 50 kGy destroyed the osteoinductive response but at 25 kGy reduced only the response with no changes in induction mechanism (Schwarz *et al.*, 1988). Wientroub and Reddi (1988) suggested doses at 30–50 kGy enhanced heterotropic bone induction leading to a high level of mineralisation. These mixed results with respect to the effects of gamma irradiation on the biological properties of bone require more studies to be conducted. In one of the recent findings, gamma irradiation at 25–31 kGy had no significant effect on the osteoinduction of human demineralised bone matrix (DBM) powder or DBM-based products with collagen carrier in paste form (Nataraj *et al.*, 2008). In addition, the study using rat model showed that gamma irradiation of the DBM did not cause an increase in inflammation. In fact, radiation at doses lower than 30 kGy reduced immunogenicity of allografts (Goldberg and Stevenson, 1992). Another finding by Alanay *et al.* (2008) reported that 50 kGy did not decrease the fusion rates in a rat spinal fusion model when using human DBM irradiated at −50°C. Therefore, irradiation up to 30 kGy at low temperature may be used to sterilise DBM without affecting osteoinductive properties.

Frozen bones irradiated at 35 and 50 kGy showed new bone formation but freeze-dried bones irradiated at the same doses showed no osteoinduction due to fast resorption (Dziedzic-Goclawska *et al.*, 1991). It seems that remodelling is better when irradiated and deep-frozen bone is used rather than irradiated and freeze-dried bone. It is obvious that type of processing and temperature during irradiation have significant role to play in bone remodelling.

Conclusion

The availability of sterile tissue graft materials is of great importance and radiation dose selected in sterilising tissues must be sufficient to inactivate or kill all microorganisms on the tissue with no detrimental effects on the tissue. Nevertheless, radiation while sterilising tissues has other effects on the physical properties of the tissues. In general, gamma irradiation at low dose will preserve the biological and biomechanical properties of tissue. Types of processing (freezing, drying and preservation) and conditions during irradiation (temperature and water content) will influence the radiation effects on tissue.

The conditions for radiation sterilisation of tissues need to be specified and controlled well. By understanding the radiation effects, we can sterilise tissues while minimising deleterious effects, maintaining native structure as well as the biochemical and biomechanical properties of tissues and hence tissue functions.

References

Alanay A, Wang JC, Shamie AN, Napoli A, Chen C, and Tsou P (2008). A novel application of high-dose (50 kGy) gamma irradiation for demineralized bone matrix: effects on fusion rate in a rat spinal fusion model. *Spine J* **8**: 789–795. Epub 2007 Aug 3.

Anderson MJ, Keyak JH, and Skinner HB (1992). Compressive mechanical properties of human cancellous bone after gamma irradiation. *J Bone Joint Surg Am* **74**(5): 747–752.

Braams R (1963). The effect of electron radiation on the tensile strength of tendon. II. *Int J Radiat Biol Relat Stud Phys Chem Med* **7**: 29–39.

Bailey AJ and Tromans WJ (1964). Effects of ionizing radiation on the ultrastructure of collagen fibrils. *Radiat Res* **23**: 145–155.

Czitrom AA (1994). Biology of bone grafting and principles of bone banking. In Weinstein SL (ed.), *The Paediatric Spine: Principle and Practice*, Raven Press Ltd., New York, pp. 1285–1298.

Dziedzic-Goclawska A (1978). Effect of radiation sterilisation on biostatic tissue grafts and their constituents. In Gaughran WRL and Goudie AJ (eds.), *Sterilisation by Ionising Radiation*, Vol. 2, Multiscience, Montreal, pp. 156–187.

Dziedzic-Goclawska A, Ostrowski K, Stachowicz W, Michalik J, and Grzesik W (1991). Effect of radiation sterilization on the osteoinductive properties and the rate of remodelling of bone implants preserved by lyophilization and deep-freezing. *Clin Orthop Relat Res* **272**: 30–37.

Feher L, Pellet S, Unger E, and Petranyi G (1975). Organisation and immunological qualities of lyophilised and radiosterilised extrasynovial dog-tendon preparation. In *Radiation Sterilisation of Medical Products*, IAEA, Vienna, pp. 225–237.

Goldberg VM and Stevenson S (1992). Biology of bone and cartilage allografts, In Czitrom AA and Gross A (eds.), *Allografts in Orthopaedic Practice*, Williams & Wilkins, Baltimore, pp. 1–14.

Hamer AJ, Stockley I, and Elson RA (1999). Changes in allograft bone irradiated at different temperatures, *J Bone Joint Surg Br* **81**(2): 342–344.

Hilmy N, Basril A, and Febrida A (1992). Effects of procurement, packaging, storage time and irradiation dose on physical properties of amnion membranes. In *Proceedings of IAEA/PFM on Radiation Sterilization of Tissue Grafts*, April 1–3, Manila, Philippines.

Kang Y-K, Jeong J-Y, Chung Y-G, Babk W-J and Rhee S-K (2005). Complications of structural allografts for malignant bone tumours. In Kennedy JF, Phillips GO and Williams PA (eds.), *Sterilisation of Tissues Using Ionising Radiations*. CRC Woodhead Publishing, England, pp. 157–162.

Klen R and Pascal J (1975). Comparison of some properties of tissue grafts sterilised by cold shock and ionising radiation. In *Radiation Sterilisation of Medical Products*, IAEA, Vienna, pp. 203–214.

Kwarecki K, Debiec H, Kurnatowski W, Nowak M, and Swiecicki W (1977). Physico-chemical properties of preserved radiation sterilised skin grafts. *Pol Przegl Chir,* 11.

Little K (1973). Biological mechanisms. In *Manual on Radiation Sterilisation of Medical and Biological Materials*. IAEA, Vienna, Tech. Rep. Series No. 149, pp. 233–243.

Nataraj C, Silveira E, Clark J, Yonchek J, and Kirk J (2008). Effect of terminal gamma sterilization on osteoinductivity. Report, RTI Biologics, Inc., Alachua, FL.

Prewett AB, O'Leary RK, and Damien CJ (1990). Investigation on the effect of low dose gamma irradiation on the collagenous and non-collagenous proteins of bone matrix. In *Proceedings of the 12th Annual Meeting of American Society for Bone and Mineral Research*, Atlanta.

Rasmussen TJ, Feder SM, Butler DL, and Noyes FR (1994). The effects of 4 Mrad of gamma irradiation on the initial mechanical properties of bone-patellar tendon-bone grafts. *Arthroscopy* **10**(2): 188–197.

Rosomoff HL and Malinin TI (1976). Freeze-dried allografts of dura mater — 20 years experience. *Transplant Proc* **8**(2 Suppl. 1): 133–138.

Schwarz N, Redl H, Schiesser A, Schlag G, Thurnher M, Lintner F, and Dinges HP (1988). Irradiation-sterilisation of rat bone matrix gelatine. *Acta Orthop Scand* **59**: 165–167.

Shiba M, Dziedzic-Goclawska A, Kaminski A, and Yamauchi M (1999). Degradation of Achilles tendon collagen irradiated in dried (lyophilised) and wet (hydrated) state. In *Proceedings of the 2nd World Congress on Tissue Banking & 8th International Conference of EATB*, 7–10 October 1999, Warsaw, Poland, p. 50.

Singh R, Purohit S, and Chacharkar MP (2007a). Effect of high doses of gamma radiation on the functional characteristics of amniotic membrane. *Radiat Phys Chem* **76**(6): 1026–1030.

Singh R, Purohit S, Chacharkar MP, Bhandari PS, and Bath AS (2007b). Microbiological safety and clinical efficacy of radiation sterilized amniotic membranes for treatment of second-degree burns. *Burns* **33**: 505–510.

Sliwowski A and Dziedzic-Goclawska A (1976). Influence of gamma radiation on the solubility of collagen-derived membranes. *Mater Med Pol* **8**: 379–381.

Strong DM (2005). Effects of radiation on the integrity and functionality of soft tissue. In Kennedy JF, Phillips GO and Williams PA (eds.), *Sterilisation of Tissues Using Ionising Radiations*. CRC Woodhead Publishing Limited, England, pp. 163–172.

Triantafyllou N and Karatzas P (1975). Effects of sterilisation doses of ionising radiation on the mechanical properties of various tissue allografts. In *Radiation Sterilisation of Medical Products*. IAEA, Vienna, pp. 215–223.

Tomford WW (2005). Effects of gamma irradiation on bone — clinical experience. In Kennedy JF, Phillips GO and Williams PA (eds.), *Sterilisation of Tissues Using Ionising Radiations*. CRC Woodhead Publishing Limited, England, pp. 133–140.

Voggenreiter G, Ascherl R, Blümel G, and Schmit-Neuerburg KP (1994). Effects of preservation and sterilization on cortical bone grafts. A scanning electron microscopic study. *Arch Orthop Trauma Surg* **113**: 294–296.

von Versen-Hoeynck F, Steinfeld AP, Becker J, Hermel M, Rath W, and Hesselbarth U (2008). Sterilization and preservation influence the biophysical properties of human amnion grafts. *Biologicals* **36**: 248–255. Epub 2008 Apr 18.

Wientroub S and Reddi AH (1988). Influence of irradiation on the osteoinductive potential of demineralized bone matrix. *Calcif Tissue Int* **42**(4): 255–260.

Yusof N (2000). Gamma irradiation for sterilising tissue grafts and for viral inactivation. *Malays J Nucl Sci* **18**(1): 23–35.

Yusof N (2007). Interaction of radiation with tissues. In Nather A, Yusof N and Hilmy N (eds.), *Radiation in Tissue Banking: Basic Science and Clinical Applications of Irradiated Tissue Allografts*, World Scientific, Singapore, pp. 99–107.

Yusof N and Hilmy N (2007). Physical and mechanical properties of radiation sterilised amnion. In Nather A, Yusof N and Hilmy N (eds.), *Radiation in Tissue Banking: Basic Science and Clinical Applications of Irradiated Tissue Allografts*, World Scientific, Singapore, pp. 155–168.

Chapter 30

Effects of Gamma Irradiation on the Biomechanics of Bone

Aziz Nather*, Ahmad Hafiz Zulkify[†] and Shu Hui Neo*

Introduction

Lyophilised bone allografts are widely used nowadays for reconstructive surgery in various clinical disciplines. However, one major concern with such transplantation is the risk of infectious disease transmission. To reduce the risk of transmission, a good method of terminal sterilisation employed currently is to irradiate the graft with ionising radiation (gamma irradiation), which dosage is dependent on the bioburden, resistance of microorganisms to the sterilisation procedure and the sterility assurance level (SAL). Gamma irradiation is generally accepted as a safe and effective method.

While gamma irradiation may be effective in reducing the bioburden of microorganisms, especially that of viruses, many controversies have emerged regarding the effect of various dosages of gamma irradiation on the biomechanics of the bone allograft, especially at high dosages where the ability of the bone allograft to play a structural role after transplantation is thought to be compromised. Currently, the recommended dose for sterilisation of bone allografts is at 25 kGy. This is practised in

*NUH Tissue Bank, Department of Orthopaedic Surgery, Yong Loo Lin School Medicine, National University of Singapore, Singapore.
†Department of Orthopaedics, Traumatology and Rehabilitation, Kulliyah of Medicine, International Islamic University Malaysia, Malaysia.

the United Kingdom, the United States and in many tissue banks around the world, with the exception of the Central Tissue Bank in Warsaw and other multi-tissue banks in Poland where a dose of 35 kGy is used. Another exception is in Australia, where 15 kGy has become the new "golden" standard.

Compressive Properties of Bone

Anderson *et al.* (1992) investigated the compressive mechanical properties of human cancellous bone after gamma irradiation at 10, 31, 51 and 60 kGy. Two-centimetre-thick sections cut from the proximal tibiae from two male cadavers were irradiated at the abovementioned dosages of gamma irradiation. Sections that were not exposed to radiation served as a control. They found that there were no significant differences in compressive failure stress between irradiated specimens and the control, except those irradiated at 60 kGy ($p = 0.03$). When compressive strain, failure stress and the elastic modulus of sections were compared, there was no significant difference in compressive strain between all irradiated and control specimens. There was no evidence that gamma irradiation at 25 kGy affects the mechanical properties of proximal tibia allografts.

In a similar study conducted by Zhang *et al.* (1994), which investigated the force at failure, compressive strength, stiffness, Young's modulus, deformation and strain of iliac crest wedges, there was no statistically significant difference among irradiated (at 20–25 kGy), non-irradiated and deep-frozen and freeze-dried tricortical iliac crest allografts. The authors also recommended using 25 kGy for the secondary sterilisation of human iliac crest wedges.

However, there has been contention to the results of Zhang *et al.* (1994) and Anderson *et al.* (1992). Cornu *et al.* (2000) investigated the biomechanical properties of bone after freeze drying and gamma-irradiation sterilisation at 25 kGy. All treated specimens were rehydrated by immersion in saline solution and frozen specimens thawed at room temperature prior to mechanical testing. Compression tests were then performed with a 100-kN screw-driven machine. Cancellous bone slices underwent *in situ* compression between two flat-nosed steel columns. Depending on the size of the bone, 8–12 compression tests were performed. The results of the study showed that after the freeze-dried bone was irradiated, ultimate stress declined sharply from −19% (without irradiation) to −43% (after irradiation). Irradiation was found to weaken the bone significantly.

Tensile Strength of Bone

In a study by Dziedzic-Goclawska (2000), whole femurs obtained from 20-week-old WAG male rats (lyophilised or fresh) were irradiated with doses of 25, 35 and 50 kGy. Non-irradiated femurs served as controls. Using an Instron universal testing machine (Instron Corp. Canton, MA), the femurs were tested in three-point bending. The strength of the bone, or the maximum force needed to break the bone, was then calculated from the respective deflection curve. The study showed that in lyophilised femurs irradiated with 25, 35 and 50 kGy, strength decreased by 43%, 57% and 64%, respectively. In fresh femurs, the decrease was signifi-cantly lower, by 17%, 25% and 34%, respectively. It seems that hydration can reduce undesirable, radiation-induced damage to the bone allograft.

Hamer *et al.* (1996) performed a similar study which showed that the bending strength of bone was reduced to 64% of control values after irra-diation with 28 kGy and that the reduction in strength was dose dependent. Femora were obtained from seven donors at post-mortem and stripped of soft tissue. The femoral diaphyses were cut into 10–12 ring sections, approximately 1.6-mm thick, perpendicular to the long axis of the femur. Within each group from a particular donor femur, alternate rings were assigned to treatment and control groups. Specimens were passed through a commercial plant twice to give a dose of 60 kGy, and once to give a dose of 28–30 kGy. A three-point bending jig was used to study the bending strength of the specimens. Each specimen was loaded and unloaded once within their linear elastic regions, and then loaded to failure. The bending strength was compromised at 28 kGy.

Another study by Triantafyllou *et al.* (1975) showed that the bending strength of bone was markedly reduced to 10–30% of controls by a com-bination of lyophilisation and radiation sterilisation at a dose of 33 kGy.

In a more recent study by Akkus and Belaney (2005), tensile monotonic tests were performed on bone tissue following gamma-irradiation sterili-sation. The results are discussed in greater detail in the following section.

Fatigue Life of Bone

Akkus and Belaney (2005) studied the extent of degradation in the fatigue life of bone tissue following gamma-radiation sterilisation at a standard dose level of 36.4 kGy. High-cycle and low-cycle tests of control and irradiated human cortical bone tissue were performed. These simulate the dynamic nature of the physiological loading conditions experienced by allografts that cannot be found through monotonic loading.

Akkus and Belaney (2005) showed that the elastic modulus and the yield stress were not affected by sterilisation at 35 kGy, but the energy to fracture, the post-yield energy and the fracture strain experienced significant reductions of 86.4%, 70.0% and 60.5%, respectively.

In the low-cycle regime, the average fatigue life of control specimens was on the order of 10^5 cycles. The low-cycle fatigue life of irradiated specimens, on the other hand, was dramatically reduced by 99.5%, to only on the order of 100 cycles to failure. In the high-cycle range, none of the control specimens failed within 300 000 cycles, while all the irradiated specimens failed within 12 240–74 400 cycles. The mean fatigue life of high cycle was reduced by 86.8% due to gamma-radiation sterilisation.

Akkus and Belaney (2005) recommended that gamma-radiation sterilisation of the graft should be at the lower end of the standard dose range, i.e. 25 kGy to minimise the impairment of mechanical and fatigue properties of the allograft.

Torsion of Bone

At a high irradiation dosage of 60 kGy, Bright and Burchardt (1983), Komender (1976) and Triantafyllou *et al.* (1975) all showed that the specimens showed a reduction in bending, compression and torsion strength. The torsion strength was decreased to 65% at a dose of 60 kGy and to 70% by a combination of irradiation at 30 kGy and freeze-drying. Up to 30 kGy though, 90% of torsion strength was maintained.

Nather *et al.* (2004) studied the biomechanical strength of freeze-dried and gamma-irradiated cortical allografts using the tibial diaphysis of the adult cat. Large allografts (two-third of the right tibial diaphysis) of 28 adult cats were procured and lyophilised to reduce the water content to

5%, and irradiated to a dose of 25 kGy. The allografts were transplanted back into the same 28 cats, four for each observation period of 4, 6, 8, 12, 16, 28 and 36 weeks. The corresponding segment of the unoperated left tibia served as the control for each cat. After the respective observation periods, the cats were sacrificed and the allografts taken and embedded in rectangular jigs or moulds of $3.2 \times 2.4 \times 3.2$ cm using quick setting dental cement, leaving the central 2-cm portion free for torsion testing. Allografts were loaded to failure with an external rotation torsional force till an oblique fracture was produced.

The study showed that the maximum torque of freeze-dried and gamma-irradiated cortical allografts were significantly weaker than deep-frozen cortical allografts. At 24 weeks, the maximum torque was only about 12% of normal strength compared to 64% for deep-frozen allografts — one-fifth the strength of deep-frozen allografts. This difference was shown to be statistically significant in all observation periods (8, 12, 16 and 24 weeks) as shown in Fig. 1.

Nather *et al.* (2004) then concluded that lyophilised and gamma-irradiated cortical bone allografts, which only possessed one-fifth the strength of large deep-frozen cortical bone allografts, were not suitable for use in massive reconstruction of the extremities especially in the lower

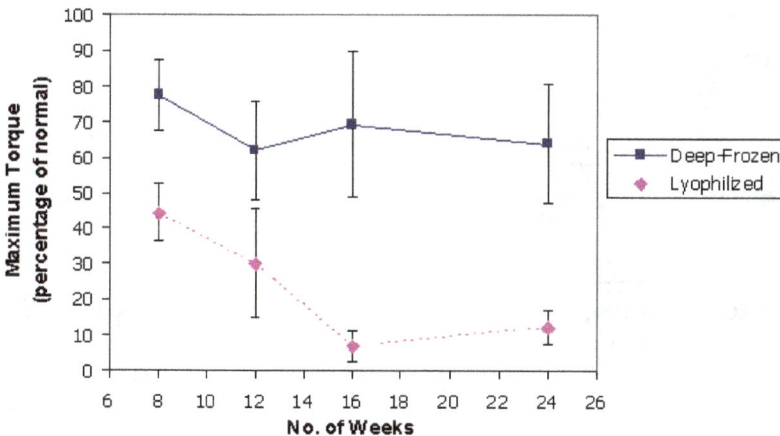

Fig. 1. Graph showing maximum torque of deep-frozen and freeze-dried cortical allografts.

limbs where immediate weight bearing was required. Instead, allografts which had been lyophilised and irradiated were useful as morsellised bone grafts for packing cavities in bones such as simple bone cysts or for packing cavities in the maxillofacial region.

Fracture Toughness

In a study investigating the fracture resistance of gamma-irradiated (27.5 kGy) cortical bone allografts by Akkus and Rimnac (2001), fracture toughness was found to be reduced and the ability of bone tissue to undergo damage in the form of microcracks and diffuse damage was significantly impaired.

This might be because gamma irradiation induces immature intramolecular cross-links between collagen molecules and scission of tropocollagen α-chains. Since allograft bone tissue are also used in various geometries such as dowels, pins and screws, which have stress concentrations present in the form of holes, ridges and screw threads, the impaired fracture resistance of bone in the presence of a stress concentration is a concern for radiation sterilised allografts.

Summary

Table 1 summarises the study by various investigators on the effect of gamma radiation on the biomechanical strength of bone.

Table 1.   Biomechanical studies of effect of gamma radiation on bone.

Dosage of gamma irradiation	Author	Findings
25–30 kGy	Anderson *et al.* (1992) Zhang *et al.* (1994) Hamer *et al.* (1996)	No effect on biomechanics of bone at 20–25 kGy. Bending strength of bone was reduced to 64% of control values at 28 kGy. Reduction was dose dependent.

(Continued)

Table 1.   (*Continued*)

Dosage of gamma irradiation	Author	Findings
	Nather *et al.* (2004)	At 25 kGy, maximum torque of freeze-dried and gamma-irradiated cortical allografts from cats was significantly weaker than deep-frozen cortical allografts.
	Dziedzic-Goclawska (2000)	Bending strength was decreased by 43% at 25 kGy.
	Akkus and Rimnac (2001)	Fracture toughness was reduced and the ability of cortical bone to undergo damage in the form of microcracks and diffuse damage was impaired at 27.5 kGy.
	Triantafyllou *et al.* (1975), Komender (1976),	Up to 30 kGy, 90% of torsion strength is maintained.
	Bright and Burchardt (1983)	Torsion strength is decreased to 70% by a combination of irradiation at 30 kGy and freeze-drying.
30–50 kGy	Triantafyllou *et al.* (1975)	Bending strength of bone was reduced to 10–30% of controls by lyophilisation and radiation sterilisation at 33 kGy.
	Akkus and Belaney (2005)	At 36.4 kGy, tensile fatigue life of cortical bone is reduced by two orders of magnitude. Fracture strain was reduced by 60.5%.
	Dziedzic-Goclawska (2000)	Bending strength was decreased by 57% and 64% at 35 and 50 kGy, respectively.
60 kGy	Triantafyllou *et al.* (1975), Komender (1976), Bright and Burchardt (1983)	At 60 kGy, torsion strength is decreased to 65%.
	Anderson *et al.* (1992)	Compressive failure stress was reduced significantly at 60 kGy.
	Hamer *et al.* (1996)	Limit of proportionality was only compromised at 60 kGy irradiation.

Conclusion

The effect of gamma irradiation on the biomechanics of bone very much dependent on the dosage employed. There has been much contention regarding the best dosage for irradiation of bone such that the biomechanics and strength of bone are preserved while also reducing the bioburden and sterilising the graft effectively so that it is safe for use.

References

Akkus O and Rimnac CM (2001). Fracture resistance of gamma radiation sterilized cortical bone allografts. *J Ortho Res* **19**: 927–934.

Akkus O and Belaney RM (2005) Sterilization by gamma radiation impairs the tensile fatigue life of cortical bone by two orders of magnitude. *J Ortho Res* **23**: 1054–1058. Epub 2005 Apr 20.

Anderson MJ, Keyak JH, and Skinner HB (1992). Compressive mechanical properties of human cancellous bone after gamma irradiation. *J Bone Joint Surg Am* **74**(5): 747–752.

Bright R and Burchardt H (1983). The biomechanical properties of preserved bone grafts. In Friedlaender GE, Mankin HJ and Sell KW (eds.), *Bone Allografts: Biology, Banking and Clinical Applications*, Little, Brown & Co., Boston, pp. 223–232.

Cornu O, Banse X, Docquier PL, Luyckx S, and Delloye C (2000). Effect of freeze-drying and gamma irradiation on the mechanical properties of human cancellous bone. *J Ortho Res* **18**: 426–431.

Dziedzic-Goclawska A (2000). The application of ionizing radiation to sterilize connective tissue allografts. In Phillips GO (ed.), *Radiation and Tissue Banking*, World Scientific Publishing Co. Pte. Ltd., Singapore, pp. 57–99.

Hamer AJ, Strachan JR, Black MM, Ibbotson CJ, Stockley I, and Elson RA (1996). Biomechanical properties of cortical allograft bone using a new method of bone strength measurement. A comparison of fresh, fresh-frozen and irradiated bone. *J Bone Joint Surg Br* **78**(3): 363–368.

Komender A (1976). Influence of preservation on some mechanical properties of human haversian bone. *Mater Med Pol* **8**: 13–17.

Nather A, Thambiah A, and Goh JCH (2004). Biomechanical strength of deep-frozen versus lyophilized large cortical allografts. *Clin Biomech* **19**: 526–533.

Triantafyllou N, Sotiropoulos E, and Triantafyllou JN (1975). The mechanical properties of lyophilized and irradiated bone grafts. *Acta Orthop Belg* **41**: 35–44.

Zhang Y, Homsi D, Gates K, Oakes K, Sutherland V, and Wolfinbarger L, Jr (1994). A comprehensive study of physical parameters, biomechanical properties, and statistical correlations of iliac crest bone wedges used in spinal fusion surgery. IV. Effect on gamma irradiation on mechanical and material properties. *Spine* **19**(3): 304–308.

Part VIII

Clinical Application

Chapter 31

Environmental Management: Control and Safety

Nazly Hilmy* and Norimah Yusof[†]

Introduction

Environmental management is a part of total quality system management that should be implemented in a tissue bank. Steps involved in procurement, processing and distribution of tissue allografts shall take place in an appropriately controlled environment of the tissue bank, whereby the area is monitored for microbiological contamination and air control. Allografts to be transplanted should be free from harmful effects to recipients. The environmental management control includes the protection of the tissue bank's personnel, ranging from managing hazard circumstances to waste treatment of the tissue bank, to prevent contamination of the environment inside and outside of the tissue bank. A well-organised and -managed tissue bank should be safe and pose minimal risk to the safety and health of the workers and the environment (TGA, AATB, APASTB and EATB Standards). This chapter explains how to implement the environmental management in tissue banking system.

*BATAN Research Tissue Bank (BRTB), Center for the Application of Isotopes and Radiation Technology, BATAN, Jakarta 12070, Indonesia.
[†]Malaysian Nuclear Agency (NM), Bangi 43000 Kajang, Selangor, Malaysia.

Environmental Control

(Refer to: IAEA, 1997; TGA, 1995; Nather *et al.*, 2007; EATB, 1997)

Environmental control in the tissue bank is a concept of how environmental conditions should be provided, maintained and validated, so that it is safe to perform procurement and processing in the tissue bank and the grafts produced are of high quality and safe for transplantation. Aseptic retrieval or procurement is the retrieval of tissues using methods that restrict or minimise contamination with microorganisms originated from donors, environment, retrieval personnel and/or equipment. Procurement and processing of tissues should be done in a clean room, i.e. a room in which the count of airborne particles is monitored and controlled within a defined specific limit. A high standard of tissue bank cleanliness and personnel hygiene is required to protect each product from contamination especially by the environment or by the operators. A standard operating procedure (SOP) should be developed to minimise contamination from microbes, chemicals and dust substances. There are four main sources of potential contamination of tissues in the tissue bank that should be controlled and monitored, as follow:

1. Environment in procurement and processing area — It can be controlled by taking air sampling regularly for microbe and particle counts. Swab test can be done for the floor, wall and ceiling. SOPs should be established and results should be documented. Cleaning equipment that generates contamination such as particles, dust or aerosols should be avoided wherever possible.
2. Personnel or worker — They should wear a protective clothing.
3. Equipment and material to be used in procurement and processing of tissues — All instruments and equipment that are directly in contact with tissues should be sterile.
4. Cross-contamination from tissues of different donor — Avoid pooling system, i.e. tissues from different donor should not mixed. The handling of tissues should be controlled so as to minimise the risk of contamination from other donor tissues.

Several requirements that should be implemented in a tissue bank in relation to the environmental monitoring are as follows:

- The areas where materials or tissues are exposed (such as the areas for procurement and processing of tissues) should be monitored regularly for microbial contamination. The result of monitoring should be the basis for assessment of air-control system.
- Temperatures where materials/tissues are held or stored should be monitored and demonstrated to be suitable for the particular materials/tissues.
- Access to environmentally controlled areas should be from corridors or other manufacturing areas. Where internal doors are barrier to avoid cross-contamination, they should be closed when not in use.
- Door that leads from processing areas directly to the outside, such as fire exist, should be secured in such a way that it may be used only as emergency exits.
- Where pest control is needed, it should be carried out in such a way to ensure that chemicals used do not contaminate materials/tissues.
- Implement good manufacturing practice (GMP) concepts in all steps of work of procurement, processing and distribution. All procedures needed should be documented.

Environmental Safety

(Refer to: Nather *et al.*, 2007; IAEA, 1997; AATB, 2002)

Each tissue bank shall provide and promote a safe work environment by developing, implementing and enforcing safety procedures. Safety precautions and procedures for maintaining a safe work environment shall include in SOP manual and shall conform to applicable international and national law or regulation.

Safe Working Environment for the Personnel

Safety aspects in a tissue bank in handling of raw materials and products are similar to clinical and biomedical laboratory as well as pharmaceutical and medical devices manufacturing facility that produce transplant or implant materials. A well-organised and -managed tissue bank should be safe and pose minimal risk to the safety and health of the worker and the environment. Personnel in tissue bank are mostly working with equipment (bone saw, deep freezer, cutter, cryopreservation, fitting, etc.), chemicals that may cause hazard, infectious agents (raw materials, blood which might be contaminated with microbes, viruses and prion) and waste of the tissue bank.

Protection of the Workers

Protection of workers can be done through:

- barrier system and
- personal habit and practice.

Barrier System

Barrier system in a tissue bank is useful to assist in implementing environmental safety including microbiological aspect. The barrier system comprises:

- *Primary barriers* around the hazardous materials, such as infectious and dangerous materials and equipment. Use correct equipment and implement GMP in all steps of work in the tissue bank.
- *Secondary barrier* around the workers. The workers should wear protective clothing when working, including mask, gowns, gloves, safety glasses and overshoes. Figures 1, 2, 3 and 4 show several types of protective clothing for workers and visitors in different tissue banks. Personal hygiene such as hand washing before and after working should be done routinely. Vaccination for hepatitides B and C should be provided at no charge for all staff of the tissue bank and the results should be documented.
- *Tertiary barrier* around the tissue bank for safe environment. This barrier includes the management of waste disposal, security and limited access for public including visitors. Barrier system in a tissue bank is illustrated in Fig. 5.

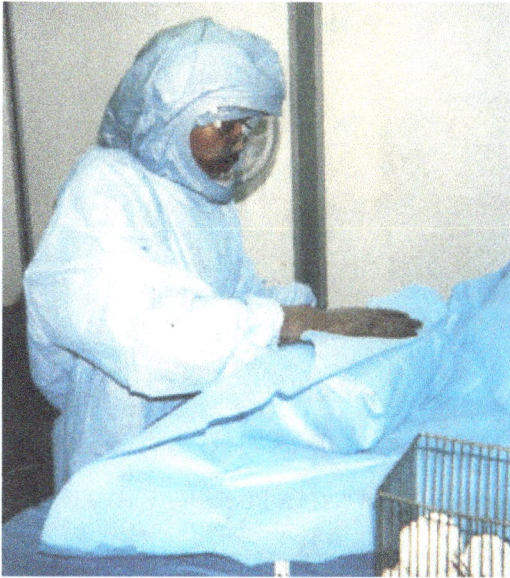

Fig. 1. Worker at North West Tissue Center, USA.

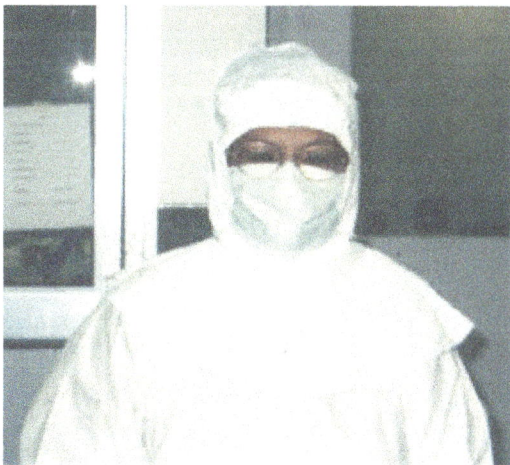

Fig. 2. Protective clothing for visitor at LifeNet Tissue Bank, USA.

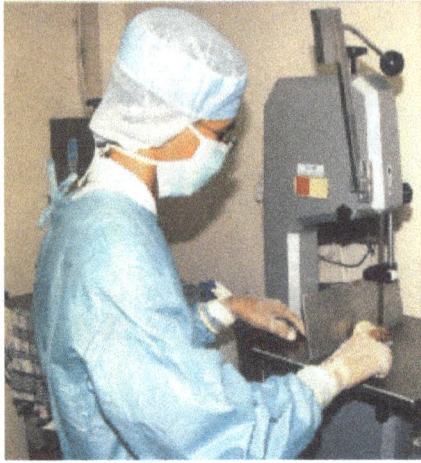

Fig. 3. Worker at BATAN Research Tissue Bank, Jakarta.

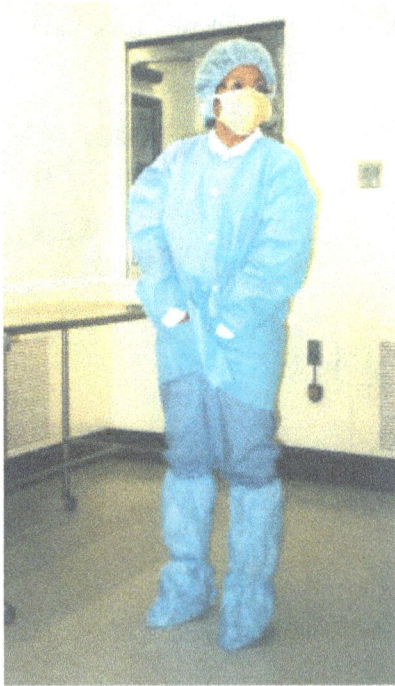

Fig. 4. Protective clothing for visitor at North West Tissue Center, USA.

TERTIARY BARRIERS AROUND TISSUE BANK
Materials, Safe Waste Disposal, Limited Access, Care of Visitors

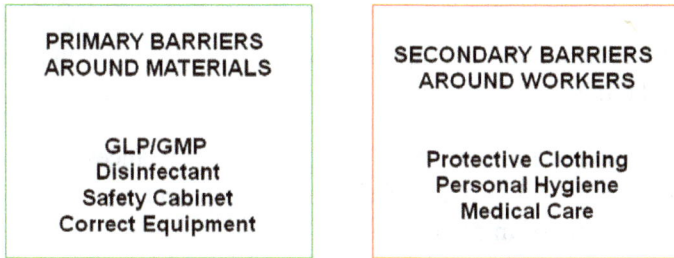

PRIMARY BARRIERS AROUND MATERIALS	**SECONDARY BARRIERS AROUND WORKERS**
GLP/GMP Disinfectant Safety Cabinet Correct Equipment	Protective Clothing Personal Hygiene Medical Care

Fig. 5. Barrier system for organising environmental safety in tissue banking.

Safe Personal Habit and Practices

- Several personal habits that should be avoided: food, beverage and smoking are not allowed in the tissue bank; beard, long hair should be covered; used gloves should not be worn outside the tissue bank.
- Personal practices that should be avoided: personal items, plants and pets should not be kept in the tissue bank; books, magazines and journals from external sources should not be taken into biohazardous area.
- All types of shoes or sandals should not be worn in laboratory. Use only laboratory shoes to reduce level of microbial contamination.
- Working alone should be avoided.

Safety Policy

- Organisation for safety policy should be implemented.
- Job description of personnel in tissue bank should be set up clearly.
- Define relation of any functional adviser, such as with fire prevention officers, radiation protection officers and safety officers, with the personnel of the tissue bank.
- Define special training for staff in safety and safety policy.

Information, Instruction and Training

- There must be adequate and appropriate staff.
- Identify need for each level of staff and develop training program accordingly.

- Integrate health and safety training with job training.
- Trained staff in safety should guide visitors and ancillary staff.

Several SOP documents on implementation of safety policy in tissue bank are listed here:

- Storage and use of hazardous materials and substances.
- Maintenance and safe use of electrical equipment.
- Safe handling of microorganisms.
- Safety of maintenance staff who might enter hazardous area.
- Safety of employees/personnel who could be affected by maintenance work.
- Arrangement for periodic environmental and other monitoring.
- Emergency procedures in the event of fire, explosion, injury, radiation, chemical mishandling, etc.
- Arrangement for working after office hours and during weekends.
- Arrangement to ensure the safety of contactor, visitor and others.
- Specific operating instruction of equipment.
- Specific safety training of staff.
- Provision of protective clothing.
- Regular inspection of equipment.
- Waste disposal.

Waste Disposal

(Refer to: Nather *et al.*, 2007; EATB and EAMST, 1997; PIC, 2000)

Human tissue and other hazardous waste items shall be disposed of in such a manner that should prevent hazardous risks to the tissue bank's personnel or the environment. Waste of the tissue bank includes tissues and related waste materials that come from procurement and processing of tissues. Contaminated waste tissues and materials can lead to transmission of diseases to the tissue bank's workers and the general public as well as to the environment.

Several Types of Waste in Tissue Bank

- Left over pieces of bone, soft tissues and muscles.
- Saw dust from bone cutting.

- Waste materials from procurement: tissue pieces, plastic pack, clothes and blood-contaminated materials.
- Waste materials from processing: contaminated materials from cleaning benches, saw, knife, vessel, etc.
- Expired tissues.
- Invalid tissues, i.e. tissues with positive result from serological and microbiological tests. These tissues can be used for research.
- Materials discarded from microbiological test such as culture media, pieces of tissues, gauze, cotton cloth, etc.

Collecting and Disposal of Waste Materials

- Waste should be collected into plastic bags immediately after procurement and processing.
- Wet waste should be wrapped in multi-layer plastic bags and labelled.
- Sharp objects must be placed in a special envelope and labelled.
- Waste can be incinerated or buried.
- All activities should be documented and special attention paid to the colour of the waste container in order to avoid mix up with other containers.

Labeling of Waste Bag/Container

The waste bag must be labelled, indicating:

- date,
- source of waste,
- name and sign of responsible personnel and
- name and address of the tissue bank.

Conclusion

Environmental management as part of quality system management should be implemented in tissue banking. The environmental management encompasses management control and management safety is a prerequisite in providing safe environment for workers and clean environment for graft processing with the main aim to produce high-quality grafts.

Most of the wastes from tissue bank are human tissues. To respect the donor and to protect the environment around the tissue bank, special attention should be taken for proper waste management system in the tissue bank. Relevant SOPs on the environmental management should be set up in each tissue bank.

References

American Association of Tissue Banks (AATB) (2002). Standards for Tissue Banking, McLean, VA, USA.

European Association of Tissue Banks (EATB) (1997). General Standards for Tissue Banking, OBIG Transplant, Vienna.

European Association of Tissue Banks (EATB) and European Association for Muscular Skeletal Tissue (EAMST) (1997). Common Standards for Muscular Skeletal Tissue Banking, OBIG Transplant, Vienna.

International Atomic Energy Agency (IAEA) (1997). Multi-Media Distance Learning Package on Tissue Banking, Modules 3, 4 and 5, IAEA/NUS, Interregional Training Centre, Singapore.

Nather A, Yusof N, Hilmy N, Kang Y-K, Gajiwala AL, Ireland L, Kumta S, and Yim C-J (2007). Asia Pacific Association of Surgical Tissue Bank (APASTB) Standards for Tissue Banking. In Nather A, Yusof N and Hilmy N (eds.), *Radiation in Tissue Banking*, World Scientific, Singapore, pp. 383–442.

Pharmaceutical Inspection Convention (PIC) (2000). Guide to Good Manu-facturing Practice (GMP) for Medicinal Products. Pharmaceutical Inspection Co-Operation Scheme-PIC/S, Geneva.

Therapeutic Goods Administration (TGA) (1995). Australian Code of GMP for Therapeutic Goods — Human Tissues, Service and Health, Australia.

Chapter 32

Clinical Applications of Bone Allografts

Aziz Nather*, Zameer Aziz* and Jia Ming Low*

Introduction

In orthopaedic surgery, gamma-irradiated and deep-frozen (−80°C) large cortical bone allografts, as well as gamma-irradiated, morselised and freeze-dried corticocancellous or cancellous bone strips/chips, remain an important option for filling large and small bone defects. Autografts are the preferred option. However, there are limitations to using autografts. This includes the size, shape, and quantity of bone needed for the reconstruction as well as the complications from harvesting autografts from iliac crest (e.g. persistent donor site pain, haematoma formation and donor site infection) (Montgomery *et al.*, 1990).

Thus, there is a need for a good tissue bank that processes high-quality bone allografts. The tissue bank must conform to standards equivalent to those of the European Association of Tissue Banks (EATB) and the American Association of Tissue Banks (AATB).

In Singapore, the NUH Tissue Bank is the only national musculoskeletal tissue bank which provides bone and soft tissue allografts to the hospitals. The NUH Tissue Bank follows the IAEA/Asia Pacific Association of Surgical Tissue Banking Standards, which is comparable to the standards of EATB and AATB.

Gamma irradiation at a dosage of 25 kGy is used as the final-end sterilisation step in its processing techniques. Femoral heads and morselised

*NUH Tissue Bank, Department of Orthopaedic Surgery, Yong Loo Lin School of Medicine, National University of Singapore, Singapore.

and lyophilised bones are irradiated in a gamma chamber in the Department of Nuclear Medicine, Singapore General Hospital. The long bones are gamma irradiated in a cobalt-60 plant at the Malaysian Nuclear Agency (NM) in Bangi by Dr Norimah Yusof (Nather and Thambiah, 1996).

Before the distribution of tissues to surgeons, informed consent is required from all recipients. Patients are informed that the products are not free of disease transmission. The risk of HIV transmission in deep-frozen allograft bone transplantation is about 1 on 1.6 million (Buck *et al.*, 1989). This is acceptable considering that the risk of HIV transmission with blood transfusion is 1 in 250000. There is no risk of hepatitis C transmission, since a 25-kGy dose of gamma irradiation inactivates the hepatitis C virus. The risk is zero for lyophilised bone allografts. Allografts are therefore an attractive option provided that they obtained from a reliable tissue bank processing high quality bone allografts.

In addition, a catalogue of available products and instructions on how the grafts should be prepared and used is provided by the tissue bank. Transplant surgeons must be trained in the proper preparation and usage of the allograft used. When new surgeons use them for the first time, the director joins the surgery team to make sure that the surgeon prepares the grafts adequately and uses them appropriately.

Gamma-Irradiated and Deep-Frozen Large Bone Allografts Preparation

Before the start of the operation, the femur or tibia in its "triple wrap" (Nather, 2000a) must be thawed for at least an hour. A separate donor team and a separate donor trolley are needed to prepare the allografts adequately. The outer layer (plastic) is opened and the bone, wrapped in inner (plastic) and middle (linen) layers, is passed by the circulating nurse to the donor team.

The bone is soaked in a large basin containing 1–2 L of normal saline with 1 g of ampicillin and 1 g of cloxacillin. The donor team starts preparing the graft by completely removing all soft tissues (including muscles and periosteum) off the bone. For intercalary grafts, the ends of the bones are then osteotomised with an oscillating saw and the medullary contents meticulously removed using a manual intramedullary reamer. All blood and bone marrow must be removed. Next, the bone is mechanically flushed with normal saline using jet lavage to make sure that all soft tissues, marrow, and blood have been completely removed, as the latter are immunogenic. The cleaned bone that has been cut to its required dimension is then soaked in a new basin containing new normal saline with 1 g of ampicillin and 1 g of cloxacillin for at least 30 minutes before it is used.

Gamma-Irradiated and Deep-Frozen Femoral Heads Preparation for Spinal Fusion

The circulating nurse opens the outer jar of the "double jar" (Nather, 2000a) to pass the inner jar to the donor team. The team opens the inner jar and soaks the femoral head in a kidney dish of saline containing antibiotics (ampicillin and cloxacillin). After thawing, all the cartilage is meticulously removed from the head. The head is then cut into about four small pieces, which are jet lavaged with saline to remove all soft tissues, marrow, and blood. The bone pieces are then passed through a bone mill to produce smaller pieces. These are mixed with autografts to provide a 50% mix (50% allograft and 50% autografts) that is ready to be placed on the prepared raw spinal bed for spinal fusion.

Gamma-Irradiated and Lyophilised Bone Allografts Preparation

The outer layer of the vacuum-packed graft is removed by the circulating nurse and passed to the donor team. The donor team removes the pieces of bone out of its inner layer. The bones are reconstituted in a small amount of saline containing antibiotics (ampicillin and cloxacillin) for 5 to 10 minutes before their use.

Appropriate Allograft Selection

For the reconstruction of large cortical bone defects, gamma-irradiated and deep-frozen cortical allografts must be used. This is crucial in the lower limb surgery where functional weight bearing is required immediately after the surgery. Nather *et al.* (2004) showed that gamma-irradiated and deep-frozen allografts exhibited 64% maximum torque strength of normal bone strength in adult rabbits *in vivo* in 24 weeks. Care must therefore be taken by the surgeon by using the strongest implant for the reconstruction, i.e. fluted interlocking nails rather than plating where possible to prevent the allograft from fracturing.

In contrast, gamma-irradiated and lyophilised cortical allografts were significantly weaker than deep-frozen allografts, with only 12% maximum torque strength of six months (one-fifth of the strength of deep-frozen allografts) (Nather *et al.*, 2004). Irradiated and lyophilised bone allografts can only be used as "fillers" for filling bone defects and not for structural functions requiring weight loading. Clinical applications include elevating calcaneal fractures, filling bone cysts, elevating bumper fractures etc.

The incidence of complications can be minimised by using the correct allografts for the various applications and high-quality bone grafts.

Clinical Applications

Between October 1988 and December 2005, a total of 854 bone and soft tissue transplantations were performed using allografts procured and processed by the NUH Tissue Bank (Nather, 2004). Of these, 238 were of soft tissue allograft transplantations. Soft tissue grafts were excluded from the following study, as the NUH Tissue Bank Protocol does not gamma irradiate soft tissue allografts.

As shown in Table 1, a total of 616 gamma-irradiated musculoskeletal bone transplantations were performed during the abovementioned period. Of these, 184 were performed for spinal surgery, 123 for hip surgery, 88 for bone trauma, 75 for malignant bone lesions, and 28 for benign bone lesions.

Spinal Surgery

Deep-frozen femoral head allografts were used for posterior spinal fusion in 132 cases. Indications included degenerative stenosis, degenerative spondylolisthesis, burst fractures, idiopathic scoliosis, congenital scoliosis, and secondary cord compression. Pure autografts were used for facet joint

Table 1. The number of gamma-irradiated bone transplantations performed.

Indication	No. of cases
Spine surgery	
• Posterior spinal fusion	132
• Anterior spinal fusion	52
Hip surgery	
• Revision total hip replacement	113
• Primary total hip replacement	10
Knee surgery	
• Revision total knee replacement	10
• Knee arthrodesis	6
Malignant bone lesion	
• Massive bone reconstruction	57
• Curettage and bone grafting	18
Benign bone lesion	28
Trauma	
• Calcaneal fracture	20
• Tibial condyle fracture	18
• Periprosthetic fracture	25
• Other fractures	25
Other bone lesions (including maxillofacial lesion)	102
Total	616

fusions. The bulk of the decorticated and freshened spinal fusion bed was then packed with allografts and used as a 50% mix with autografts (Nather, 2000b). Figure 1 shows a posterior spinal fusion using allografts for a burst fracture at lumbar 3 vertebra. Of the 132 cases, nine (6.82%) encountered complications: two with deep infections, two with superficial infections, and five with pseudarthrosis with implant failure (Nather, 2004).

Anterior spinal fusions using allografts with lyophilised femoral cortical rings were performed in 52 cases. Indications included burst fractures, osteoporotic burst fractures and secondary spinal cord compression.

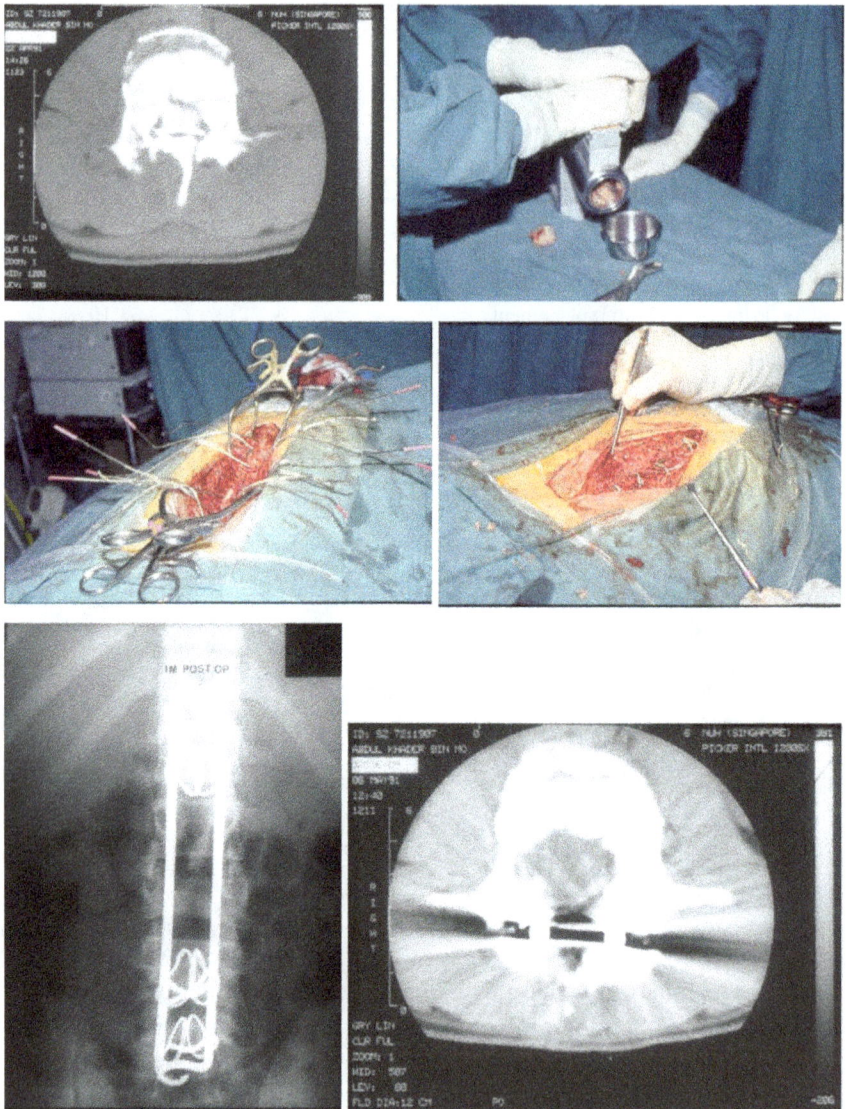

Fig. 1. A posterior spinal fusion using femoral head allografts for a burst fracture at lumbar 3 vertebra.

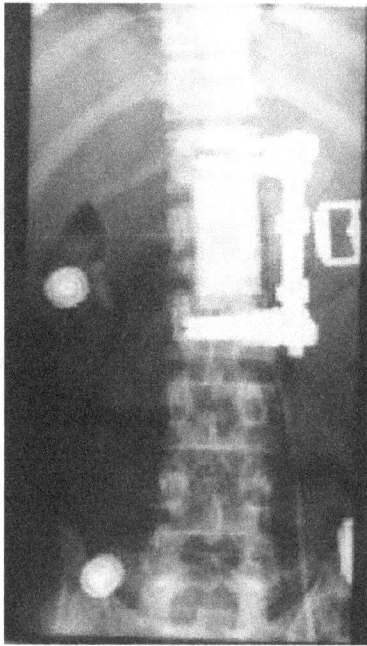

Fig. 2. A case of anterior reconstruction for metastasis to the spine secondary to renal cell carcinoma at lumbar 2 vertebra using a lyophilised femoral cortical ring and Kaneda instrumentation.

Figure 2 shows a case of anterior reconstruction for metastasis to the spine secondary to renal cell carcinoma at lumbar 2 vertebra. No infection was seen in all the 52 cases. However, one case developed persistent serous discharge from the thoracic wound following reconstruction for a burst fracture of the first lumbar vertebra. The discharge settled after two weeks (Nather, 2004).

Hip Surgery

Gamma-irradiated and lyophilised cortical or corticocancellous allografts were used in 123 cases, of which 113 were used for revision total hip replacement.

Figure 3 shows a cortical onlay strut allograft being used for revision total hip surgery. No complications were encountered with hip surgery.

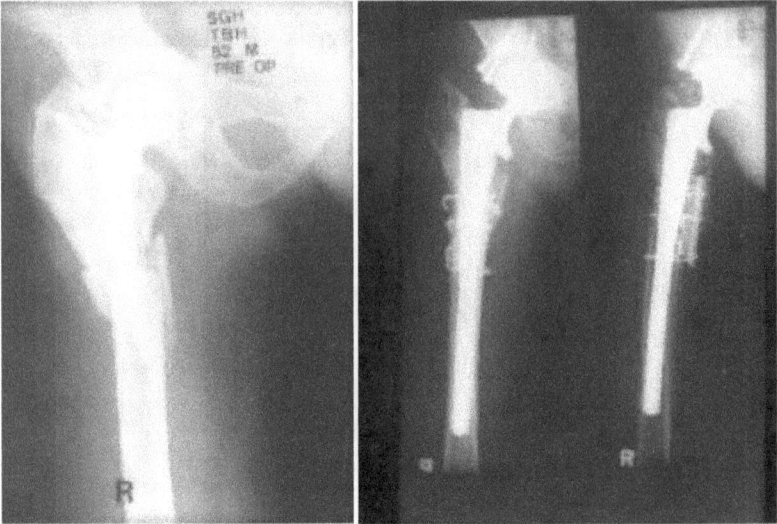

Fig. 3.　A cortical onlay strut allograft being used for revision total hip surgery.

Malignant Bone Lesions

Out of 57 cases of massive bone reconstruction for limb salvage surgery, six (10.5%) complications were encountered. These included two with non-salvageable deep infections requiring above-knee amputations and four superficial infections that were successfully treated (Nather, 2004).

Benign Bone Lesions

Curettage and bone grafting were performed in 28 cases. Figure 4 shows allografts being used for a simple bone cyst.

Trauma Surgery

Gamma-irradiated and lyophilised corticocancellous allografts were commonly used to elevate depressed calcaneal fractures. Figures 5 and 6 show their use in depressed tibial condyle fractures and depressed calcaneal fractures, respectively.

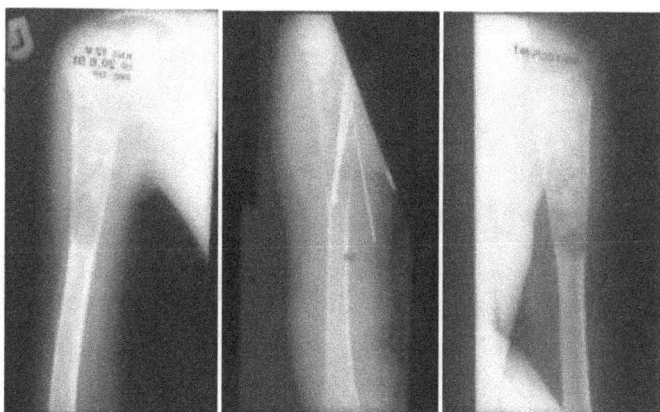

Fig. 4. Femoral head allografts being used for a simple bone cyst.

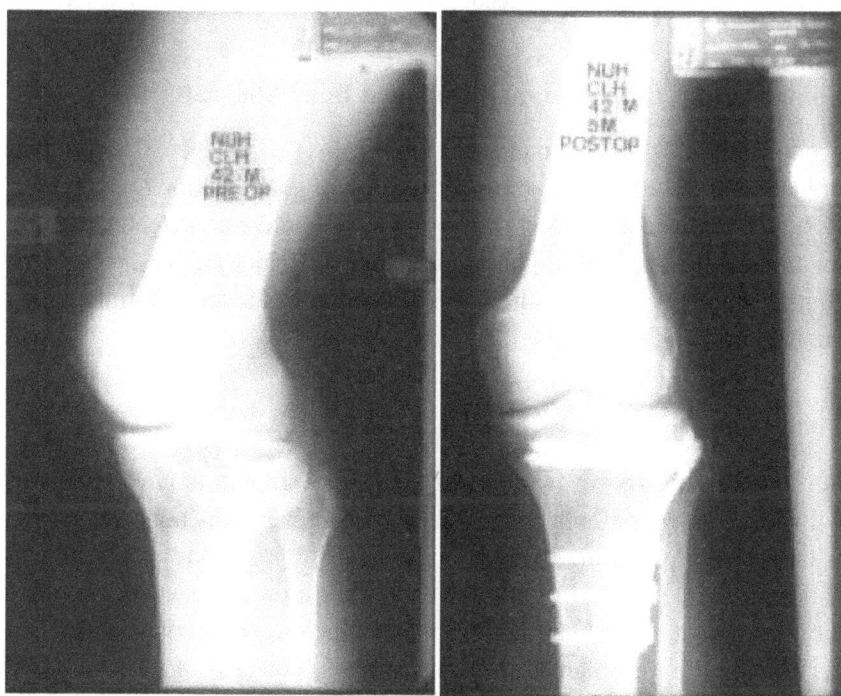

Fig. 5. The use of lyophilised corticocancellous allografts in depressed tibial condyle fractures.

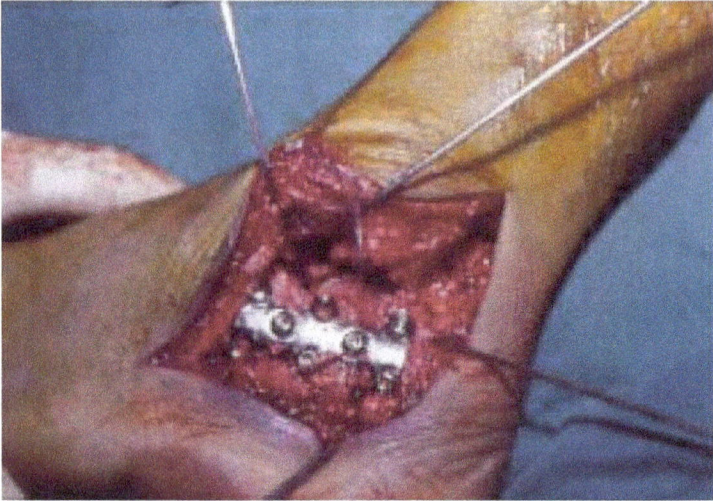

Fig. 6. The use of corticocancellous allografts in depressed calcaneal fractures.

Complications from Gamma-Irradiated Bone Allografts

Complications were encountered in 19 (3.1%) of the 616 gamma-irradiated bone allografts transplanted — a good outcome indeed. A higher (10.5%) complication was found with massive bone reconstruction for tumours in the limbs. This is expected for major surgery, where the complication rate (even without using gamma-irradiated allografts) is in the range of 10–20%.

Conclusion

In orthopaedic surgery, gamma-irradiated bone allografts have an essential role to play for both deep-frozen and freeze-dried bone allografts. A tissue bank providing high-quality tissue allografts and experience in choosing the correct allografts for each indication are needed to reduce the number of complications. The technical expertise of the surgeon performing the transplantation is crucial to the success of the operation. The indications for the use of such grafts are diverse. They range from spine surgery, hip surgery, knee surgery and trauma surgery to many other conditions in orthopaedic Surgery.

References

Buck BE, Malinin TI, and Brown MD (1989). Bone transplantation and human immunodeficiency virus. An estimate of risk of acquired immunodeficiency syndrome (AIDS). *Clin Orthop Relat Res* **240**: 129–136.

Montgomery DM, Aronson DD, Lee CL, and LaMont RL (1990). Posterior spinal fusion: allograft versus autograft bone. *J Spinal Disord* **3**: 370–375.

Nather A (2000a). Procurement systems and availability. In Phillips GO (ed.), *Radiation and Tissue Banking*, World Scientific, Singapore, pp. 263–288.

Nather A (2000b). Allografts and bone substitutes in orthopaedics and mandibular reconstruction. In Phillips GO, Strong DM, von Versen R and Nather A (eds.), *Advances in Tissue Banking*, Vol. 4, World Scientific, Singapore, pp. 149–167.

Nather A (2004). Musculoskeletal tissue banking in Singapore: 15 years of experience 1988–2003). *J Orthop Surg (Hong Kong)* **12**: 184–190.

Nather A and Thambiah J (1996). Allografts for spinal surgery. In Czitrom AA and Winkler H (eds.), *Orthopaedic Allograft Surgery*, Springer-Verlag, Wien, pp. 203–210.

Nather A, Thambyah A, and Goh JC (2004). Biomechanical strength of deep-frozen versus lyophilized large cortical allografts. *Clin Biomech (Bristol, Avon)* **19**: 526–533.

Chapter 33

Clinical Applications of Irradiated Amnion Grafts

Menkher Manjas*, Petrus Tarusaraya[†] and Nazly Hilmy[‡]

Introduction

Amnion is a collagen-rich, thin, transparent and tough membrane, lining the chorion laeve and placenta that produces the amniotic fluid at the earliest period of foetus. It is the innermost layer of foetal membranes. The function is to protect the foetus from unwanted material during intrauterine development.

The amniotic membrane consists of a thick basement membrane and an avascular stroma, with thickness of 0.02–0.5 mm. Basement membrane contains type IV collagen, type VII collagen, laminin 1, laminin 5, fibronectin, allantoin, lysozyme, transferrin, progesterone and several kinds of growth factors. The sub-chain of collagen IV and collagen VII is identical to that of conjunctiva and laminins, which facilitate corneal epithelial cell adhesion, and play an important role in ophthalmic surgery. Avascular stroma contains growth factors and anti-inflammatory proteins, as well as acts as natural inhibitor to various proteases (Kamardi *et al.*, 1993; Panakova and Koller, 1997; Koller and Panakova, 1998).

---

*Dr. M. Djamil Hospital Tissue Bank, Department of Surgery, Faculty of Medicine, Andalas University, Padang, Indonesia.
[†]Sitinala Leprosy Hospital, Tangerang, Indonesia.
[‡]BATAN Research Tissue Bank (BRTB), Center for the Application of Isotopes and Radiation Technology, BATAN, Jakarta 12070, Indonesia.

The angiogenetic capability of amniotic membrane stimulates neovascularisation and induces the development of new blood vessels. The content of allantoin, lysozyme, transferrin and progesterone plays an important role in ulcer healing mechanism, which serves as bacteriostat and bactericide (Rao and Chandrasekharam, 1981; Nursal, 1993). According to Gruss and Jirsch (1978), the allantoin serves as antibody generator, and high concentration lysozyme constitutes an enzyme that acts as bacteriostatic and bacteriolytic agent. Meanwhile, progesterone is bacteriostatic in nature against Gram-positive bacteria. In all the layers in the amnion, there is no identifiable structure of blood vessels, lymphatic vessels or nervous tissues. From immunologic viewpoint, amnion is a suitable transplant material since it does not express HLA-A, HLA-B or HLA-DR antigens and therefore rejection does not occur (Farazdaghi *et al.*, 2001).

Due to amniotic membrane functions such as promotion of wound healing, tight adherence to the wound surface, soft and easy to shape wound surface, satisfactory adhesive properties, good elasticity and sufficient transparency, wound control becomes easier without secondary redressing. As amniotic membrane also allows increase in mobility of wound area, diminishes pain, prepares the skin area for closure, and stimulates neovascularisation. It can be used as a wound covering as well as for ulcer healing (Hilmy *et al.*, 1987, 1994; Menkher and Helmi, 2001).

Fresh amniotic membranes have been used as biological dressing since 1913 (Stern, 1913) and used in ophthalmic surgery since 1940 (DeRoth, 1940). Since 1989, preserved amnion grafts, i.e. the lyophilised and radiation-sterilised amnion (ALS radiation) and air-dried and radiation-sterilised amnion (AAS radiation) produced by several tissue banks such as BATAN Research Tissue Bank, M. Djamil Hospital Tissue Bank, Soetomo Hospital Tissue Bank and Sitanala Hospital Tissue Bank in Indonesia have been applied as wound dressing for burn wound, open wound, after-surgical wound and diabetic and leprous ulcers. The preserved amnion grafts can be kept at room temperature for more than two years. ALS radiation was used as amnion grafts for ophthalmic and dental surgery and AAS radiation mostly for dressing for all kinds of wound. In 2002, those tissue banks produced more than 8000 pieces of amnion grafts per year and used them as dressing for all kinds of wounds and ulcers, as well as for ophthalmic and maxillofacial surgeries in more than 50 hospitals in Indonesia. The healing time of the wound using amniotic membranes is reduced up to 50% compared to the healing time using conventional wound dressing (Hilmy *et al.*, 1987; Kamardi *et al.*, 1993; Tarusaraya and Hilmy, 1998). Indira *et al.* (2000) reported that there is no difference in healing time between fresh/frozen amnion and lyophilised amnion/ALS radiation for ophthalmic surgery.

This chapter describes the application of amnion grafts as wound dressing for leprous ulcer, burn wound, diabetic ulcer and dressing for post-surgical wound including post-skin graft.

Benefits and Types of Amnion Grafts

The procurement, processing, packaging and radiation sterilisation of amniotic membrane to be used as amnion grafts can be found at Chapter 20 of this book.

Type of Amnion Grafts

The type of amnion grafts used for clinical applications can be divided into:

- viable/fresh (hypothermical preservation),
- short-term storage in saline or combined with antibiotic and then store at 4°C in refrigerator, mostly can be stored for up to 14 days,
- long-term storage by freezing at –85°C in DMSO (dimethylsulphoxide) or antibiotic solution,
- long-term storage by cryopreservation at –70°C and
- glycerolised (amnion in 85% glycerol) or other method.

The amnion grafts processed, radiation sterilised or others method include:

- freeze- or air-dried/oven-dried and frozen grafts and then sterilised by radiation or others method.

Benefits of Amniotic Membrane

The healing effect or promotion of healing of amnion as dressing mostly is based on its chemical and biomechanical properties such as:

- antibacterial and angiogenetic effects,
- accelerated and protected the epithelisation and granulation and stimulation of neovascularisation,
- ability to decrease pain,
- scarless healing,
- no case of rejection/avoid immunological reaction and
- tight adherence to the wound surface, increase in mobility and elasticity, translucence, semi-permeability and biodegradability (Matthews, 1981; Panakova and Koller, 1997; Koller, 2001; Menkher and Helmi, 2001).

Application of ALS and AAS Radiations on Several Kinds of Wound and Ulcer as well as for Dressing onto After-Surgical Wound in Indonesia

Some applications such as skin defect dressing of amnion graft for several kinds of wound are as follows:

- post-skin grafting and donor site (Hanafiah, 1989),
- shallow clean burn and second grade and post-burn deformities (Quinby *et al.*, 1982; Brown and Barot, 1986; Thalut *et al.*, 1993; Nursal, 1993; Sjaifuddin, 2001),
- chronic ulcerative defects (Troensegaard-Hansen, 1950; Ward *et al.*, 1989; Henky, 2004),
- leprous wound/ulcer (Tarusaraya and Halim, 1994; Tarusaraya and Hilmy, 1998; Bari and Begum, 1999),
- wound covering after Caesarean section (Menkher and Doddy, 2001),
- post-circumcision (Menkher and Helmi, 2001) and
- clean open wound in daily operation (Hanafiah, 1989).

Other applications of amnion grafts in surgical operation include:

- vagina reconstruction (Dhall, 1984; Nisolle and Donnez, 1992; Paraton, 2001),

- as a moulding for demineralised bone powder, it is biodegradable and compatible with oral tissues, because both are ectodermal in origin and
- ophthalmic operation for conjunctival surface reconstruction and corneal defects: pterygium removal, tumour removal and symblepharon lyses (Indira *et al.*, 2000; Djiwatmo, 2001; Getry, 2005).

The things to be taken care of before clinical application of ALS and AAS radiations are:

- before using, immerse graft in sterile saline for about five minutes, except for ophthalmic surgery.
- Amnion side onto the defects, proven to be more effective than chorion side for clinical use as wound dressing (Thalut *et al.*, 1993) and
- chorion side onto the defects, proven to be more effective than the amnion side for clinical application at ophthalmic surgery (Indira *et al.*, 2000; Getry, 2005).

Leprous Ulcer

Amnion as dressing, using fresh amniotic membrane, for leprous ulcer has been applied for a long time (Sabella, 1913); the latest publication was done by Bari and Begum (1999). During 1997–1998, Sintanala Hospital in Tangerang, Indonesia had observed and cured the leprous wound of 85 patients (male: 65 and female: 20) between the age of 12 and 60 years, by using AAS radiation as dressing. Types of wound were reaction (38 cases) and simple ulcer (60 cases). Locations of wound were at the legs (52 cases) and arms (33 cases), with the average area of the wound between 0.5 and 36 cm^2. The area to be covered with AAS radiation firstly was debrided and cleansed. The grafts were regularly replaced every three to four days until complete healing is achieved (Fig. 1) (Tarusaraya and Hilmy, 1998).

Type of dressing used was amnion/AAS radiation and ZnO ointment with the parameters observed were:

- interaction of age and dressing type on length of healing (days) (Table 1),

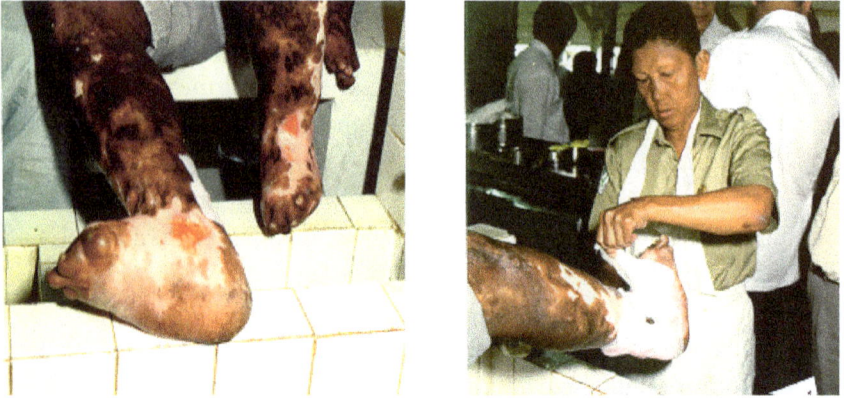

Fig. 1. Application of AAS radiation onto leprous wound at Sitanala Hospital in Tangerang, Indonesia. The wound healing time reduced up to 50% compared to using conventional dressing.

Table 1. Interaction of age and dressing type on length of wound healing (days).

Age (year)	Amniotic membrane	ZnO ointment
<20	39.8 ± 15.6	80.2 ± 15.6
20–40	23.8 ± 9.12	52.1 ± 9.1
>40	27.5 ± 11.6	60.4 ± 11.6

Table 2. Interaction of dressing and type of second-grade wound on length of healing (days).

Types of wound	Amniotic membrane	ZnO ointment
Reaction	21.3 ± 13.6	63.0 ± 6.2
Simple ulcer	39.4 ± 4.6	82.6 ± 4.6

- interaction of dressing and type of wounds on length of healing (Table 2) and
- effects of location of wound on length of healing using amnion dressing.

It can be seen that the healing of the wound to the younger patients takes longer time compared to the old patients; the conditions are probably caused by the movement of the younger patient, since most of the locations of the wound are at the leg. The condition is the same for both dressing type. The healing time of the wound using amniotic membrane as dressing for all age groups is shorter, about 50%, compared to using ZnO ointment.

Both kinds of wounds, i.e. reaction and simple ulcer, can be cured by AAS radiation and the healing time depends on the type of wound.

Tarusaraya and Halim (1994) also described the speed of the healing of leprous wound using several kinds of dressing on simple plantar ulcers as follows:

- ZnO ointment: 0.14 ± 0.10 cm^2/day.
- MSGA (magnesium sulphate glycerin acriflavine mixture): 0.10 ± 0.06 cm^2/day.
- Amnion sterile/AAS radiation: 0.29 ± 0.22 cm^2/day.

It can be cleared that the application of amniotic membrane as leprous wound dressing has several benefits as follows: the average reduction of the healing time by using amnion is up to 50% compared to that of using conventional dressing and the application of amnion is more comfortable for patients. There was no difference on healing time using AAS radiation and preserved amnion with 85% glycerol.

Burn Wound

The application of ALS radiation and AAS radiation began in Djamil Hospital in early 1990. The advantages of such applications, especially on the second-grade burn wound, are:

- able to cover all surfaces of burn wound,
- reduce pain,
- prevent further infection,
- reduce evaporation of wound and stimulate epithelisation and granulation and
- less expensive (Koller, 2001; Menkher and Helmi, 2001).

Thalut *et al.* (1993) carried out research on comparative study on 39 patients using amniotic membranes/AAS radiation and antibiotic dressing/medicated dressing as burn wound covering on wound grade 2 and grade 3 caused by fire, electricity and hot water. Parameters applied were formation time of epithelisation and granulation tissues as well as angioneogenic effect of the wound. Results showed that epithelisation formation time of tissue was faster by using AAS radiation compared to antibiotic dressing, i.e. from average of 7 days when using antibiotic dressing to average of 5 days using AAS radiation. Granulation formation time was also decreased from average of 8 days to 6.5 days consecutively. Comparative study on angiogenetic effect (that induce epithelisation and granulation) using amniotic membrane and antibiotic dressing was done, with parameter on calculation of total number of neovascularisation, using 30 samples for each kind of dressing. Total number of neovascular was increased two times higher in using AAS radiation compared to that of using antibiotic dressing as described by Nursal (1993). The wound healing time by using AAS radiation reduced up to 45% compared to that of the healing time using antibiotic wound dressing (Thalut *et al.*, 1993). The same results were also obtained by several authors as described by Koller and Panakova (1998).

During the period of 2000–2003, there were 63 hospitalised patients with cases of second- and third-grade burn wounds at Dr. M. Djamil Hospital, Padang, Indonesia. Among the patients 76.3% men and 70.9% were above 50 years old. All the wounds were treated with AAS radiation as wound covering. The average healing time of those wounds is 21–26 days, and no complications and rejections occurred.

The same results were also described by Sjaifuddin (2001) from Dr. Soetomo Hospital, Surabaya, where AAS radiation has been used successfully for clinical applications as biological skin substitutes and skin covering since 2000.

Diabetic Ulcer

At least 15% of diabetic patients from all over the world had ulcer and 40–70% of the cases were amputated (Armstrong and Lavery, 1998). The incidence was estimated that the situation will be increased by two times by 2025. The costs of local management of each diabetic ulcer in Indonesia are relatively high. It was estimated to be \$15 per day.

During the period of 1997–2000, 37 cases of diabetic ulcer were hospitalised at Dr. M. Djamil Hospital, Padang, Indonesia of which 62.3% women and 50.9% among the patients were above 50 years old. All the wounds were covered by ALS radiation. Location of the wounds is 92% on the leg with the average area between 2 and 18 cm^2. The average of their healing time is between 21 and 26 days (Dona, 1996).

In 2004, a comparative study on diabetic ulcer grade 2 (according to Wagner classification) of 11 patients using AAS radiation (Group 1) and using conventional/medicated bandage (Group 2) as wound covering was done and the parameters observed were histopathology examination and length of wound healing. Results showed that histopathological examination grade of peri healing diabetic ulcer after using AAS radiation gave higher grade than that of using conventional bandage. Based on the research findings, it can be seen that AAS radiation stimulates the increase of cell number from replacement tissues microscopically (Table 3). The average rate of the healing time of all the wounds was 25.36 ± 4.843 days (i.e. from 20 to 32 days) while the healing time of using ALS radiation (Group 1) was 19.56 ± 2.035 days and the healing time of using conventional bandage (Group 11) as control was 27.18 ± 2.897 days (Fig. 2) (Henky, 2004).

The study proves that the healing time of the ulcer by using ALS radiation as dressing is faster than that of using conventional sterile bandage, because the fast response of the substance released by the amniotic membrane used.

Table 3. The average rate of parameters of histological examination between the two groups (Group I and Group II).

Histopathology description	Group 1, N = 11	Group 2, N = 11	t-Test (p)
Granulation tissue	16.73 ± 3.197	4.18 ± 2.786	0.000
Epithelial cell	10.73 ± 2.328	3.82 ± 1.662	0.000
Neovascular cell	8.64 ± 2.420	2.73 ± 2.054	0.000
Lymphocytic cell	11.73 ± 3.197	8.73 ± 1.954	0.015
PMN cell	10.45 ± 2.423	7.73 ± 1.902	0.008

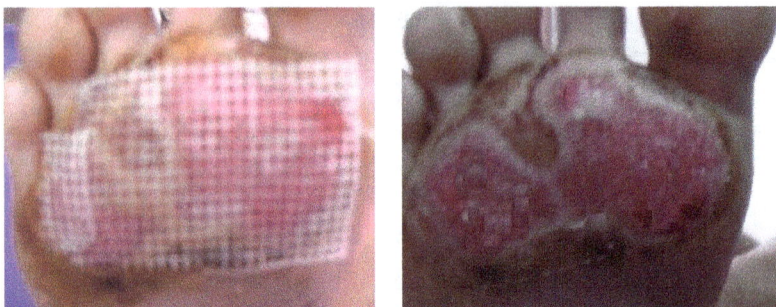

Fig. 2. Diabetic ulcer: application of ALS radiation as wound covering for foot ulcer at Djamil Hospital in Padang. First, the wound will be covered by ALS radiation and then it will heal without scar after about 15 days.

The sign of histological healing of skin ulcer can be detected by the increase of number of granulation tissues, epithelial cells, neovascularisation, lymphocyte and polymorphonuclear cell/PMN (Bose, 1979; Gruss and Jirsch, 1978). The increase in the number of healing cell/tissue, in biomolecular way, has been explained definitely in Table 3. Therefore, to find out the biomolecular mechanism and the substance found in the ALS radiation, which can stimulate the growth of healing cell/tissue on the ulcer, further research is required (Menkher *et al.*, will be published).

Post-Wound Surgery

Several types of post-wound surgeries were applied amnion as wound dressing, such as post-Caesarean operation, post-skin grafting, post-circumcision and clean wound are used in daily operation. During 1998–2002, there are 250 types of wound that were closed by AAS radiation, such as post-Caesarian operation (58 cases), circumcision (64 cases) and clean wound operation.

Post-Caesarean Operation

The wound covering usually used is medicated gauze, followed by abdominal strapping applied for three to four days. Sometimes, the wounds heal by secondary healing and left scar in abdominal wall, which makes women unpleasant. The procedural operation was done with Pfannenstiel incision. Abdominal strapping was used for flabby of abdominal wall. Radiation-sterilised and lyophilised amniotic membranes/ALS radiation produced by Dr. M. Djamil Hospital Tissue Bank were used for this study. The size of the amnion depends on the area of the wound but mostly covers 5×20 cm^2. Evaluation of all cases was followed up to eight days.

Result showed that this technique is good in achieving the best results of the wound healing and technically, it is not difficult. Both of the operation types, whether emergency or elective and dirty or clean, gave the same results of wound healing.

As there were no complications such as severe wound infection, secondary closing of wound, dehiscence, allergic reaction, as well as bad appearance of wound, it can be concluded that amnion should be recommended as wound covering for post-section Caesarean (Menkher and Helmi, 2001).

Wound Covering after Circumcision

The aim of using AAS radiation is to achieve a good wound healing, which is free from infection, pain as well as earlier use of underpants than usual.

Dorsal circumcision technique had been done in 165 boys between the age of 6 and 10 years. The works were done at Djamil Hospital for 58 boys (35.15%), while for 107 boys (64.85%), the works were carried out outside of the hospital such as at health centre or at other places such as at mass dorsal circumcision.

After suturing the wound, amnion was used as wound covering including penis glen to protect the glen from external irritation. Without using any gauze and bandage, the patients were allowed for early mobilisation. Antibiotics were given as prophylactic.

In all cases of work, the duration of wound healing, which was carried out inside and outside of hospital, was less than six days. There is no difference in duration of wound healing, which was done inside and outside of hospital ($p > 0.05$) (Menkher and Doddy, 2001).

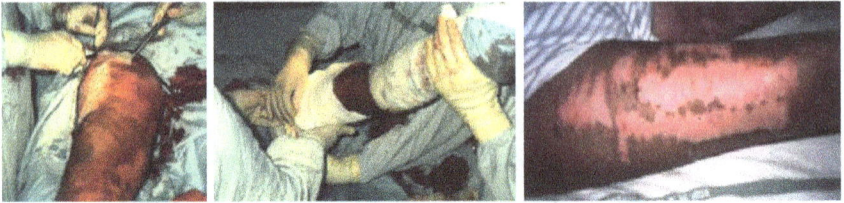

Fig. 3. Post-skin grafting: application of AAS radiation as wound covering for skin defect at donor site at Djamil Hospital in Padang. First, the wound will be covered by AAS radiation, and then by bandage. The wound will heal without scar faster than that of using medicated covering.

Post-Skin Grafting

Application of AAS radiation as wound covering for skin defect at donor site can be seen at Fig. 3. The average wound healing without scar using AAS radiation is nine days and by using medicated dressing is 11 days (Hanafiah, 1989).

From those studies, we can conclude that AAS radiation redressing gives better result if compared to original wound dressing especially in rate of infection, rate of wound healing and rate of complications.

Conclusions

Since the ALS radiation and AAS radiation can be widely used on several kinds of wounds, from dirty to clean wound and from fresh to old wound, several type of ulcer as well as post-operative wounds; therefore, we can conclude that the biologic dressing is applicable for all types of wound and gives better result than that of non-biologic dressing.

To find out the biomolecular mechanism and the substance found in the ALS radiation which can stimulate the growth of healing cell/tissue on the ulcer, further research is required.

References

Armstrong DG and Lavery LA (1998). *Diabetic Foot Ulcers: Prevention, Diagnosis and Classification*. University of Texas Health Science Center at San Antonio and Diabetic Foot Research Group, San Antonio, Texas.

Bari MM and Begum R (1999). Use of radiation-sterilised amniotic membrane grafts as temporary biological dressings for the treatment of leprotic ulcer. In Phillips GO, Kearney JN, Strong DM, von Versen R and Nather A (eds.), *Advances in Tissue Banking*, Vol. 3, World Scientific, Singapore, pp. 477–483.

Bose B (1979). Burn wound dressing with human amniotic membranes. *Ann R Coll Surg Engl* **61**(6): 444–447.

Brown AS and Barot LR (1986). Biological dressing and skin substitute. *Clin Plast Surg* **13**(1): 69–74.

De Roth A (1940). Plastic repair of conjunctival defects with fetal membrane. *Arch Opthalmology* **23**: 522–525.

Dhall K (1984). Amnion graft for treatment of congenital absence of the vagina. *Br J Obstet Gynaecol* **91**(3): 279–282.

Djiwatmo (2001). Transplantasi membran amnion. The 1st Indonesian Tissue Bank Scientific Meeting and Workshop on Biomaterial Application, Surabaya, 22–23 September, pp. 69–75.

Dona A (1996). Diabetic foot of NIDDM patient in the Internal Medicine Department Dr. M. Djamil Hospital 1990–1994. *Acta Med Indones* **XXVII**: 1341–1346.

Farazdaghi M, Adler J, and Farazdaghi SM (2001). Electron microscopy of human amniotic membrane. In Phillips GO and Nather A (eds.), *Advance in Tissue Banking*, Vol. 5, *The Scientific Basis of Tissue Transplantation*, World Scientific, Singapore, pp. 149–171.

Getry SIS (2005). Multilayered amniotic membrane transplantation for the treatment of corneal epithelial defects and ulcers, Department of Ophthalmology, Faculty of Medicine, Andalas University/M. Djamil Hospital Padang, Indonesia, will be published.

Gruss JS and Jirsch DW (1978), Human amniotic membrane: a versatile wound dressing. *Can Med Assoc J* **118**(10): 1237–1246.

Hanafiah D (1989). Clinical studies on application of sterile lyophilization amnio-chorion membrane on open wound. Presented at RCA Meeting on Radiation and Nuclear Teach for Sterilized and Clinical Quality Control of Tissue Graft, Bangkok.

Henky J (2004). The use of freeze dried radiation sterilized amniotic membrane as wound covering for diabetic ulcer, Faculty of Medicine, Andalas University, Padang, Indonesia.

Hilmy N, Siddik S, Gentur S, and Gulardi W (1987). Physical and chemical properties of freeze dried amnion membranes sterilized by irradiation. *J Atom Indones* **13**(2): 1–14.

Hilmy N, Basril A, and Febrida A (1994). The effects of procurement, packaging materials, storage and irradiation dose on physical properties of lyophilized amnion membranes. In *Proc. of IAEA Meeting of Radiation Sterilization of Tissue Grafts*, Manila, Philippines.

Indira S, Laksmi T, and Bambang S (2000). Freeze dried and fresh amniotic membranes with limbal stem cell transplantation in severe conjunctival tumor and corneal defect. In *The 8th International APASTB Conference on Tissue Banking*, Bali, 20–24 October, p. 98.

Kamardi T, Nursal H, and Nazly H (1993). Clinical studies on application of sterile irradiated freeze-dried amnio-chorio membranes on burn wound treatment. Presented at *Seminar on Wound Treatment*, Padang, Indonesia, pp. 5–24.

Koller J (2001). Healing of skin and amnion grafts. In Phillips GO and Nather A (eds.), *Advance in Tissue Banking*, Vol. 5, *The Scientific Basis of Tissue Transplantation*, World Scientific, Singapore, pp. 398–417.

Koller J and Panakova E (1998). Experiences in the use of foetal membranes for the treatment of burns and other skin defects. In Phillips GO, Strong DM, von Versen R and Nather A (eds.), *Advance in Tissue Banking*, Vol. 2, World Scientific, Singapore, pp. 353–359.

Matthews RN (1981). Wound healing using amniotic membranes. *Br J Plastic Surg* **34**(1): 76–78.

Menkher M and Doddy E (2001). Experience of using amniotic membrane after circumcision. In *Proc. Scientific Meeting Research and Development on Application of Isotopes and Radiation*, BATAN, Jakarta, pp. 165–168.

Menkher M and Helmi H (2001). Using amniotic membrane as wound covering after caesarian section operation. In *Proc. Scientific Meeting Research and Development on Application of Isotopes and Radiation*, BATAN, Jakarta, pp. 169–173.

Nisolle M and Donnez J (1992). Vaginoplasty using amniotic membranes in cases of vaginal agenesis or after vaginectomy. *J Gynaecol Surg* **8**(1): 25–30.

Nursal H (1993). The angioneogenic effect and decrease population of bacteria by amniotic membrane conservation for burn wound in Dr. M. Djamil Hospital. In *Seminar on Wound Treatment*, Padang, Indonesia, pp. 25–32.

Panakova E and Koller J (1997). Utilisation of foetal membranes on the treatment of burns and other skin defects. In Phillips GO, von Versen R, Strong DM and Nather A (ed.), *Advance in Tissue Banking*, Vol. 1, World Scientific, Singapore, pp. 165–173.

Paraton H (2001). Management vaginal agenesis. In *The 1st Indonesian Tissue Bank Scientific Meeting and Workshop on Biomaterial Application*, Surabaya, 22–23 September, pp. 77–83.

Quinby WC, Jr, Hoover HC, Scheflan M, Walters PT, Slavin SA, and Bondoc CC (1982). Clinical trials of amniotic membranes in burn wound care. *Plast Reconstr Surg* **70**(6): 711–717.

Rao TV and Chandrasekharam V (1981). Use of dry human bovine amnion as a biological dressing. *Arch Surg* **116**(7): 891–896.

Sabella N (1913). Use of foetal membranes in skin grafting. *Med Rec NY* **83**: 478–480.

Sjaifuddin N (2001). Clinical applications of biomaterial for plastic surgery. In *The 1st Indonesian Tissue Bank Scientific Meeting & Workshop on Biomaterial Application*, Surabaya, 22–23 September, pp. 57–64.

Stern W (1913). The grafting of preserved amniotic membrane to burned and ulcerated skin surface substituting skin grafts. *JAMA* **1**: 973–974.

Tarusaraya P and Halim PW (1994). Comparison study using amnion membrane, ZnO ointment, acriflavin glyserine magnesium sulphate on ulcus plantar of leprosy patients. *Med J Indones* **44**: 25–29.

Tarusaraya P and Hilmy N (1998). Comparison study using freeze dried amnion membrane and ZnO ointment as wound covering for leprosy ulcer. Presented at *The 7th International APASTB Conference on Tissue Banking*, Kuala Lumpur.

Troensegaard-Hansen E (1950). Amniotic grafts in chronic skin ulceration. *Lancet* **1**(6610): 859–860.

Ward DJ, Bennett JP, Burgos H, and Fabre J (1989). The healing of chronic venous leg ulcers with prepared human amnion. *Br J Plast Surg* **42**(4): 463–467.

Chapter 34

Clinical Applications of Radiation-Sterilised Grafts in Periodontal Surgery

Retno Dwijartini Tantin*, Paramita Pandansari[†],
Basril Abbas[†] and Nazly Hilmy[†]

Introduction

Periodontal surgery is the surgical manipulation of periodontal tissue and bone. It is performed for periodontitis cases by excising diseased gingival tissue and removing necrotic bone (Lindhe, 1995).

The periodontal disease is one of the most prevalent disorders of the dentition, which has until recently been treated mainly by symptomatic (reparative) measures. At the present time, however, important aspects of the etiology and pathogenesis of the disease are properly understood and therapeutic methods are available by which the causative factors can be eradicated or controlled and their recurrence prevented.

Pocket elimination then became the main objective of periodontal therapy. Traditionally, increased pocket depth has been the main criterion in determining whether or not periodontal surgery should be performed. However, pocket depth is no longer as unequivocal of concept as it used to be.

---

*BATAN Dental Clinic, Centre for Technology of Radiation Safety and Metrology, BATAN, Jakarta 12440, Indonesia.
[†]Batan Research Tissue Bank (BRTB), Center for the Application of Isotopes and Radiation Technology, BATAN, Jakarta 12070, Indonesia.

Freeze-dried bone allografts were introduced to periodontics in early 1970s, but had been used in orthopaedic medicine since 1950. The development of periodontal bone allografts as an alternative source of graft material was benefitted by limited number of autogenous bone. The disadvantages of fresh frozen bone allografts are the possibility of disease transfer, immunogenicity and the need for cross-matching (Robert and Pamela, 1993; Mellonig, 1994a,b), which made freeze-dried bone allografts even more attractive because they are devoid of these limitations.

Radiation-sterilised, demineralised and freeze-dried bone allograft (DFDBA), periosteum xenograft as well as the amniotic membrane produced by BATAN Research Tissue Bank were used routinely for periodontal cases, including pigmentation of gingiva in our dental clinic (Tantin *et al.*, 2008). The quality of these local products is the same as those imported ones that had been used previously. No rejection occurred in all of these cases.

This chapter explains the advantages of using radiation-sterilised bone allograft, bone xenograft, and amniotic membrane for periodontal cases.

Clinical Application of DFDBA, Periosteum Xenograft and Amniotic Membrane

Preparation of Graft Before Using (Tantin *et al.*, 2008; Tantin, 2001; Wilson and Kornman, 1992; Macphee and Cowley, 1981).

The patients with chronic periodontitis should be free from any major systemic complications, and have an acceptable level of plaque control, positive attitude towards therapy, as well as committed to live long with periodontal maintenance therapy. The sites selected should have deep probing depth, the bone loss may be horizontal and or vertical with minimal soft tissue recession. Radiographic evidence of bone loss is needed. BATAN Research Tissue Bank (BRTB)'s products, such as radiation-sterilised, demineralised and freeze-dried bone allografts (DFDBA), periosteum xenograft, as well as amniotic membrane were used for this purpose.

Previous cases that were conducted include:

- Increasing depth of periodontal pocket (50 cases).
- Post-odontectomy (10 cases). A comparison study was done with and without DFDBA in post-odontectomy on 10 cases. Healing time and clinical appearance were used as parameters.
- Pigmentation of gingival (15 cases).

The steps involved in increasing depth of periodontal pocket and post-odontectomy are:

- *Anaesthesia and incision* — Block or infiltration anaesthesia was used. Good haemostasis is important for visualisation of the bone defect and root surface to control bleeding at the defect site. Sulcular incisions to bone are made on both facial and lingual surfaces.
- *Flap reflection* (Fig. 1) — Once the incisions are made, the interproximal papillae are gently elevated to ensure they are freely movable.
- *Soft tissue debridement* — All of the soft and granulation tissues are removed and the bleeding will rapidly subside with ultrasonic instrumentation followed by curets.
- Root planning — Removal of bacterial plaque and gross debris is done via ultrasonic instrumentation. Hand instrument is used to plane the root surface.

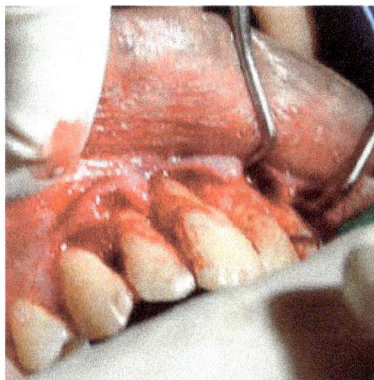

Fig. 1.   Flap operation.

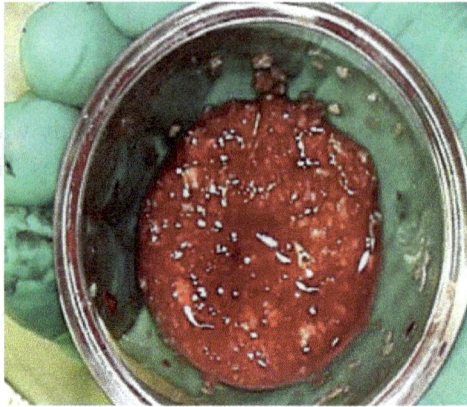

Fig. 2. Mixture of DFDBA with patient's blood.

- *Visualisation of defect site* — The surgical site is inspected to make sure that all of the soft tissues have been removed from the defect.
- *Bone graft preparation* (Fig. 2) — Radiation-sterilised DFDBA is placed in a sterile dish and then mixed with the patients blood or sterile saline.
- *Graft placement* — The mixture is packed firmly into the defect in a systematic fashion until the entire defect is filled with bone mixture. For the cases with lost of periosteum, the implantation of periosteum membrane is needed. Ten cases of periosteum implantation were done.
- *Flap closure* — Completed flap closure is important for successful result.
- *Suturing* — An interrupted vertical suturing technique is used.
- *Immediate post-surgical management* — A periodontal dressing is placed to protect the wound.
- *Supportive periodontal care* — The patient is placed in a periodontal programme commensurate for the maintenance of probing depth reduction and clinical attachment.

Pigmented Gingival

Pigmentation has been associated with a variety of endogenous and exogenous etiologic factors. It can happen in both normal and/or abnormal discolouration. It has a specific characteristic and most frequently happen at intraoral tissues. The most common location is the attached gingival, followed by the papillary gingival and then alveolar mucosa. Prevalence of gingival pigmentation was higher on the labial part of the gingival than that of palatal/lingual parts.

Classification of Pigmentation

- Localised pigmentation or amalgam tattoo.
- Multiple or general pigmentation caused by genetics, drugs (smoking and exposure to heavy metals), HIV, and nutritional deficiencies.

The treatment can be done by manual technique, electric cauterisation or combined; however, it could not improve the thickness of gingiva (Ciçek and Ertaş, 2003).

In this case, the treatment was done by using manual technique combined with radiation sterilisation amniotic membrane to improve the thickness of gingival.

The steps in the treatment are as follows:

- Scrapping by using blade until pigmentation fade away.
- Smoothing the surface of gingival.
- Irrigation with normal saline, followed by amniotic membrane on the surface of gingiva and then cover with a periodontal dressing.

Results

Chronic Periodontitis

From 50 cases of chronic periodontitis with absolute or infrabony pocket which have been done from 2006 to 2008, the growth of bone happened in all of those cases. Evaluation was done after 3–12 months of implantation by using probing depth and X-ray examination. The result of the X-ray examination can be seen at Figs. 3 and 4.

It can be seen that the growth of bone exist significantly in the defect site.

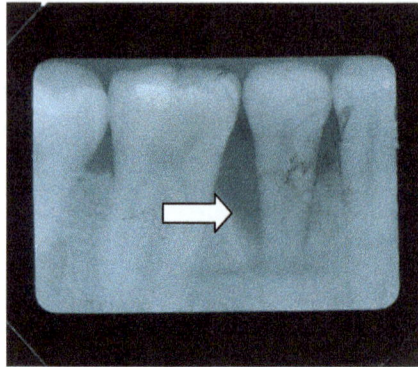

Fig. 3. Before implantation of DFDBA.

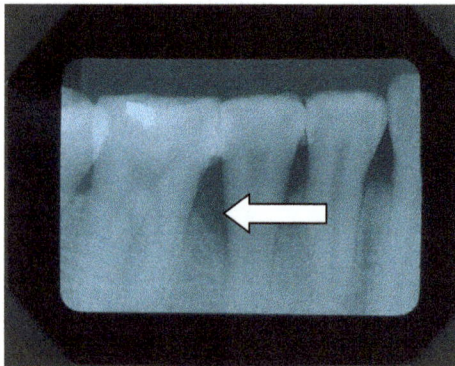

Fig. 4. Six months after implantation.

Post-Odontectomy/Post-Extraction

Ten cases of post-odontectomy were evaluated from 2007 to 2008. The parameters used are clinical appearance and healing time. Results show that implemented DFDBA improves the attachment level in cases of chronics periodontitis and post-odontectomy as well as the healing time. The healing time using DFDBA is about two times faster than that without DFDBA. The evaluation was done after one month of implantation.

Pigmentation of Gingival

Fifteen cases of pigmented gingival that were mostly caused by drugs and nutritional deficiencies were done in this work. Results showed that all of the gingival cases improved, the colour disappeared and the thickness of gingival increased significantly by the implantation of amnion membrane (Figs. 5 and 6).

It can be seen that after two weeks of treatment with amniotic membrane the pigment disappeared and the thickness of gingival increasing.

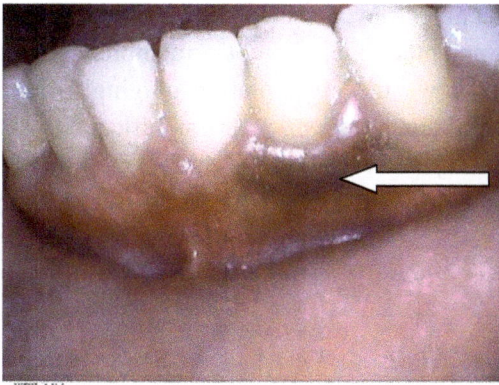

Fig. 5.   Before treatment with amnion.

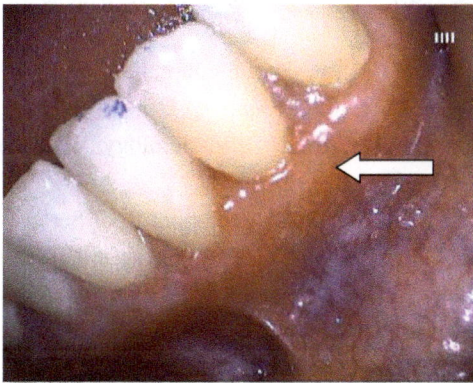

Fig. 6.   Two weeks after treatment with amnion.

This work was done on 15 cases of pigmented gingival during January to August 2009.

Discussions

The results of this study were evaluated on clinical and radiological bases. New bone formation and growth of connective tissue were demonstrated in the defect sites. They were detected by radiological technique to compare before and after treatment by DFDBA. It shows that DFDBA implantation increases the pocket depth and the healing time of the defect, and also the attachment level in cases of chronic periodontitis and post-odontectomy. Implantation of amniotic membrane in pigmentation of gingival could remove the pigmentation of gingival and revert to the normal colour and increase the thickness significantly.

There was no rejection in all of the cases done using DFDBA, periosteum and amniotic membrane. Radiation-sterilised grafts produce by BRTB can be used for periodontal defect to replace the imported grafts. The price of imported grafts is about 10 times higher than that of locally produced grafts.

References

Brunsvold AM and Mellonig JT (1992). Bone graft and periodontal regeneration. In Carton JG (ed.), *Periodontology 2000*, Munksgaard, Copenhagen, pp. 80–81.

Ciçek Y and Ertaş U (2003). The normal and pathological pigmentation of oral mucous membrane: A review. *J Contemp Dent Pract* **4**(3): 76–86.

Lindhe J (1995). *Texbook of Clinical* Periodontology, W.B. Saunders Company, pp. 219–224, 353–357.

Lowenguth AR (1993). Periodontal Regeneration: Root Surface Demineralization, In Caron JG (ed.), *Periodontology 2000*, Munksgaard, Copenhagen, pp. 53–54.

Macphee T and Cowley G (1981). *Essentials of Periodontology and Periodontics.* Blackwell Scientific Publications, London, 3rd ed. pp. 261–263.

Mellonig JT (1994a). Osseous grafts and periodontal regeneration. In Polson AM (ed.), *Periodontal Regeneration, Current Status and Direction*, Quintessence Publishing Co. Inc., pp. 71–102.

Mellonig JT (1994b). Severe chronic adult periodontitis, Today, freeze dried bone allografts are used routinely in periodontal therapy. In Wilson TG *et al.* (eds.), *Advances in Periodontics*, Quintessence Publishing Co. Inc., pp. 181–192.

Robert SG and Pamela MK (1993). Periodontal regeneration using combined technique. *Periodontology 2000*, **1**: 109–117.

Tantin RD (2001). *Kesembuhan Klinis Jaringan Periodonsium Setelah Terapi Awal Pada Kasus Chronic Adult Periodontitis Dengan Indikasi Bedah Flap*, Fakulty of Dentistry, University of Indonesia, pp. 10–11.

Tantin RD, Pandansari P, Nazly H, and Abbas B (2008). Clinical application of radiation sterilization DFDBA and DFDBA-bovine periosteum membrane in periodontal defect. Presented in 5th World Congress on Tissue Banking and 12th International Conference of APASTB held in Kuala Lumpur, Malaysia.

Wilson TG and Kornman K (1992). Diagnosis of periodontal disease and conditions using a traditional approach. In *Advances in Periodontics*, Qintessence Publishing Co. Inc., pp. 75–77.

Chapter 35

Complications of Bone Allografts

Aziz Nather* and Li Min Tay*

Introduction

The reported complication rate of tissue grafts produced by any tissue bank is always lower than the actual complication rate. This is because whilst all doctors using tissue grafts were required to report complications to the tissue bank, this was not always done. It was also difficult to verify the occurrences of such complications. It is important that tissue bankers understand that the tissue grafts they produced can cause potential complications. They should also know what these complications are. Only in this way can they produce better quality allografts to prevent more complications.

The demand for bone allograft, particularly for revision arthroplasty, is likely to increase in the future. It is important that appropriate safeguards must be taken to ensure that the safety of both donor and recipient is not compromised. Screening must be undertaken to exclude donors with potentially serious transmissible diseases (e.g. hepatitis B, hepatitis C, HIV and syphilis), malignancy and systemic disorders that could compromise the biological or biomechanical integrity of the skeleton (Palmer *et al.*, 1999).

The main complications of bone allografts include infection, immune rejection, host–graft junction non-union, allograft resorption, fracture of allograft and transmission of biological diseases such as HIV, hepatitis B, hepatitis C and syphilis.

*NUH Tissue Bank, Department of Orthopaedic Surgery, Yong Loo Lin School of Medicine, National University of Singapore, Singapore.

In Singapore, allografts produced by the tissue bank were used by surgeons from general hospitals in various parts of the island — National University Hospital, Singapore General Hospital, Tan Tock Seng Hospital, Alexandra Hospital, Changi General Hospital, National Eye Centre, National Cancer Centre and private hospitals including Mt Elizabeth Hospital, Gleneagles Hospitals, East Shore Hospital and Camdem Medical Centre. In a recent review of 718 large allografts, the complication rate was 46% (Mankin *et al.*, 1996). Rates increased with the size of the graft and the complexity of the procedure. They also reflected the success of graft incorporation. In general, if large grafts are incorporated poorly, fracture, non-union and infection could be expected in 19%, 17% and 11% of procedures, respectively (Unlinked Anonymous HIV Serosurveys Steering Group, 1995).

Graft incorporation may depend on the patient's immune response (Stevenson *et al.*, 1992). The immune response was directed mainly against cellular debris in the graft, but the bone matrix itself was also antigenic. It stimulated resorption of the graft, which could lead to rapid dissolution of the graft (Berrey *et al.*, 1990).

Surgeons who managed patients after multiple revision knee arthroplasties might have to contend with large osseous defects. Such defects may not be amenable to the use of metallic steps and wedges that are an integral part of modern knee revision systems. The use of structural allografts is a viable alternative for the treatment of massive bone loss (Clatworthy *et al.*, 2001).

Infection in Bone Allografts

Failure of a bone allograft was usually due to infection or fracture or both. Reported rates of infection of 11% to more than 15% were considerably higher than might be expected and were far higher than the rates of infection for any other operative procedure (Mankin, 1983).

According to Lord *et al.* (1988), out of 283 patients who had a massive allograft of bone, infection developed in 33 patients (11.7%).

Factors that influenced the treatment of infection included type of infection, type of pathogen and its antibiotic sensitivities, integrity of the soft tissues and the anatomical location of the infection.

At early assessment, 27 (81.8%) out of the 33 infected allografts were considered to be a failure because an amputation or a resection of the graft was needed. However, nine allografts were salvaged by removing the graft, administering antibiotics and inserting another allograft.

Hence, Lord *et al.* (1988) clearly demonstrated that infection was the principal complication of implantation of a fresh-frozen cadaver allograft in patients. The rate of infection was high. The allograft, possibly because of its potential as an antigen or because of the host-versus-graft response, appeared to have virtually no defences against pathogens that were implanted at the time of operation or those that could arrive at the allograft site post-operatively via the blood stream.

Immune Rejection

It is useful and important to remember that autografts are the standard to which all other types of grafts must be compared (Burchardt, 1983). Perhaps the most vexing and recurrent question raised in association with osteochondral allografts over the past 40 years had been the role of the immune system in determining the biological fate of these transplanted tissues (Friedlaender and Goldberg, 1991).

Immunological rejection of bone occurred when a biological failure of incorporation was the result of a specific immune response (Friedlaender, 1991). Clinical success had been reported to be considerably lower than 90% for patients who had a massive osteochondral allograft.

In evaluation of failures, it was important to understand the biology of bone graft, host site and systemic origin that precluded or interfered with success (Friedlaender, 1987). Graft incorporation was most successful biologically and biomechanically when histocompatibility differences were minimised by matching of tissue type or by application of preservation techniques to the graft.

However, rejection was more difficult to identify and quantify in transplants of musculoskeletal tissue than in transplants of organs such as kidneys in which there were readily identifiable markers of function. Although clinical evidence of rejection of such transplants had been indirect, it had been indicated that cells of these tissues have antigens that the host mounted an immune response to them, and that the incorporation of a graft was influenced by its immunogenicity (Stevenson and Horowitz, 1992).

Non-union of the Allograft–Host Junction

The rate at which a graft was incorporated into the host skeleton was one indication of whether a graft was accepted or rejected. Successful incorporation included the formation of callus around the host–graft junctions, as well as in the internal repair of the graft (Burchardt, 1983).

For at least 65 years, bone grafting had been used to fuse joints and to repair skeletal defects. Unfortunately, non-union or fatigue failure still occurred as frequently as in the past (Burchardt, 1983). Non-union of the host–allograft junction after massive allograft transplantation for patients with malignant bone tumours was a common complication (Brien *et al.*, 1994). According to Hornicek *et al.* (2001), 163 out of 945 patients who underwent allograft transplantation (17.3%) experienced non-union. Of the 163 patients, 108 had a successful outcome with union at the junction site after reoperations on their limbs. Overall, the event of allograft–host junction non-union corresponded to a worse allograft outcome.

Approximately 50% of the patients required at least one additional surgical procedure to treat the non-union. The percentage of failure increased with more surgical procedures performed. Despite treatment, 49 patients (30%) had failure of the graft and required metallic prosthesis replacement, repeat allograft or amputation to achieve a stable limb.

Previous studies have shown that complications such as infection or fracture after allograft transplantation had not only prolonged morbidity but also reduced the proportion of patients with satisfactory functional outcome (Berrey *et al.*, 1990). Many patients with infection and fracture of the allografts had to undergo removal of the allografts. From these data, it is evident that non-union in the absence of infection of fractures was a less serious problem. Approximately 70% of these patients did reasonably well regardless of allograft type.

Allograft Resorption

Allograft resorption was a problem seen increasingly in long-term follow-up (Gross *et al.*, 1995). Possible factors contributing to allograft resorption were immune response, stress shielding, mechanical disuse caused by distal cement fixation and absence of host-bone wrapping around the graft (Haddad *et al.*, 2000). There has been concern that allograft resorption, which occasionally has led to early failure (Chandler *et al.*, 1994), may be a greater problem with longer follow-up. Early reports of resorption have the potential for subsequent graft failure (Haddad *et al.*, 2000). From the study conducted by Wang and Wang (2004), 15 patients who had undergone hip reconstruction using a proximal femoral allograft were reviewed. It was noted that allograft resorption was seen in three (20%) out of the 15 patients; there was mild resorption in one hip and severe resorption in two hips.

Gross *et al.* (1995) reviewed another 168 proximal femoral allografts and reported one significant and six minor resorptions. Masri *et al.* (1995) reported four mild and ten severe resorptions in a review of 39 patients undergoing proximal femoral allograft replacement. Haddad *et al.* (1995) studied 40 proximal femoral allografts that demonstrated mild, moderate and severe resorption, giving an overall resorption rate of 50%.

Based on a study by Clatworthy *et al.* (2001), 25 patients, with 27 intact knee replacements, who had not had a revision had 35 grafts that could be evaluated radiographically for resorption. One proximal tibial graft exhibited moderate resorption, and two demonstrated mild resorption. Five grafts had progressive resorption, resulting in implant migration and loosening.

In another experiment conducted by Blackley *et al.* (2001), 63 total hip arthroplasties in 60 patients were revised with a proximal femoral allograft-prosthesis construct. Peripheral allograft resorption in some form was seen in 13 (27%) out of the 48 hips in living patients. Mild resorption was seen in 10 hips (21%) and moderate resorption in two (4%). The resorption took several years to appear. One failure after revisions was due to resorption.

The probability of increasing rates of allograft resorption over time raised grave concerns for the use of allografts, and thus, continuous and prolonged observation was strongly recommended for these patients (Wang and Wang, 2004). There was concern that allografts might resorb over time because of revascularisation by creeping substitution (Clatworthy et al., 2001).

Fracture of Allograft

The use of frozen allografts was a very old procedure, but it was also the most controversial because the results have been unpredictable and it had been associated with a high rate of complications (especially infection and fracture) (Mankin, 1983). One of the major complications in the use of a massive frozen allograft was fracture of the allograft.

Based on the study conducted by Berrey et al. (1990), 43 patients who had been treated with allograft had subsequently fractured them. The overall incidence of fracture was almost 16%. After treatment, 33 of the 43 patients (77%) were considered to have successful (excellent or good) result and only five (12%) had failure. These data suggested that fracture of an allograft was a less severe complication than was originally thought.

However, it is still believed that fractures were a major and uncontrollable complication of the procedure. This observation was supported by the failure to find any major distinguishing features between the group that had a fracture of an allograft and the rest of the patients who had received an allograft. The most important finding was that a fracture of an allograft could be treated effectively and that, when this was done, continuity and useful function could be restored. Thus, this complication was unlike infection of an allograft, which was usually disastrous (Lord et al., 1988).

Transmission of Biological Diseases

Approximately 150 000 musculoskeletal allografts, including bone, tendon, and cartilage allografts were used by orthopaedic surgeons in the United States annually. Despite the popularity of the use of these tissues, the transmission of disease through transplantation of allografts remained a concern. Because allografts were procured from humans, there was always a risk of the transmission of disease from the donor to the recipient. However, by studying the cases in which diseases have been transmitted, an understanding of this process might be established and lead to the development of strategies to reduce the risk of transmission (Tomford, 1995).

It had been established from case reports that HIV, as well as hepatitis B virus (HBV) and hepatitis C virus (HCV) could be transmitted through musculoskeletal allografts (Simonds *et al.*, 1992; Pereira *et al.*, 1991; Schratt *et al.*, 1996) but the risk may be considered remote with the current standard of tissue banking (Lemaire and Mason, 2000).

HIV

The risk of HIV transmission from deep-frozen allograft was one in a million, while the risk for lyophilised allograft is negligible.

Simonds *et al.* (1992) reported the transmission of HIV1 from a seronegative organ and tissue donor. In his case study, a 22-year-old HIV1 seronegative man died from gunshot in head in October 1985. Four solid organs and 54 tissues were procured by Lifenet Transplant Services. Seven out of 54 recipients of the tissues were infected with HIV1, while all the four recipients of organs were infected. In fact, three out of four unprocessed frozen bones were infected.

However, all tissues processed in some way tested negative. Two receiving corneas, three receiving lyophilised soft tissue, 25 receiving ethanol-treated bone, three receiving irradiated dura mater and one receiving marrow-evacuated frozen bone were tested negative for HIV1.

Therefore, it was concluded from this incident that all tissue grafts transplanted should be processed. Even frozen bones should be gamma irradiated. Lyophilised tissues should be used wherever possible as it was safer to use than deep-frozen bones. However, deep-frozen bones were recommended when biomechanical strength could not be provided by lyophilised bones.

In assessing the risk of transmission of viral disease through musculoskeletal allografts, it was important to note that the most recent reported case of transmission of the HIV through the use of such a graft occurred in 1985 (Simonds *et al.*, 1992). Since that case, at least 500 000 musculoskeletal allografts have been transplanted in the United States, with no reports of infections due to the transmission of the HIV. This suggested that, with a combination of donor screening and blood testing, the risk of viral transmission through transplantation of a musculoskeletal allograft was low.

HCV

Conrad *et al.* (1995) reported transmission of HCV by tissue transplantation. In the first case study, a woman with a history of having sexual partners with hepatitis C, committed suicide; 23 grafts were procured — six frozen bones, 12 freeze-dried tissues, three soft tissue grafts and two corneas. All grafts except corneas were irradiated. It was discovered that one recipient was reactive to the hepatitis C antibody — irradiated and freeze-dried bone for spinal fusion.

In the second case study, 14 grafts were procured from a man who died of multiple injuries. These included four frozen bones, eight cryopreserved soft tissues and two corneas. Nine out of 12 of these grafts were transplanted in recipients, and it resulted in four recipients being reactive to the hepatitis C antibody.

Hence, it was seen from the above cases that HCV could be transmitted by bone, ligament and tendon allografts. However, gamma irradiation of 17 kGy might inactivate HCV in tissue. All tissue grafts including deep-frozen tissue grafts should be gamma irradiated to eliminate the risk of hepatitis C transmission. Deep-frozen soft tissues could be irradiated at lower dosage, perhaps 15 kGy.

HBV

The detection of HBV in the blood sample of a prospective organ donor had historically limited the utilisation of organs recovered from such individuals. The transmission of this blood-borne pathogen via organ transplantations was hazardous to allograft recipients for two reasons: first, HBV had a remarkably high infectivity rate and second, HBV replication was enhanced by immunosuppression (Kidd-Ljunggren and Simonsen, 1999).

By the late 1970s, 13–18% of surgeons had been infected with the HBV compared with 3–5% of the general population. The annual rate of infection in healthcare workers was found to range from 0.5% to 5.0%, compared with 0.1% in the general population of the USA (Shapiro, 1995).

In 1954, Shutkin reported a case of hepatitis B transmitted through transplantation of a bone allograft. The graft consisted of cancellous and cortical chips, which had been stored frozen, and had been used to elevate the depressed plateau. The allograft had been obtained from the proximal tibial metaphysis from an above-the-knee amputation specimen.

The presence of the HBV was determined with an antigen test (hepatitis B surface antigen), but the antigen might not be detectable for as long as a month after exposure to the virus.

Syphilis

There has been no report in literature of syphilis being transmitted by tissue allografts, either bone or soft tissue allografts.

Conclusion

It is therefore very important that tissue bankers must recognise how unsafe allografts could be and be committed to produce allografts of the highest quality to provide safe tissue transplantation practice.

To do this, measures must be taken at every step of the procurement and processing of the allografts including:

- strict and meticulous donor selection criteria,
- sterile or clean procurement,
- comprehensive laboratory screening tests,
- highest standards of processing under laminar flow conditions,
- end-processing gamma irradiation sterilisation,
- performing quality control tests,
- proper release procedures for clinical use of allografts,
- good understanding on how to use the allografts and proper surgical preparation of the allografts before transplantation,
- proper transplantation of allografts,
- good understanding of whether allografts could withstand the biological and biomechanical functions they were reconstructed for and

- transplant surgeons must take preventive measures to prevent complications, including the use of cyclosporine C or other antibiotics post-operatively for a period of time.

References

Burchardt H (1983). The biology of bone graft repair. *Clin Orthop Relat Res* **174**: 28–42.

Berrey BH Jr, Lord CF, Gebhardt MC, and Mankin HJ (1990). Fractures of allografts. Frequency, treatment, and end-results. *J Bone Joint Surg Am* **72**(6): 825–833.

Blackeey HRL, Davis AM, Hutchinson CR, and Gross AE (2001). Proximal femoral allografts for reconstruction of bone stock in revision arthroplasty of the hip. A nine to fifteen-year follow-up. *J Bone Joint Surg Am* **83**(3): 346–354.

Brien EW, Terek RM, Healey JH, and Lane JM (1994). Allograft reconstruction after proximal tibial resection for bone tumors. An analysis of function and outcome comparing allograft and prosthetic reconstructions. *Clin Orthop Relat Res* **303**: 116–127.

Chandler H, Clark J, Murphy S, *et al.* (1994). Reconstruction of major segmental loss of the proximal femur in revision total hip arthroplasty. *Clin Orthop Relat Res* **298**: 67–74.

Chung RT, Feng S, and Delmonico F (2001). Approach to the management of allograft recipients following the detection of hepatitis B virus in the prospective organ donor. *Am J Transplant* **1**(2): 185–191.

Clatworthy MG, Ballance J, Brick GW, Chandler HP, and Gross AE (2001). The use of structural allograft for uncontained defects in revision total knee arthroplasty. A minimum five-year review. *J Bone Joint Surg Am* **83**: 404–411.

Conrad EU, Gretch DR, Obermeyer KR, Moogk MS, Sayers M, Wilson JJ, and Strong DM (1995). Transmission of the hepatitis-C virus by tissue transplantation. *J Bone Joint Surg Am* **77**: 214–224.

Friedlaender GE (1987). Bone grafts. The basic science rationale for clinical applications. *J Bone Joint Surg Am* **69**: 786–790.

Friedlaender GE (1991). Bone allografts: the biological consequences of immunological events. *J Bone Joint Surg Am* **73**(8): 1119–1122.

Friedlaender GE and Goldberg VM (1991). *Bone and Cartilage Allografts: Biology and Clinical Applications*, The American Academy of Orthopaedic Surgeons, Park Ridge, Illinois.

Gross AE, Hutchinson CR, Alexeeff M, Mahomed N, Leitch K, and Morsi E (1995). Proximal femoral allografts for reconstruction of bone stock in revision arthroplasty of the hip. *Clin Orthop Relat Res* **319**: 151–158.

Haddad FS, Garbuz DS, Masri BA, *et al.* (2000). Structural proximal femoral allografts for failed total hip replacements: A minimum review of five years. *J Bone Joint Surg Br* **82**: 830–836.

Haddad FS, Garbuz DS, Masri BA, Duncan CP, Hutchison CR, and Gross AE (2000). Femoral bone loss in patients managed with revision hip replacement: results of circumferential allograft replacement. *Instr Course Lect* **49**: 747–762.

Horniek FJ, Gebhardt MC, Tomford WW, Sorger JI, Zavatta M, Menzer JP, and Mankin HJ (2001). Factors affecting nonunion of the allograft-host junction. *Clin Orthop Relat Res* **382**: 87–98.

Kidd-Ljunggren K and Simonsen O (1999). Reappearance of hepatitis B 10 years after kidney transplantation. *New Engl J Med* **341**: 127–128.

Lemaire R and Masson JB (2000). Risk of transmission of blood-borne viral infection in orthopaedic and trauma surgery. *J Bone Joint Surg Br* **82**(3): 313–323.

Lord CF, Gebhardt MC, Tomford WW, and Mankin HJ (1988). Infections in bone allografts. Incidence, nature, and treatment. *J Bone Joint Surg Am* **70**(3): 369–376.

Mankin HJ (1983). Complications of allograft surgery. In Friedlander GE, Mankin HJ and Sell KW (eds.), *Osteochondral Allografts. Biology, Banking and Clinical Applications*, Boston, Little, Brown, pp. 259–274.

Mankin HJ, Gebhardt MC, Jennings LC, Springfield DS, and Tomford WW (1996). Long-term results of allograft replacement in the management of bone tumors. *Clin Orthop Relat Res* **324**: 86–97.

Masri BA, Spangehl MJ, Duncan CP, Beauchamp CP, and Myerthal SL (1995). Proximal femoral allografts in revision total hip arthroplasty: a critical review. *J Bone Joint Surg Br* **77**(Supp. 3): 306–307.

Palmer SH, Gibbons CL, and Athanasou NA (1999). The pathology of bone allograft. *J Bone Joint Surg Br* **81**(2): 333–335.

Pereira BJ, Milford EL, Kirkman RL, and Levey AS (1991). Transmission of hepatitis C virus by organ transplantation. *N Engl J Med* **325**: 454–460.

Schratt HE, Regel G, Kiesewetter B, and Tscherne H (1996). HIV infection caused by cold preserved bone transplants. *Unfallchirurg* **99**(9):679–684.

Shapiro CN (1995). Occupational risk of infection with hepatitis B and hepatitis C virus. *Surg Clin North Am* **75**(6): 1047–1056.

Shutkin NM (1954). Homologous-serum hepatitis following the use of refrigerated bone-bank bone. *J Bone Joint Surg Am* **36**(7): 160–162.

Simonds RJ, Holmberg SD, Hurwitz RL, Coleman TR, Bottenfield S, Conley LJ, Kohlenberg SH, Castro KG, Dahan BA, Schable CA, *et al.* (1992). Transmission of human immunodeficiency virus Type 1 from a seronegative organ and tissue donor. *N Engl J Med* **326**: 726–732.

Stevenson S and Horowitz M (1992). The response to bone allografts. *J Bone Joint Surg Am* **74**(6): 939–950.

Tomford WW (1995). Transmission of disease through transplantation of musculoskeletal allografts. *J Bone Joint Surg Am* **77**: 1742–1754.

Wang JW and Wang CJ (2004). Proximal femoral allografts for bone deficiencies in revision hip arthroplasty: a medium-term follow-up study. *J Arthroplasty* **19**(7): 845–852.

Chapter 36

Allograft Engineering

Aziz Nather* and Yan Yi Han*

Introduction

Deep-frozen bone allograft is commonly used in the repair of large bone defects. However, it exhibits poor biological healing and incorporation, remaining largely inert following transplantation.

Healing of cortical bone allograft occurs via "creeping substitution" (Burchardt, 1993). This involves initial resorption of the transplanted bone allograft, followed by replacement of the bone allograft with bone and bone cells from the host (Nather *et al.*, 1990). However, the extent of "creeping substitution" that occurs is limited by a weak inflammatory response produced upon transplantation of the bone allograft into the recipient. As a result, host vasculature is ineffective in penetrating the graft medullary canal (Enneking and Mindel, 1991) and trabecular resorption of bone is restricted to a few millimeters of allograft bone at the host–bone junction (Kumta *et al.*, 2001).

In addition, the periosteal surface of bone has a thin layer of unmineralised bone matrix that prevents osteoclastic bone resorption (Chambers and Fuller, 1985). Thus, new bone laid down by the periosteum of the host envelopes the allograft for a short extent, usually a few centimeters (Kumta *et al.*, 2001). This envelope is important to the successful amalgamation of the host bone interface (Kumta *et al.*, 2001).

*NUH Tissue Bank, Department of Orthopaedic Surgery, Yong Loo Lin School of Medicine, National University of Singapore, Singapore.

Unlike normal bone, these large masses of mineralised but acellular bone are incapable of responding to stress by remodelling.

Pioneering Techniques

Knowing that massive, deep-frozen bone allograft does not undergo much biological remodelling or osseous integration, they are best not used alone.

A large cortical allograft can instead be used in combinations as suggested:

- "allograft–autograft composite" and
- "allograft–vascularised fibula composite".

Other techniques that have been introduced to improve host–bone union and biological incorporation of bone allograft included the step-cut technique. A step cut to approximate allograft with host bone resulted in stronger union and more stable host–bone interface, due to the larger surface contact area than that across conventional end-to-end apposition (Fig. 1).

Contact area between graft and host bone could also be increased by telescoping the host bone into the allograft (Fig. 2) (Kumta *et al.*, 1998). By telescoping the bone ends, junctional stability is enhanced as the

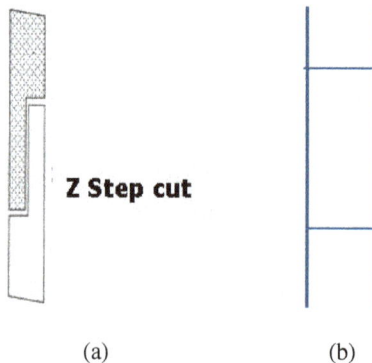

(a)   (b)

Fig. 1.   (a) Z step cut increases allograft–host contact area and stability. (b) Conventional end-to-end apposition.

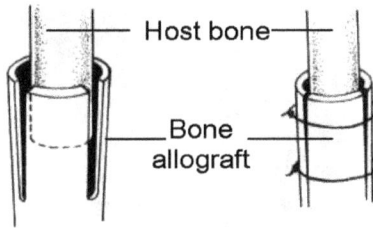

Fig. 2. Technique of telescoping allograft and host bone.

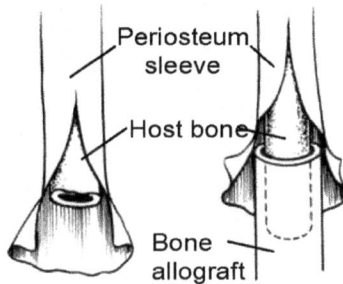

Fig. 3. Additional sleeve of periosteum used to wrap around the allograft.

available contact area was almost six times greater than with conventional end-to-end apposition. An additional sleeve of periosteum from the host bone could be retained and used to envelope the allograft (Fig. 3). Excellent union was observed in this series with the use of this technique, which required minimal additional internal stabilisation (Fig. 4).

Allograft–Vascularised Fibula Composite

To augment the healing of large cortical bone allograft, Capanna *et al.* (1993) placed a vascularised bone graft (fibula) within the shell of a bone allograft to create the "allograft–vascularised fibula composite" or "shell technique" (Fig. 5).

The vascularised fibula was able to remodel quickly under stress, while the outer allograft shell provided secure mechanical support that allowed for weight bearing. These properties resulted in better biological

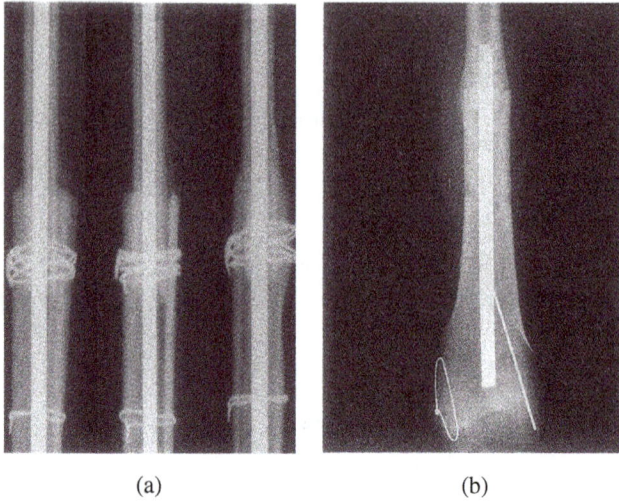

Fig. 4. (a) Telescoped junction showing progressive union through an enveloping callus. (b) Telescoped allograft with intramedullary nail and minimal internal stabilisation at the distal metaphyseal end.

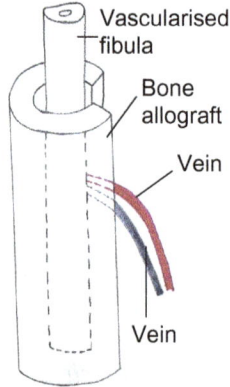

Fig. 5. Vascularised bone graft (fibula) placed within the shell of an allograft.

healing of the bone allograft. Rapid hypertrophy and cellular repopulation of allograft were observed. The living bone and allograft were fused to each other and amalgamated within a short period (Fig. 6). Vascularised fibula induced new bone formation within the allograft and enabled the otherwise inert allograft to remodel under loading. At the metaphyseal

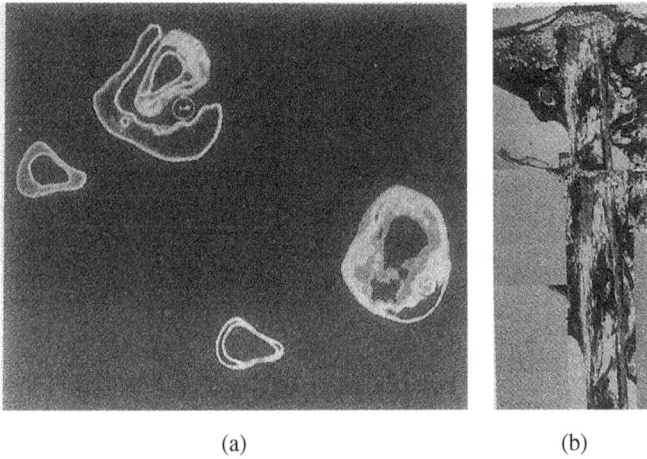

(a)   (b)

Fig. 6. (a) Combined allograft vascularised bone graft reconstruction. Vascularised bone graft within shell of allograft. Rapid incorporation and fusion within 12 months. (b) Hypertrophy and amalgamation of host bone with allograft.

Fig. 7. Histological section shows extensive new bone formation around original osteonal arrangement without resorption having occurred — formative remodelling.

ends, where allograft bone was less dense, "formative remodelling" occurred. This is a process that bypasses allograft resorption, allowing for rapid cellular repopulation of the allograft (Fig. 7). Juxtaposing of vascularised bone and allograft may yield these benefits as well (Fig. 8).

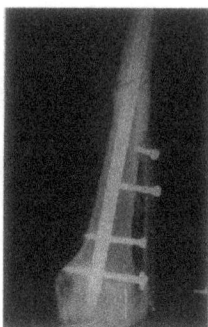

Fig. 8. Vascularised fibula juxtaposed with distal femoral allograft. Excellent healing of host–allograft junction is observed, which is enhanced by contribution from vascularised fibula.

It is evident that a hybrid graft had superior mechanical and biological properties to normal allograft. The combination of living bone or vascular tissue enhanced the biological potential of bone allograft and overcame most of the disadvantages associated with its use.

Autograft–Allograft Composite

Since bone allograft had little osteogenic potential, large cortical allograft is best used in combination with autograft in posterior spinal fusion, by packing the posterolateral spinal fusion bed with a mixture of deep-frozen bone allografts and autografts (Nather, 1999).

Allograft Engineering

Allograft engineering improved the biological healing and incorporation of large, deep-frozen cortical bone allograft. Allograft played an important role as the scaffold for implantation of mesenchymal stem cells (MSCs) or platelet-rich plasma (PRP).

An experimental study was conducted, which compared the biological healing of autograft, allograft and allograft into which MSC or PRP had been implanted. A defect measuring 1.2 cm large was created in the mid-tibia of New Zealand white rabbits (Fig. 9). This defect was repaired with

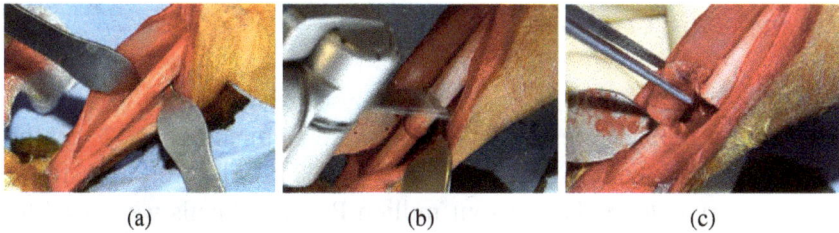

(a)   (b)   (c)

Fig. 9. (a) Exposing the rabbit tibia. (b) Proximal and distal osteotomies are performed. (c) 1.2 cm-large segment of the rabbit tibia is excised.

Table 1. Experimental plan showing number of rabbits examined per group at fixed time intervals.

Observation period	Group 1 (autograft)	Group 2 (allograft)	Group 3 (allograft–MSC)	Group 4 (allograft–PRP)
12	4	4	4	4
16	4	4	4	4
24	4	4	4	4
Total	12	12	12	12

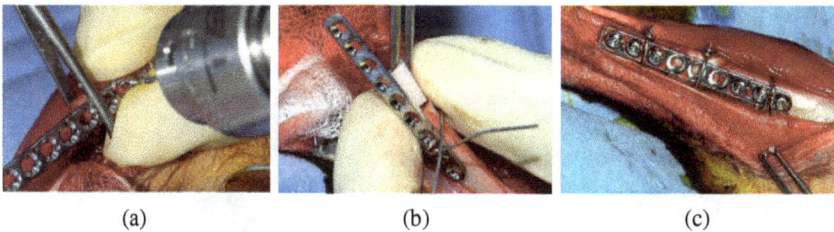

(a)   (b)   (c)

Fig. 10. (a) Nine-hole mini plate. (b) Fixation of tibial defect with nine-hole mini plate and cerclage wire. (c) Fixated tibial defect.

an autograft segment (Group 1), an allograft segment (Group 2), an "allograft–MSC composite" (Group 3) or an "allograft–PRP composite" (Group 4) (Table 1). The tibial defect was then fixed with a nine-hole mini plate and two cerclage wires (Fig. 10).

Allograft–MSC Composite

Methodology

MSCs from the iliac crest bone marrow were cultured until 80% confluence at primary culture was achieved (Fig. 11). Cells were transferred
to secondary culture. Two to four million Passage 2 cells were used for
transplantation into bone allograft. Human fibrinogen from the TISSEEL
kit was seeded with the cells and mixed with human thrombin to form
MSC gel (Fig. 12a, b). The cell pellet formed was then packed into the

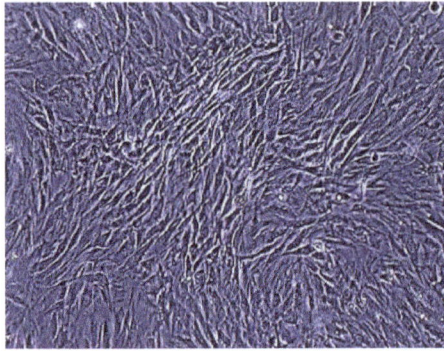

Fig. 11.   Bone marrow MSCs at 80% confluence.

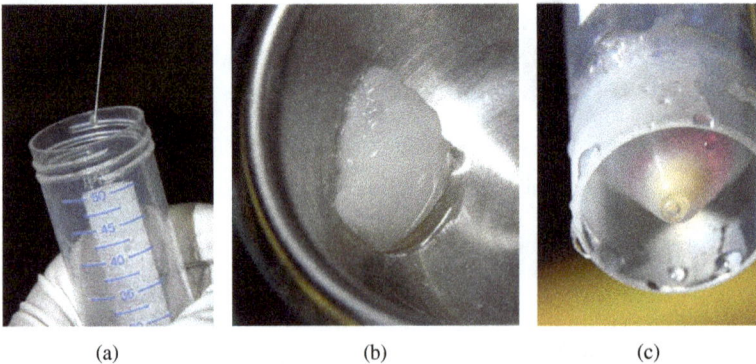

(a)   (b)   (c)

Fig. 12.   (a) Fibrin glue is added, one drop at a time, to fibrinogen seeded with MSCs,
until MSC gel is formed. (b) MSC gel. (c) MSC pellet.

medullary cavity and both ends of the host–graft junctions in Group 3 specimens (Fig. 12c).

Results

The union of host–graft junctions was compared and found to occur in all Group 1 (autograft) and Group 3 ("allograft–MSC composite") specimens at 12 weeks (Figs. 13a, c). For Group 2 (allograft) specimens, union had not occurred at 12 weeks (Fig. 13b). At 16 weeks, only 50% of Group 2 specimens showed union and at 24 weeks, all Group 2 specimens showed union.

Resorption cavities were more prominent in Group 3 specimens than in Group 2 specimens (Fig. 14). Figure 15 shows that the resorption index (RI) was significantly higher in Group 3 specimens than in Group 2 specimens for all three periods. The RI for Group 1 and Group 3 specimens did not differ significantly for all three periods.

New bone formation was more abundant in Group 3 specimens than in Group 2 specimens (Fig. 16). Figure 17 shows that the new bone formation index (NBFI) was significantly higher in Group 3 specimens than in Group 2 specimens for all three periods. The NBFI for Group 1 and Group 3 specimens did not differ significantly for all three periods.

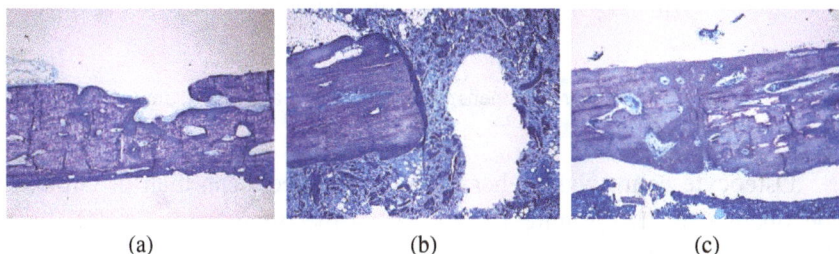

(a)   (b)   (c)

Fig. 13. (a) Autograft specimen showing union at host bone–graft junctions at 12 week, under a magnification of 40×. (b) Allograft specimen showing non-union at host bone–graft junctions at 12 weeks, under a magnification of 40×. (c) "Allograft–MSC composite" specimen showing union at host bone–graft junctions at 12 weeks, under a magnification of 40×.

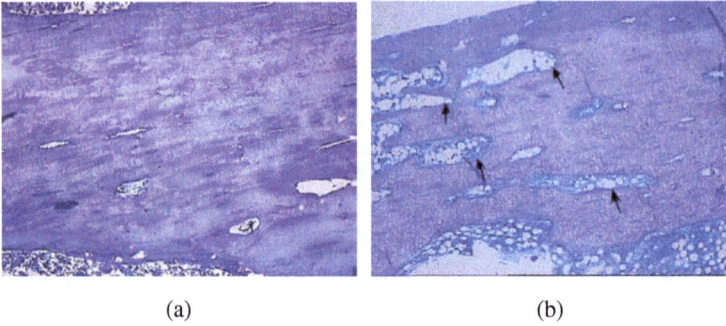

(a)   (b)

Fig. 14. (a) Allograft specimen showing small resorption cavities at 12 weeks, under a magnification of 40×. (b) "Allograft–MSC composite" specimen showing large resorption cavities at 12 weeks, under a magnification of 40×.

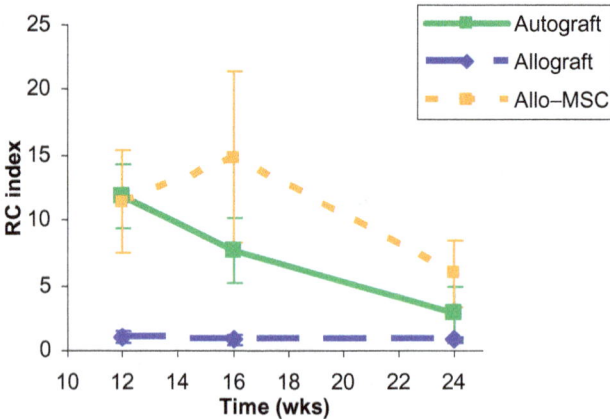

Fig. 15. RI for specimens from Groups 1, 2 and 3 over time.

Osteocyte count was higher in Group 3 specimens than in Group 2 specimens (Fig. 18). Figure 19 shows that the osteocyte index (OI) was significantly higher in Group 3 specimens than in Group 1 and Group 2 specimens for all three periods.

Conclusion

The use of "allograft–MSC composite" in the reconstruction of a large bone defect significantly improved the biological healing of bone

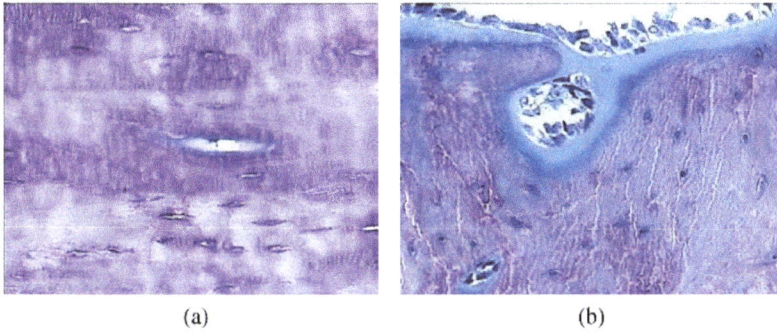

(a) (b)

Fig. 16. (a) Allograft specimen showing little new bone formation at 12 weeks, under a magnification of 40×. (b) "Allograft–MSC composite" specimen showing abundant new bone formation at 12 weeks, under a magnification of 40×.

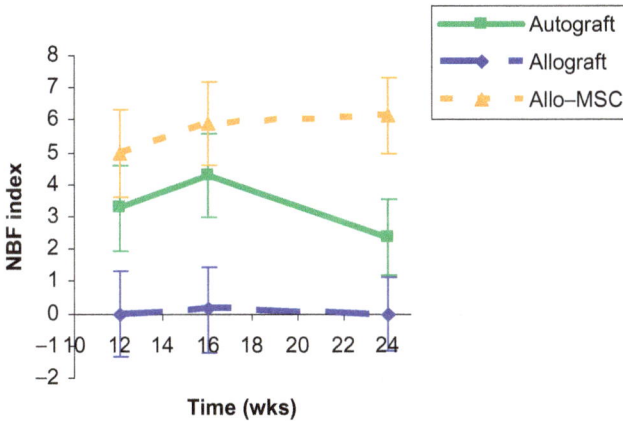

Fig. 17. NBFI for specimens from Groups 1, 2 and 3 over time.

allograft as shown by better union at host–graft junctions, increased resorption, increased new bone formation and increased osteocyte count of the allograft when allograft–MSC composite was used.

Allograft–PRP Composite

Autologous PRP contained growth factors such as PDGF, which stimulated mitosis and angiogenesis and TGF-β, which promoted new bone formation.

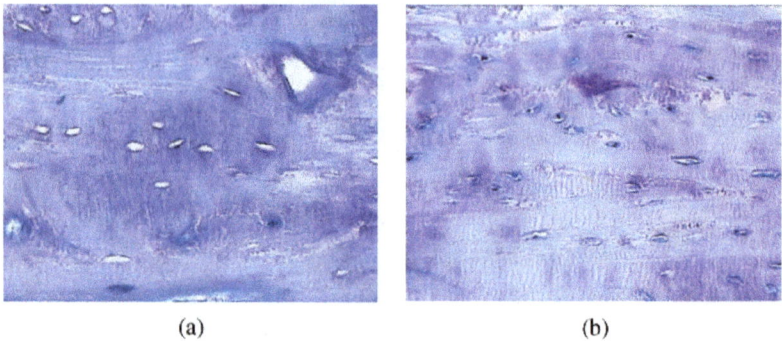

(a)   (b)

Fig. 18. (a) Allograft specimen showing few osteocytes at 12 weeks, under a magnification of 40×. (b) "Allograft–MSC composite" specimen showing an increased number of osteocytes at 12 weeks, under a magnification of 40×.

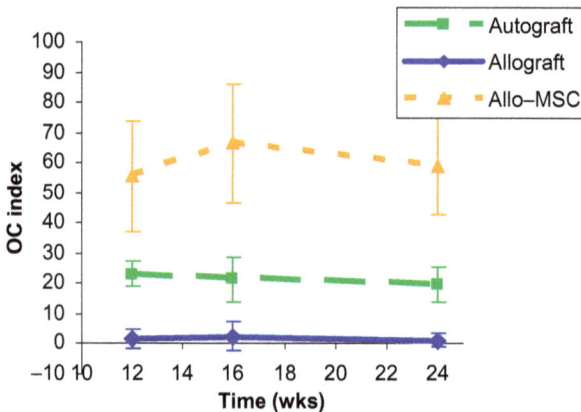

Fig. 19. OI for specimens from Groups 1, 2 and 3 over time.

PRP is affordable, at about SGD700 per application. It is efficient and safe to use; immunogenicity is not an issue since it is autologous.

Methodology

PRP is harvested from blood collected from a human donor using the aseptic method. Blood is loaded into a collection set and centrifuged at 3000 rpm for 3 minutes and 45 seconds. This is known as the soft spin (Fig. 20a). To harvest PRP, the supernatant is subjected to hard spin or centrifugation at 3 000 rpm for 13 minutes (Fig. 20b). PRP is harvested from the pellet obtained from hard spin and seeded with human fibrinogen from a TIS-SEEL kit. Human fibrin is added to obtain PRP gel (Fig. 20c), which is implanted in bone allograft to create the "allograft–PRP composite". This composite was used to repair the tibial defects of Group 4 specimens.

Results

The union of host–graft junctions was compared and found to occur in all Group 1 (autograft) and Group 4 ("allograft–PRP composite") specimens at 12 weeks (Fig. 21a, c). For Group 2 (allograft) specimens, union had not

(a)   (b)   (c)

Fig. 20. (a) After soft spin. (b) After hard spin. (c) Human fibrin is added to fibrinogen seeded with PRP to form PRP gel.

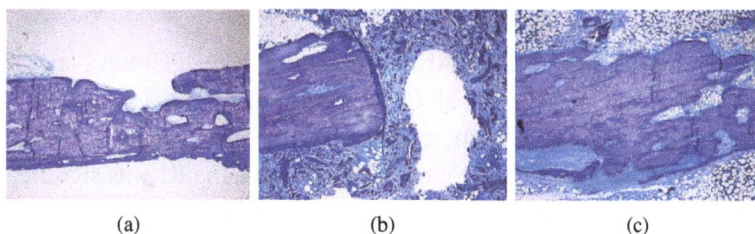

(a)   (b)   (c)

Fig. 21. (a) Autograft specimen showing union at host bone–graft junctions at 12 weeks, under a magnification of 40×. (b) Allograft specimen showing non-union at host bone–graft junctions at 12 weeks, under a magnification of 40×. (c) "Allograft–PRP composite" specimen showing union at host bone–graft junctions at 12 weeks, under a magnification of 40×.

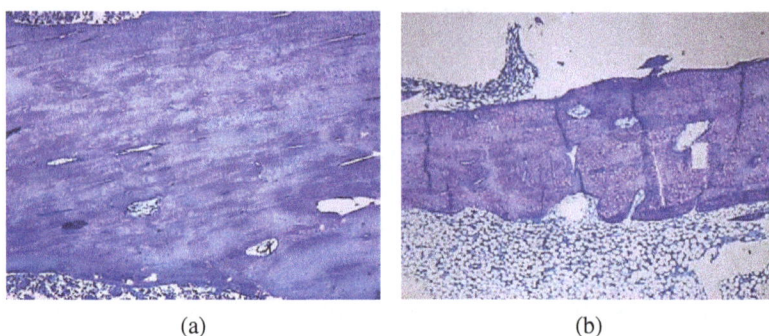

(a)　　　　　　　　　　　　　　(b)

Fig. 22. (a) Allograft specimen showing small resorption cavities at 12 weeks, under a magnification of 40×. (b) "Allograft–PRP composite" specimen showing large resorption cavities at 12 weeks, under a magnification of 40×.

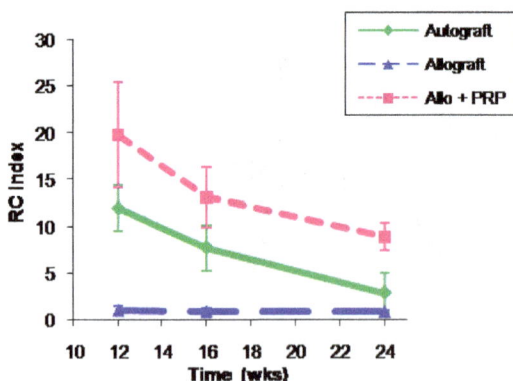

Fig. 23. RI for specimens from Groups 1, 2 and 4 over time.

occurred at 12 weeks (Fig. 21b). At 16 weeks, only 50% of Group 2 specimens showed union and at 24 weeks, all Group 2 specimens showed union.

Resorption cavities were more prominent in Group 4 specimens than in Group 2 specimens (Fig. 22). Figure 23 shows that the RI was significantly higher in Group 4 specimens than in Group 2 specimens for all three periods.

At 12 weeks, Group 2 and Group 4 specimens experienced little new bone formation (Fig. 24). Figure 25 shows that the NBFI was significantly higher for Group 4 specimens as compared to Group 1 specimens only after 12 weeks.

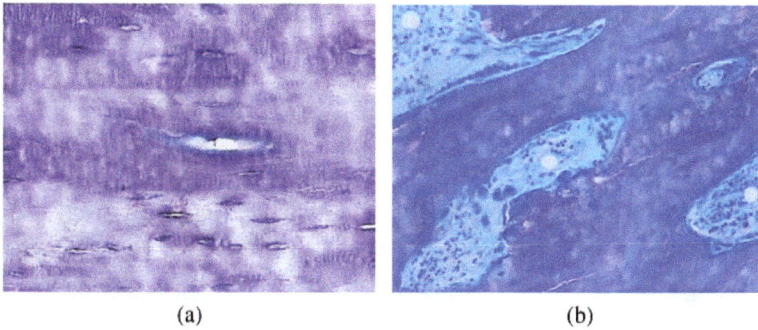

(a)   (b)

Fig. 24. (a) Allograft specimen showing little new bone formation at 12 weeks, under a magnification of 40×. (b) "Allograft–PRP composite" specimen showing little new bone formation at 12 weeks, under a magnification of 40×.

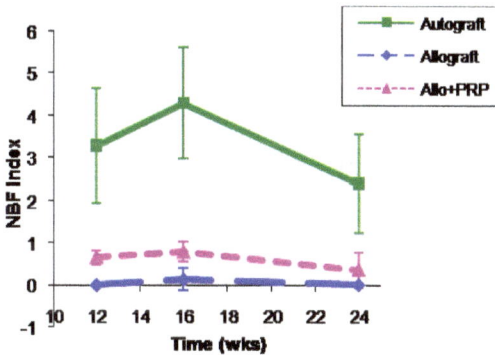

Fig. 25. NBFI for specimens for Groups 1, 2 and 4 over time.

At 12 weeks, few osteocytes were seen in Group 2 specimens, while a fair number of osteocytes were seen in Group 4 specimens (Fig. 26). The OI was significantly higher for Group 4 specimens as compared to Group 2 specimens only after 24 weeks (Fig. 27).

Conclusion

The use of "allograft–PRP composite" in the reconstruction of a large bone defect has improved the biological healing of bone allograft as

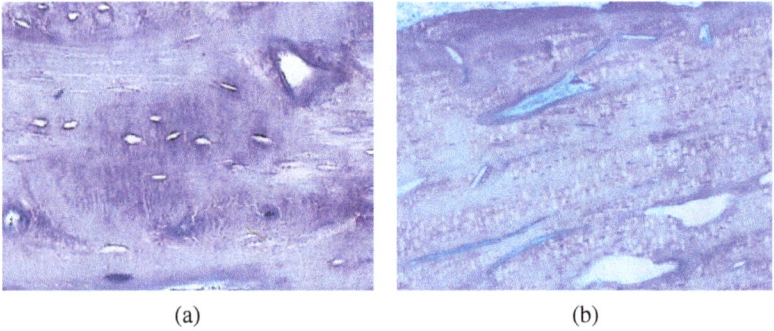

(a)   (b)

Fig. 26. (a) Allograft specimen showing few osteocytes at 12 weeks, under a magnification of 40×. (b) "Allograft–PRP composite" specimen showing a fair number of osteocytes at 12 weeks, under a magnification of 40×.

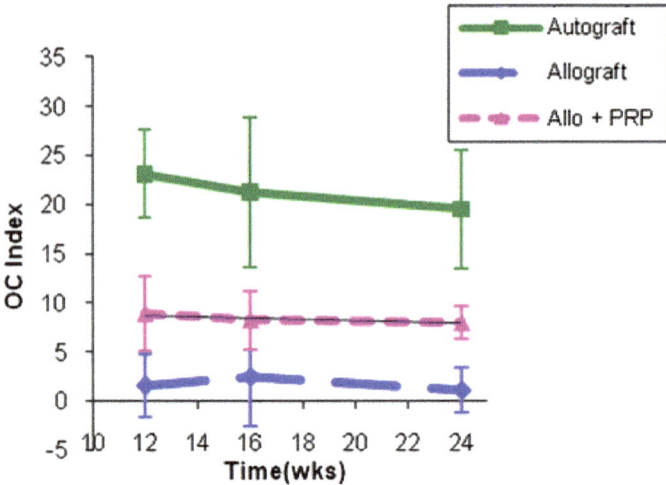

Fig. 27. OI for specimens of Groups 1, 2 and 4 over time.

shown by better union at host–graft junctions and increased resorption. However, the "allograft–PRP composite" did not perform as well as the "allograft–MSC composite" did, as the former achieved no significant increase in new bone formation and osteocyte count when the composite was used, as compared to when allograft was used alone.

Tissue Engineered Bone Substitute

The search is on for a bone substitute, whose scaffold can biodegrade and bioresorb, leaving behind osteocytes to form new bone that is able to integrate with host bone and remodel with time. Scaffolds available included calcium phosphate, polycaprolactone, coral and hydroxyapatite.

PCL-TCP–MSC Composite

Methodology

Polycaprolactone-tricalcium phosphate (PCL-TCP) was used as a scaffold for reconstruction of tibial defects created in 24 adult New Zealand white rabbits.

As before, a defect measuring 1.2 cm large was created in the mid-tibia of New Zealand white rabbits. This defect was repaired using an autograft segment (Group 1), a deep-frozen bone allograft (Group 2), PCL-TCP scaffold (Group 5) or a "PCL-TCP–MSC composite" (Group 6) (Table 2).

In Group 5 specimens, PCL-TCP scaffold alone was used to repair the tibial defect (Fig. 28). In Group 4 specimens, the "PCL-TCP–MSC composite" was used. This is a PCL-TCP scaffold that has been seeded with autologous bone marrow derived mesenchymal precursor cells, using fibrinogen from a TISSEEL kit as the carrier (Fig. 29).

Results

In Group 5 (PCL-TCP) specimens, there was non-union at the host bone–scaffold junctions in all specimens even at 24 weeks (Fig. 30a). On

Table 2. Experimental plan showing number of rabbits examined per group at fixed time intervals.

Observation period	Group 5 (PCL-TCP)	Group 6 (PCL-TCP–MSC)
12	4	4
16	4	4
24	4	4
Total	12	12

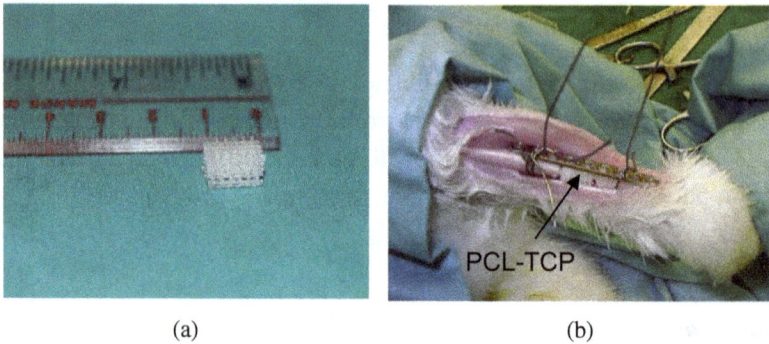

Fig. 28. (a) PCL-TCP scaffold. (b) Reconstruction with PCL-TCP scaffold.

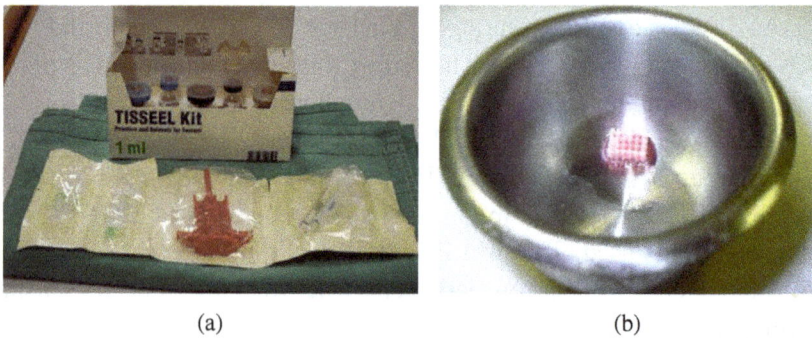

Fig. 29. (a) TISSEEL kit by Baxter, using T500 for rapid solidification. (b) Embedding MSCs in PCL-TCP using fibrinogen.

X-ray, no osseous union was seen in the defect (Fig. 30b). PCL-TCP scaffold biodegraded at 24 weeks, causing the gap to narrow (Fig. 31). The gap was bridged by layers of collagen fibres, with fibrocytes in between the fibres (Fig. 32).

In Group 6 (PCL-TCP–MSC) specimens, there was non-union at the host bone–scaffold junctions in all specimens even at 24 weeks (Fig. 33a). On X-ray, no osseous union was seen in the defect (Fig. 33b). PCL-TCP scaffold biodegraded at 24 weeks, causing the gap to narrow (Fig. 34). The gap was bridged by layers of collagen fibres, with fibrocytes in between the fibres (Fig. 35).

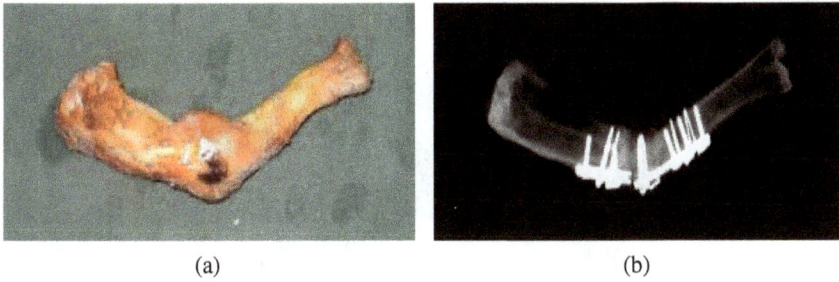

(a)   (b)

Fig. 30. (a) Mobility felt at 24 weeks in Group 5 specimen (T5-7). (b) No osseous union in Group 3 specimen (T5-7) at 24 weeks.

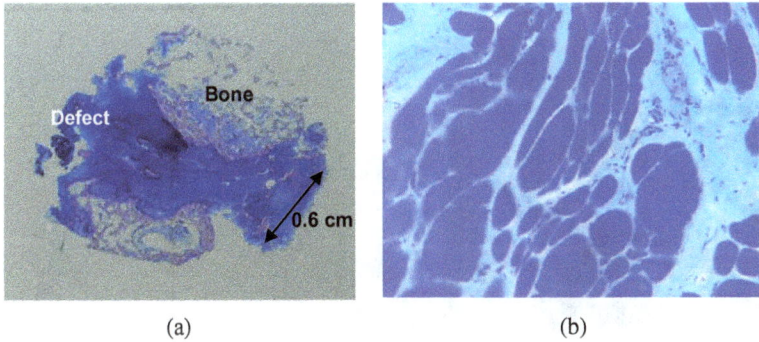

(a)   (b)

Fig. 31. (a) Histological section of narrowed bone defect due to degradation of scaffold in Group 5 specimen (T5-7). Gap has been reduced from 1.2 cm to 0.6 cm. (b) Remnants of PCL-TCP scaffold after it has biodegraded in Group 3 specimen, at a magnification of 200×.

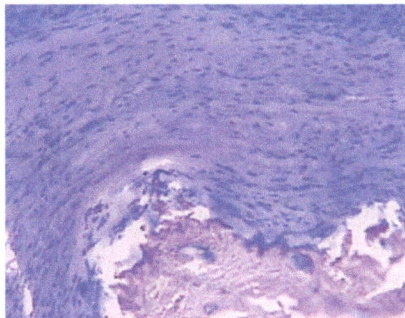

Fig. 32. Histological section of Group 5 specimen (T5-7) at 24 weeks showing no osseous union, at a magnification of 200×. Fibrous union is seen with collagen fibres and fibrocytes.

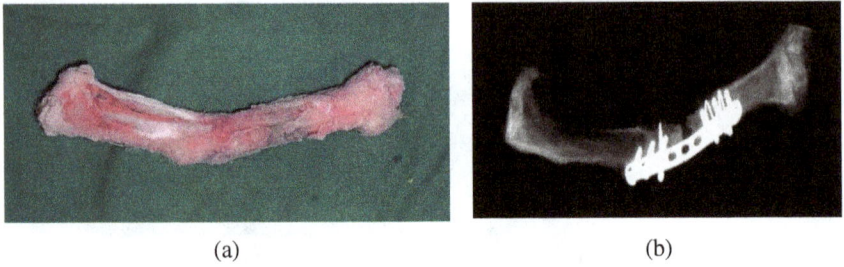

(a)   (b)

Fig. 33. (a) Mobility felt at 24 weeks in Group 6 specimen (M2). (b) No osseous union in Group 6 specimen (M2) at 24 weeks.

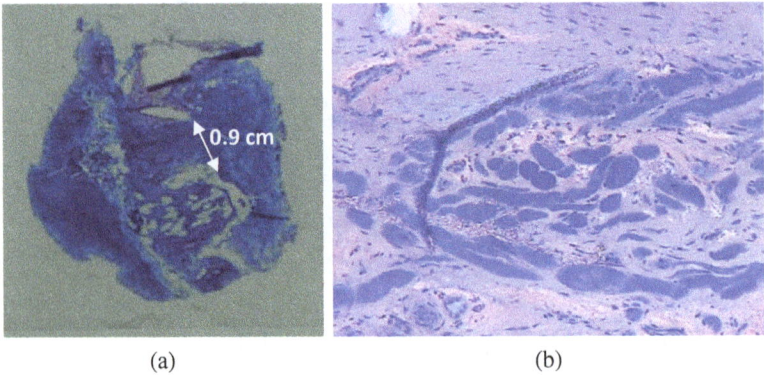

(a)   (b)

Fig. 34. (a) Histological section of narrowed bone defect due to degradation of scaffold in Group 6 specimen (M2). Gap has been reduced from 1.2 cm to 0.9 cm. (b) Remnants of PCL-TCP scaffold after it has biodegraded in Group 6 specimen (M2), at a magnification of 200×.

Conclusion

Results have shown that PCL-TCP scaffold is too weak for weight bearing. This caused the bone defect to narrow in both Group 5 and Group 6 specimens. Fibrous union instead of osteoid union bridged the gap.

In addition, no bone cells, chondrocytes and stem cells were observed in Group 6 specimens. The MSCs implanted in the PCL-TCP scaffold did not reconstitute new bone.

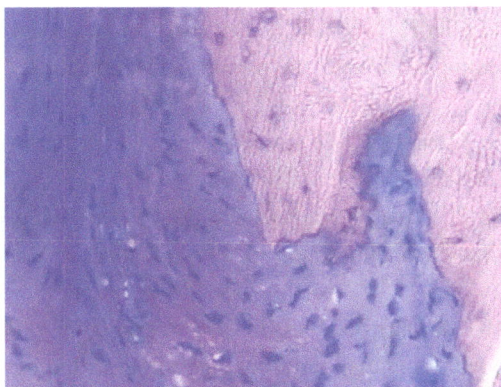

Fig. 35. Histological section of Group 6 specimen (M2) at 24 weeks showing no osseous union, at a magnification of 200×. Fibrous union is seen with collagen fibres and fibrocytes.

PCL-TCP, with cancellous bone strength, was not a good scaffold. The search for a third-generation scaffold with cortical bone strength to produce a "tissue engineered bone substitute" is still on. Bone allograft remained the best scaffold at the moment.

Conclusion

Bone allograft remained the best option for reconstruction of massive bone defects. It was therefore important that allograft engineering must improve the disadvantages associated with bone allograft, such as poor osseointegration of the allograft itself and delayed union occurring at the host–bone allograft junctions. The implantation of MSCs and/or PRP into allograft could improve the healing properties of bone allograft, so that it could become a more favourable option in the repair of large bone defects.

References

Bruchardt H (1983). The biology of bone graft repair. *Clin Orthop* **174**: 28.
Capanna R, Buffalini C, and Campanacci M (1993). A new technique for reconstructions of large metadiaphyseal bone defects — A combined graft (allograft shell plus vascularised fibula). *Orthop Traumatol* **2**(3): 159–177.

Chambers TJ and Fuller K (1985). Bone cells predispose bone surfaces to resorption by exposure of mineral to osteoclastic contact. *J Cell Science* **76:** 155–165.

Enneking WF and Mindel ER (1991). Observations on massive retrieved bone allografts. *J Bone J Surg* **73**: 1123–1142.

Kumta SM, Leung PC, Griffith JF, Roebuck DJ, Chow LTC, and Li CK (1998). A technique for enhancing union of allograft to host bone. *J Bone J Surg* **80B**: 994–998.

Kumta SM, Chow LTC, Griffith J, Fu LLK, and Leung PC (2001). The biology of massive bone allografts: understanding allograft biology and adapting it toward successful clinical application. In Nather A (ed.), *The Scientific Basis of Tissue Transplantation*, Singapore, pp. 456.

Nather A, Balasubramaniam P, and Bose K (1990). Healing of non-vascularised diaphyseal bone transplants: an experimental study. *J Bone J Surg [Br]* **79**: 830–834.

Nather A (1999). Use of allografts in spinal surgery. *Ann Transpl* **4**: 19–22.

Index